Water and Wastewater Management

Global Problems and Measures

Series Editors

Müfit Bahadir, Institut für Ökologische und Nachhaltige Chemie, Technische Universität Braunschweig, Braunschweig, Germany

Andreas Haarstrick, Leichtweiss-Institut für Wasserbau, Exceed, Technische Universität Braunschweig, Braunschweig, Germany

Water and wastewater management are among the greatest challenges of our century and the challenges posed by climate change will become even greater. Unfortunately, however, most efforts, especially in developing countries but also in the so-called developed countries, have been less than optimal or not optimal at all. In particular, there are still too many people who have to live without clean water and decent sanitation. Today, 2.2 billion people lack access to safely managed drinking water and wastewater, and 4.2 billion people lack safely managed sanitation services. The question, why this is the case - especially in developing countries - as well as other urgent water and wastewater management issues, are discussed in this book series. Contributions therein present in more detail critical reviews, discussions, and analysis of the water and wastewater situation and management aspects in different regions and countries worldwide.

Ali Müfit Bahadir · Andreas Haarstrick ·
I. Ethem Karadirek ·
Mehmet Emin Aydin ·
Serife Yurdagül Kumcu ·
Amitava Bandyopadhyay
Editors

Hydrology and Urban Water Supply

Springer

Editors
Ali Müfit Bahadir
Water and Wastewater Management
Braunschweig, Niedersachsen, Germany

I. Ethem Karadirek
Department of Environmental Engineering
Akdeniz University
Antalya, Türkiye

Serife Yurdagül Kumcu
Necmettin Erbakan University
Selçuklu, Türkiye

Andreas Haarstrick
Leichtweiß Institute
Technical University
Braunschweig, Germany

Mehmet Emin Aydin
Department of Civil Engineering
Necmettin Erbakan University
Konya, Türkiye

Amitava Bandyopadhyay
Department of Chemical Engineering
University of Calcutta
Kolkata, West Bengal, India

ISSN 2731-3166   ISSN 2731-3174 (electronic)
Water and Wastewater Management
Global Problems and Measures
ISBN 978-3-031-72588-3   ISBN 978-3-031-72589-0 (eBook)
https://doi.org/10.1007/978-3-031-72589-0

This Springer imprint is published by the registered company Springer Nature Switzerland AG
The registered company address is: Gewerbestrasse 11, 6330 Cham, Switzerland

Foreword

In our rapidly changing world, the availability and sustainable management of water resources have become critical global challenges. Decades of misuse, poor management, over-extraction of groundwater, and contamination of freshwater resources have intensified the water stress. In addition, countries are facing growing challenges linked to degraded aquatic ecosystems, water scarcity caused by climate change, malpractices in the management of water and sanitation, and insufficient cooperation on transboundary waters. Thus, it is crucial to understand the most complex interdependency between these factors and to develop innovative strategies to ensure sustainability of water for present and future generations. In 2020, only 74% of the world population had access to safely managed drinking water services. Still, 2 billion people live without safely managed drinking water and 1.2 billion people lack even a basic level of sanitation services. Access to safe water, sanitation, and hygiene is the most basic human need for health and well-being. Clean water and safe sanitation together are simply a basic human right.

Water scarcity is a growing problem in many parts of the world, and conflicts and climate change are exacerbating the issue. To address these problems, in 2015 the United Nations set 17 Sustainable Development Goals (SDGs) to guide the policy efforts until 2030, among which SDG 6 particularly is dedicated to clean water and sanitation for all. It focuses on ensuring the availability and sustainable management of freshwater and wastewater treatment. This goal emphasizes the importance of universal access to safe drinking water, improved water resources management, appropriate wastewater treatment, and well water-use efficiency. Achieving universal coverage by 2030 will require a 6-fold increase in current global rates of progress on drinking water, a 5-fold increase in sanitation, and an 8-fold increase in hygiene. Boosting infrastructure investment, improving cross-sectoral coordination, and addressing climate change are key to setting SDG 6 on the track. The mutual influence between hydrology, climate change, SDG 6, and urban water supply has emerged as a complex and pressing topic of concern. As cities expand, their demand for water increases, placing immense strain on existing supply systems. In an era defined by ever-increasing urban populations, cities face mounting pressure to secure water resources that can meet the demands of their residents, industries, and ecosystems. Balancing the needs of diverse stakeholders while protecting water quality, maintaining ecological integrity, and adapting to climate change presents a complex puzzle to solve.

The involved disciplines behind the study of water in all its forms have emerged as a guiding force in understanding the delicate balance between human needs and the natural environment. Hydrology is the scientific discipline of the movement, distribution, and quality of water in the Earth's system, and it is inextricably linked to climate patterns. Hydrology and climate change are two of the most significant and interlinked issues facing our global community today. Water is the essence of life on our planet. However, the global climate is changing at an unprecedented rate, and this is having a profound impact on the water cycle and the availability of freshwater resources. The interaction between hydrology and climate change requires a comprehensive understanding of both in order to address the challenges and opportunities that lie ahead. The world is currently experiencing a global climate crisis, with rising temperatures, more frequent extreme weather events, and changing precipitation patterns affecting communities and ecosystems around the globe. One of the most crucial systems impacted by these changes is the hydrologic cycle.

The field of hydrology has long been recognized as a vital module of environmental sciences, with significant implications for water resource management, land-use planning, and ecosystem health. However, the impact of climate change on the hydrologic cycle has introduced new challenges and complexities, which require innovative approaches and interdisciplinary collaboration. As temperatures continue to rise and precipitation patterns shift, hydrologic systems are being tested like never before. We are seeing the disappearance of glaciers, permafrost, and snowpack; increases in sea level and coastal flooding; and changes in the timing and quantity of river flows. These have social, economic, and environmental impacts that are already being felt around the world and will further intensify in the coming years.

This book is a comprehensive exploration of the impacts of climate change on the hydrologic cycle and the complex relationship between water and the environment that will impair future urban water supply and sustainable wastewater management for reuse of reclaimed water. Drawing on the recent scientific research, it examines the ways in which changing weather patterns are affecting the amount, quality, and distribution of water resources around the world. This book is structured to provide an introduction to the basics of hydrologic science, and the complex systems that drive the hydrologic cycle. It dives into the impacts of climate change on global hydrology, including changes in precipitation, evaporation, and runoff patterns. The book further explores the role of hydrologic science in mitigating the impacts of climate change through innovative water management practices and adaptation strategies for human consumption in urban areas and agricultural crop production. Through the use of case studies, scientific analysis, and personal investigations, it sheds light on the many interconnected factors that shape the hydrologic cycle and the ways in which water resources are being adapted to respond to the changing climate. From the role of technology in managing water scarcity to the importance of community engagement in climate adaptation strategies, the book explores a wide range of approaches to responding to the challenges of climate change.

The overall aim is to provide a comprehensive overview of the interaction between hydrology and climate change, highlighting the latest research findings, practical applications, and future directions. Key aspects of hydrology and climate change in urban water supply are addressed, inter alia: (i) climate change and its impact on surface water management; (ii) methods of modelling climate change on watershed hydrology; (iii) implications of climate change for water resources management, including adaptation and mitigation for drought and flood management strategies; (iv) role of water pollution with heavy metals, pesticides and pharmaceuticals, and microplastics in ecosystem and public health risks; and (v) controlling water losses and water pricing for improved urban water management.

The book also explores the crucial role that hydrologic science and management can play in adapting to these changes and building more resilient communities. While climate change represents a significant challenge, there are also opportunities to rethink our water management practices, to invest in more sustainable infrastructure, and to collaborate across related disciplines and sectors.

Hydrology and climate change are interconnected topics, and this book aims to provide a valuable resource for students, researchers, policymakers, and practitioners, who are interested in understanding the complex, interlinked issues of hydrology and climate change. The book offers new insights and perspectives to inspire dialogues and collaborations, which is crucial to effectively addressing the challenges of urban water supply under changing climate conditions. As the world faces unprecedented environmental challenges, including droughts, floods, and rising sea levels, the need for a deeper understanding of the water cycle and its response towards resilient communities has never been greater.

Prof. Dr. Muzaffer Şeker
President of TÜBA
Turkish Academy of Sciences
Ankara, Türkiye

Prof. Dr. Dr. Dr. h.c. Ali Müfit Bahadir
Technical University Braunschweig
Braunschweig, Germany

Preface

Water is a precious resource that sustains all forms of existence on our planet. As our world advances with leaps and bounds into the urban era, the need for comprehensive hydrology and efficient urban water supply systems has become more crucial. The delicate balance between meeting the demands of growing urban populations and safeguarding the sustainability of our water resources requires careful assessment, strategic planning, and good management. Hydrology which is the science dealing with the occurrence, circulation, distribution, and properties of the waters of the earth and its atmosphere, plays a pivotal role in understanding the distribution, movement, and availability of water resources. It encompasses a vast array of interrelated disciplines, including meteorology, geology, ecology, and engineering. Scientists/engineers analyse the relationship between rainfall, surface flow, and groundwater flow and provide effective use and management of water resources.

Water supply becomes a formidable challenge by increasing urban activities. Ingeniously designed and diligently maintained water supply systems play a vital role in the sustainability of water supply systems.

Reliable access to clean water is not only essential for daily living but also for driving economic activities, sustaining industrial facilities, and nurturing a thriving urban landscape. The importance of hydrology and urban water supply is further amplified in the context of climate change. Increasing temperature, altering precipitation patterns with time and place, and extreme weather events pose unprecedented challenges to water management. In order to ensure water security for future generations, scientists extra force to fortify the cities against the impacts of climate change by harnessing the principles of hydrology and by adopting resilient urban water supply practices.

In striving to challenge hydrology and supply of urban water, we must leverage innovation and research. Cutting-edge technologies, such as remote sensing, data analytics, and artificial intelligence, empower scientists to collect, analyse, and interpret vast amounts of hydrological data. By embracing innovation, we can devise novel solutions and make informed decisions to optimize water resource allocation and management.

This book delves into the intricacies of hydrology and urban water supply, delving into the science, practices, and policies that underpin these vital domains. Through the collective wisdom of experts and practitioners, it illuminates the challenges faced by cities worldwide and offers pathways to build sustainable, resilient, and equitable water systems for the urban centres of currents and futures.

Antalya, Türkiye Assoc. Prof. Dr. I. Ethem Karadirek
Konya, Türkiye Prof. Dr. Mehmet Emin Aydin
Konya, Türkiye Prof. Dr. Serife Yurdagül Kumcu
Kolkata, India Prof. Dr. Amitava Bandyopadhyay

About This Book

This special edition titled *Hydrology and Urban Water Supply* is a concise and informative book that explores the intricate relationship between hydrology and urban water provision. Authored by experts in the field, this book offers a comprehensive exploration of the challenges and solutions associated with urban water supply management in the context of hydrology. It covers topics such as water sources, treatment technologies, distribution systems, and sustainable water management practices. With its meticulous analysis and practical insights, this book equips professionals, researchers, and policymakers with the knowledge necessary to address the growing demands of urban water supply in an era of climate change and urbanization.

This book is part of a book series (special editions) based on the publication of the book *Water and Wastewater Management*, published by Springer in 2022 (ISBN 978-3-030-9528-7). The part about *Wastewater Management* (Technologies) edited in this book will be deepened with this first special edition in terms of technological topics.

Contents

About the Editors

Ali Müfit Bahadir studied chemistry at the Free University Berlin and Bonn University in Germany, received his Ph.D. 1975 from Bonn University and his Associate Professor 1988 from Munich Technical University. He became a full Professor of Environmental and Sustainable Chemistry at the Technische Braunschweig in 1989. Since 1997, he is an honorary doctor and visiting professor at the Selcuk University Konya in Turkey. His research fields cover environmental chemistry and analyses, environmental pollution through industrial processes and products, pesticide chemistry and metabolism in soil and water, ecotoxicology, sustainable chemistry, renewable feed stocks, and biodiesel and bio-lubricants. During the last 20 years, his research focused on R&D of sustainable water and wastewater management in developing countries. He retired in 2016.

Andreas Haarstrick studied Chemistry at the Technische Universität Braunschweig (1983–1989) and received his doctorate in 1992 in bioengineering. Since 2006, he is Professor for Bioprocess Engineering at the TU Braunschweig. His teaching and research cover modelling biological and chemical processes in heterogeneous systems, development of models predicting pollutant reduction in and emission behaviour of landfills, growth kinetics at low substrate concentrations under changing environmental conditions, Advanced Oxidation Processes (AOP), and groundwater management. Since 2012, he is the managing director of the DAAD exceed-Swindon Project dealing with sustainable water management in developing countries.

I. Ethem Karadirek graduated from Department of Environmental Engineering at Akdeniz University in 2005. He received his doctorate degree from Akdeniz University in 2005 in the field of urban water supply systems. He had research activities at University of Cincinnati in the USA (2012–2013) in the field of hydraulic and water quality modelling of water distribution systems. In 2020, he received the title of Associate Professor of Environmental Engineering at Akdeniz University and has been working in the same department ever since. His teaching and research cover hydraulics, fluid mechanics, water supply, design of wastewater collecting systems, water losses, water resources engineering, and environmental systems analysis. He has ongoing national research projects on water loss management in water distribution systems. He has published many research papers in international peer-reviewed journals and proceedings volumes. He is currently the Vice Director of the Institute of Natural and Applied Sciences at Akdeniz University (2021–ongoing), and Director of Environmental Problems Research and Application Center at Akdeniz University (2023–ongoing).

Mehmet Emin Aydin studied Civil engineering at Selçuk University in Konya, Turkey (1980–1985) and received his Ph.D. in 1993 in the field of drinking water treatment at Civil Engineering Department of Loughborough University of Technology in England. He is appointed as Associate Professor Dr. in Environmental Engineering of Selçuk University in 1997. Since 2003, he is full Professor in Environmental Engineering initially at Selçuk University and then since 2012 at Necmettin Erbakan University in Konya, Turkey. Since 2018, he has been a full professor of civil engineering at Necmettin Erbakan University. His teaching and research cover water supply, wastewater disposal, water pollution, water and wastewater treatment, ecotoxicology, PAHs, PCBs, pesticides, pharmaceuticals, and illicit drugs in the environment. Professor Dr. Aydın published 232 papers in international peer-reviewed journals and proceedings volumes; 82 presentations in National symposiums; 14 international book chapters, six book editorships, and co-editorships.

Serife Yurdagül Kumcu graduated in 1992 at Selçuk University of Konya, Turkey. She worked on labyrinth spillways and scour measurements around the bridge abutments in Middle East Technical University (METU) during her Master's and Ph.D. studies, respectively. After that, she worked for State Hydraulic Works as a research engineer in the Hydraulic Laboratory of the Technical Research and Quality Control Department. Later, she accomplished a comparative study with University of Mississippi in US on flood control and management, as well as sedimentation problems using NCCHE computer model and physical scale models as a Scientific and Technological Research Council of Turkey Foundation Scholar. Currently, she has been working for Necmettin Erbakan University as a Professor and she is Head of Institute of Science. Her research fields cover hydraulics, fluid mechanics, hydrology, sediment transport, water resources and sustainability, physical model studies, and similarity.

Amitava Bandyopadhyay studied Chemical Technology at the University of Calcutta and Chemical Engineering at the Indian Institute of Technology (IIT), Kharagpur, received his Ph.D. from the IIT, Kharagpur in 1996. Currently, he is a Full Professor of Chemical Engineering at the University College of Science and Technology, University of Calcutta. Prior to his university position for the last 15 years, he spent more than 10 years in a senior position at the West Bengal Pollution Control Board, the state regulatory agency responsible for implementing various environmental rules and regulations. He also spent eight years working with other academic institutions and organizations addressing environmental engineering and process formulations. He has been a Visiting Faculty Member of School of Environment, Resources and Development, Asian Institute of Technology (AIT), Bangkok, Thailand under the Faculty Secondment Programme nominated by the MHRD, Government of India. His research fields cover air pollution control, air and water quality management, environmental impact assessment, air emission dispersion studies, solid waste management, process plant safety, wastewater treatment, CO_2 capture and environmental jurisprudence under Indian regulatory regime. Besides, characterizing gas-liquid absorbers forms a part of his research areas.

Surface Water

Integrated Hydrodynamic and Water Quality Models in Surface Waters: Focus on Assessment of the Impacts of Climate Change

Meltem Kacikoc

Abstract

The pressure elements which occur as a result of human activities have intensified in recent years due to climate change and population growth. The negative effects of climate change have increased the urgency of managing the limited water resources. Water quality models are indispensable tools for effective management of water resources. Climate change and water resources, which are in mutual interaction, both have a specific and complex structure separately. For this reason, it is vital to use the appropriate model for the effective management of water resources. The aim of this work is to provide an overview of the sustainable management of water resources under the impact of climate change. To this end, this study also emphasizes the significance of integrated use of hydrodynamic and water quality models.

Keywords

Climate change · Eutrophication · Hydrodynamic · Modelling · Water quality

1 Introduction

Climate change is gaining momentum due to anthropogenic activities such as the use of fossil fuels, agricultural activities and changes in land use. Since the start of the Industrial Revolution in 1750, greenhouse gases have seen an unprecedented increase, which was not the case for 800,000 years before [1]. Between 1750 and 2020, the amount of atmospheric CO_2 has increased from 280 ppm to 413.2 ppm, CH_4 from 715 to 1889 ppb, and N_2O from 227 ppb to 333.2 ppb. During COVID-19 pandemic, due to the restrictions, there was an approximately 5.6% decrease in CO_2 emissions caused by fossil fuels. Accordingly, even though the CO_2 increase between 2019 and 2020 was relatively lower than between 2018 and 2019, for the other greenhouse gases (CH_4 and N_2O), the increase rates were higher than the average increase rate for the last 10 years [2]. Greenhouse gas concentrations continue to rise to this day. Climate model estimates show that atmospheric CO_2 concentration will continue to rise and reach 800 ppm by the end of this century unless steps are taken to reduce emissions [3, 4]. Greenhouse gas emissions due to anthropogenic factors have caused post-industrial global warming. As of 2017, global temperatures have increased by about 1 °C compared to pre-industrial revolution levels. Unless effective measures are taken to reduce greenhouse gases, the global temperature is

M. Kacikoc (✉)
Environmental Engineering Department, Suleyman Demirel University, Isparta, Turkey
e-mail: meltemkacikoc@sdu.edu.tr

A. M. Bahadir et al. (eds.), *Hydrology and Urban Water Supply*, Water and Wastewater Management,
https://doi.org/10.1007/978-3-031-72589-0_1

predicted to rise as much as 4.8 °C by the end of the twenty-first century [5]. The existing water resources are basic necessities for the survival of human beings, the welfare of society, as well as ecologic sustainability. Today, more than 2 billion people do not have adequate access to clean drinking water and are deprived of well-conducted sanitation services [6, 7]. UNESCO has noted that it is expected for the global water consumption to increase gradually due to the population increase and global change, and for the water studies to intensify due to global change [8, 9]. This situation highlights that water is not renewable, it needs to be managed well, and how important water reclamation is. Climate change is threatening water security worldwide. Global water security includes the satisfaction of the following conditions: (i) adequate water availability, (ii) acceptable water quality, (iii) sustainable water accessibility [10]. Climate change is directly affecting all these aspects of water security. As a result, it would be difficult especially for developing countries to reach Sustainable Development Goals (SDGs) in many dry areas due to the increased water scarcity [11]. For this reason, works that aim to ensure the security of water resources both in terms of quantity and the quality by taking into consideration the pressure of climate change.

Water quality models have become increasingly popular around the world for determining water quality and quantity, as well as sustainable adaptation activities under the effects of climate change. Models integrating hydrodynamics and water quality have been widely developed and used for the assessment of many aspects of lakes, rivers, reservoirs, ponds, estuaries and coastal waters. [12, 13]. This chapter discusses the effects of climate change on water resources and the importance of using models in the sustainable management of water resources.

2 The Effects of Climate Change on Water Resources

The impact of climate change on water resources can be observed in two areas: water accessibility and water quality [14]. Continued greenhouse gas emissions and climate change lead to intensified hydrologic cycles, and significant changes in precipitation, temperature and evaporation [15–17]. As the hydrological cycle intensifies under climate change, it is generally expected that wet regions will become wetter and dry regions drier [11, 18]. However, intensification of extreme events has been detected in both humid and arid regions worldwide [19, 20]. The effects of climate change are experienced more severely and more often in the form of droughts and floods in terms of the availability of water resources [21]. Increase in air temperature and change in precipitation (amount, timing, shape, and density) due to climate change have the potential to alter water flows [22]. Climate change can alter precipitation distribution based in both time and space and contribute to changing streamflow patterns [23]. Rising temperatures and increased precipitation contribute to increased river discharges and floods during the wet seasons and prolonged droughts during the dry seasons [23, 24]. Increasing annual average temperatures impact evapotranspiration and water flow balance and thus water yield [23, 25, 26]. Snowmelt is the most important source of river runoff in spring and early summer, and during low-flow periods it replenishes groundwater [27, 28]. Higher average temperatures due to climate change can create a snow-to-rain shift during the snowmelt season or type of precipitation and reduce water storage in the snowpack [28, 29]. The presence of water is defined by the difference between precipitation and evapotranspiration [30]. However, basin water retention is a factor that influences the flow slowing, which plays a critical role in determining hydrological drought risks [31]. If water seepage capacity is exceeded, more intense precipitation events will increase runoff [31, 32]. Different basins respond differently to the same change in climatic factors, depending on the geographical and hydrogeological characteristics of the basin and the amount of lake or groundwater storage in it [33]. More intense precipitation events will lead to a decrease in infiltration, as dry periods reduce the water absorption capacity of the soil. In combination with increased evapotranspiration, this would ultimately reduce water availability in the aquifer area [21]. Surface water resources, which are currently threatened by influences such as population growth, rapid industrialisation, industrial agricultural activities, are very sensitive to climate change. Between 1984 and 2015, about 90,000 km^2 of permanent surface water was lost [34]. Surface waters under such pressure will experience a significant reduction in net availability with a tendency towards further deterioration [33].

The Fourth National Climate Assessment reported a decline in surface water quality due to the increases in water temperature and in the frequency of high-intensity precipitation [35, 36]. Increased pollutant discharge due to flooding, increased temperature and dissolved oxygen depletion during drought, as well as changes in water ecology are some of the possible effects of climate change on water quality [14, 37]. Climate change causes direct and indirect negative effects on the physical, chemical, and biological properties of surface water resources [38]. For example, global warming directly affects the water temperature, while physical and chemical processes related to temperature in the water column change [39]. The direct and indirect impacts of climate change on water quality are generally of a similar magnitude. However, in some cases the indirect impacts may exceed the direct impacts, such as increases in eutrophying nutrients, chemical contaminants from industrial wastewater, sewage treatment plants in surface waters as well as biogeochemical and biological processes [40–42].

Drought-induced low flows and water levels in rivers increase the residence time of soluble and particulate matter and reduce the flushing rate or dilution in surface waters and their oxygen levels [43]. Higher temperatures will increase oxygen demand in water systems where photosynthesis is not dominant, thereby increasing the occurrence of anoxic and low oxygen conditions. However, this situation will gradually become permanent in the hypolimnion layer in lakes [40]. The amount of pollutants (organic, inorganic) leached from industrially used agricultural soils as well as the overflow from the sewage system into surface waters will increase with the frequency of flooding [42]. Water bodies are also strongly affected by processes in their basins. The hydrology, biology and chemistry as well as hydraulic processes in water catchment areas will be permanently changed by climate change [40, 44]. The nutrient load of surface waters increases will increase, creating conditions that favour rapid algal growth [45].

2.1 Temperature

In the case of eutrophication, a mass increase of phytoplankton algae and especially cyanobacteria can be observed [46]. Cyanobacteria, which are harmful algae, become more dominant under climate change conditions as they show maximum growth rates at relatively high temperatures compared to other phytoplankton groups (diatoms, green algae) [47]. Increases in water temperature, earlier onset of thermal stratification, as well as longer and more intense lake thermal stratification are consequences of climate warming [48, 49]. If the climate continues to change as predicted by global climate models, stratification in lakes is expected to be permanent [50]. Stratification makes it more advantageous for cyanobacteria as well, which can swim independently in the water surface layer on a higher water temperature for photosynthesis [51]. Also, when the hypolimnion becomes anoxic due to stratification, there will be a significant release of nutrients (especially phosphorus) from the sediment. These "internal nutrient sources" can be significantly larger than the external nutrient loads during the summer months and can be the main drivers of primary production in a lake [52].

2.2 Solar Radiation

Solar radiation is an important power source for photosynthesis, the basic life activity of algae, and a limiting factor for the growth of phytoplankton [53]. Under normal conditions, adequate sunlight provides good conditions for algae growth, leading to an increase in water temperature, thus further affecting phytoplankton growth and increasing the risk of water eutrophication [53, 54]. But for algae, there is a maximum growth rate beyond a certain threshold at which the growth rate will decrease [53, 55]. This condition results in Harmful Algal Blooms (HABs), an increase in cyanobacteria that have a strong tolerance to UV radiation [56]. In the absence of light, plants take oxygen from the water and dissolved oxygen is depleted. Under anaerobic conditions, phosphorus release from sediment can cause water eutrophication [53, 55, 57].

2.3 Wind

Wind speed is a very important driver in the eutrophication mechanism due to its effect on thermal stratification, sediment/nutrient upwelling, and horizontal mixing [58]. Climate model projections show that wind speed increases in some areas and decreases in others [59, 60]. Decreased wind speed will increase stratification [61, 62] and this will increase the risk of having low oxygen concentrations [63]. Low wind speed will accelerate the lake's thermal responses to warming, stratification times will be longer [50], and thus these conditions would raise the possibility of dominance of toxic cyanobacteria [63]. Increased wind stress could potentially prevent increase in thermal stratification [64, 65]. However, the shear stress caused by the wind on the lake surface can be efficiently transferred downstream to the lakebed via the shallow water column, causing resuspension of nutrients from the sediment into the water column [65]. This will again trigger the eutrophication mechanism.

2.4 Precipitation

Net precipitation is expected to increase in large parts of the world, but decrease in parts of the subtropics and in limited areas of the tropics and the Mediterranean. [63, 66]. In areas with higher net precipitation, higher run-off will lead to higher primary production in other surface waters, and higher biomass will possibly accelerate hypoxia [67–69]. Nutrient concentrations in the water can remain high, as longer residence times increase the exchange between sediment and water column and thus the internal nutrient cycle [63, 70]. However, there is an increased frequency of heavy rain events even in regions with less annual precipitation [71, 72]. Extreme precipitation will cause sudden increases in nutrient loads in surface waters and will lead to nutrient enrichment of sediments, which are an important source of internal nutrients.

Studies have shown that climate change and eutrophication are not one-way interactions, but can influence each other [73]. It will be difficult to determine the ultimate responses of ecosystems to climate change and to make possible predictions [69].

3 Assessing the Effects of Climate Change with Models

Conceptual models (such as the HBV model [74, 75], the HYMOD model [76]), have been widely used in climate change impact studies, and distributed hydrological models (such as MIKE-SHE [77], SWAT [78, 79]) has begun to be implemented [80]. In recent years, however, certain concerns have been raised regarding the impacts of climate change on water eutrophication on a global scale [55, 81, 82]. Under natural conditions, surface water sources move from the oligotrophic phase to the mesotrophic, eutrophic, and swampy phases, respectively. And eventually, water sources completely disappear, evolving from the swamp phase to land [83]. While in nature this is a slow process, the effects of climate change make water resources vulnerable to eutrophication and accelerate it [44, 84]. The most important aspect of reducing and controlling the eutrophication problem is to understand the complex hydrodynamic and convection processes of water resources. To simulate these processes hydrodynamic models are used [85]. Hydrodynamic models can provide important information about eutrophication models, including water velocities and circulation patterns, mixing and distribution, water temperature and density stratification, and sediment resuspension [86]. Since water movement and mixing processes are closely related to biochemical kinetics in the water column, the ability to combine multidimensional hydrodynamic and mass transport simulations with algorithms to predict component kinetics may be crucial for many water quality models [87].

3.1 Hydrodynamic Model

Changes in temperature, solar radiation, wind, precipitation etc. due to climate change will primarily affect the hydrodynamic structure of water resources. Therefore, in order to develop adaptation strategies, it is necessary to know how future climate change will affect the hydrodynamic conditions of water resources [88]. Eutrophication management should be based on a comprehensive understanding of hydrodynamics. Because algal bloom occurs mainly under favourable hydrodynamic and meteorological conditions, and algal growth reduction is determined by pollution loading and hydrodynamic conditions [89]. There is consensus that the substantial change in hydrodynamic state is a key factor in controlling algal blooms by influencing thermal stratification, nutrient transport, and the relationship between mixing depth and euphotic depth [90, 91].

Water quality is not only dependent on internal and external loads, but also closely related to the complex hydrodynamic conditions of water bodies [92, 93]. Hydrodynamic properties such as residence time of water, surface circulation pattern, and turbulent mixing have been reported as important features for maintaining the quality of surface water bodies [94]. Therefore, to better understand surface water flow and pollutant transport patterns, researchers and managers should focus on hydrodynamic studies of water bodies [95]. The hydrodynamic conditions of water resources are directly related to the transport, diffusion and mixing of pollutants as well as exchange and self-cleaning capacity [93]. In recent years, various engineering techniques like water diversion, water transfer, and inflow regulation etc. have been utilized to improve hydrodynamic conditions [93]. It is vital to analyse the effects of the measures planned to be implemented in this direction on the hydrodynamics of the water source before they are implemented. After adequate calibration and verification, numerical modelling is a potent tool for quantitatively examining hydrodynamic parameters since it can be used to assess a variety of options, including reservoir operations, discharge management, bathymetry changes, etc. [89]. Hydrodynamic simulation has also been utilized in numerous recent research to evaluate how climate change would affect water resources [95–98].

Numerous numerical models have been used to simulate the temperature, hydrodynamics, and associated biogeochemical and biological processes in lakes and reservoirs [99]. All models have identical mechanical explanations for the underlying physical processes, but they vary in their dimensionality, or the number of spatial dimensions they take into account [99]. High-dimensional numerical simulations of hydrodynamic models have become useful for comprehending the complex roles played by various driving processes [95]. Some studies using hydrodynamic models have shown that changes in hydrodynamics are the main cause of algal blooms in water bodies [100]. Similarly, the impacts of shifting wind direction were found to have a higher impact on the distribution of sediment concentration in lakes with directionally varying strong winds than advection and diffusion [95, 101]. Results from water quality models are very dependent on the hydrodynamic framework due to advection and dependence on variables like temperature or radiation [102, 103], so a robust assessment of the hydrodynamic component is crucial for coupled models [104].

3.2 Water Quality Model

Water quality, including physical, chemical, biological and other properties, is predominantly controlled by hydrodynamic processes and various mechanisms that govern the fate and transport of pollutants [12].

We can collect water quality models under 2 groups as (i) receiving water quality models and (ii) watershed water quality models. The majority of watershed models are either developed for general watershed and resource management or for hydrological modelling to help soil and water management. Most current watershed models have limited or non-existent water quality modelling capabilities [105]. Receiving water quality models frequently incorporate hydrodynamic models. As they are created with representations of various temporal and spatial variables, receiving water body types, pollutants of relevance, and processes, they differ greatly in their ability to predict water quality [105, 106].

Especially under the effects of climate change, the protection of water quality, especially eutrophication, which has a very complex mechanism, poses significant difficulties in the management of surface water bodies. To cope with these challenges, it is necessary to use process-based models that are strongly based on scientific theory against changes in flow regime, morphology, and radiation balance [107]. The predictive ability of these process-based models including mathematical descriptions of surface water hydrodynamic physics [107], provides the only real way to screen for numerous management alternatives [94, 108]. There are numerous software and tool types that vary in complexity and ability from 1 to 2D and 3D formulations [107].

In the last years, there have been advances in the application of three-dimensional (3D) models from one-dimensional (1D) models thanks to the integration of hydrodynamic model and water quality models, while also simulating more complex interactions of physical, chemical and biological processes [109]. Basic three-dimensional well-established eutrophication models such as WASP [110], CE-QUAL-ICM [111], Delft3D-WAQ [112], EFDC [113], MIKE3_WQ [114] and ECOM/RCA [115] have been used to simulate eutrophication dynamics in rivers, lakes, estuaries, and coastal waters. One of the crucial components of the eutrophication dynamics is the interaction between the sediment–water column, therefore, processes that define the interactions between lake sediment and water column should be taken into consideration in water quality studies [116]. These models have been improved over time regarding their performance, including the addition of interaction between the water column and the sediment layer [117]. Generally, the basic physical, chemical and biological processes of aquatic ecosystems can be simulated by these models but only a few models such as WASP, CE-QUAL-ICM, Delft3D-WAQ and EFDC are capable of simulating the effects of sediment on water quality [116, 118]. Apart from the interaction between the water column and the sediment, the interaction between the water body and the basin is also very important. To demonstrate the interaction between the water body and basin to be modelled, combining hydrodynamic models with hydrological models is widely used [119].

The models help to understand the responses of lake ecosystems to the reduction of nutrient loads and water transfer, as well as to quantitatively evaluate the eutrophication status of these water resources [120]. Numerical model applications play an important role in understanding the eutrophication mechanism and the causes of water pollution. Better water control and resource management policies can be created to improve or maintain the quality of aquatic ecosystems by using these tools [121].

4 Integrated Usage of Hydrodynamic and Water Quality Models: Case Studies

4.1 WASP

Thanks to its flexible structure, the WASP model, which can combine several interacting systems consisting of biochemical oxygen demand (BOD), organic nitrogen, DO, ammonia, nitrates, bacteria, phytoplankton, phosphates, pH, and solids, has several modules for organic toxicants, simple toxicants, eutrophication, mercury and non-ionic organic toxicants [122]. Eutrophication module simulates water quality in terms of nutrients, bacterial decay, DO and reactive pollutants in water [123]. The model includes six transport areas: the water column, water in the sediment spaces, sedimentation and resuspension rates in water for the three sediment groups, and transport due to precipitation/evaporation [124, 125]. The WASP model system includes a hydrodynamics sub-model (DYNHYD), which simulates water transport while the water quality model models the transport and interaction of contaminants in water [123]. Other hydrodynamic models like EFDC, EPD-RIV1, and SWMM can also be integrated with WASP [110]. Alam et al. [126] evaluated the effects of increased temperature and solar radiation due to climate change on the water quality of the Sitalakhya River using the WASP model. The model predicted that ammonia–nitrogen and BOD5 would gradually decrease with increasing temperature due to climate change, and the orthophosphate concentration would decrease due to higher orthophosphate uptake by phytoplankton in the upper part of the river. In the area under the influence of wastewater treatment plant discharge, the increase in orthophosphate was also predicted by the model, as a large amount of organic phosphorus would be converted into orthophosphate with the increase in temperature. Another study using the WASP model evaluated the potential future effects of climate change on the water quality of Lake Chungju [127]. The results of the study showed that Chungju Lake will pass from the mesotrophic level to the eutrophic level in terms of total phosphorus and chlorophyll-a. This study confirmed that climate change will cause more eutrophic lake conditions, further promoting algae growth and altering aquatic ecosystems.

4.2 CE-QUAL-ICM

CE-QUAL-ICM is a 3D eutrophication model that includes a hydrodynamic model and a sediment diagenesis model [111, 121]. The model calculates mass transport, kinetic conversions, and mass balances for 36 water quality condition variables, including a large number of algae, zooplankton, carbon, nitrogen, phosphorus, silica, and dissolved oxygen [128]. The unstructured grid feature allows it to link with a variety of hydrodynamic models, notably CH3D and EFDC [128]. In a study conducted in the Chesapeake Bay, the effect of sea level rise and warming on circulation and water quality was simulated with the predicted climatic conditions in 2050, with the CE-QUAL-ICM model [129]. In a warming condition projected in 2050, using a set of global circulation models (GCMs) that predict a 1.5–1.9 ° C increase in monthly air temperatures, the water temperature is estimated to increase by 0.8–1 °C and the summer average anoxic volume to increase by 1.4 percent. Besides the effect of algae growth in the warming state, increased anoxia is also associated with increases in stratification. As the temperature of the surface water rises faster than the temperature at higher depths, the surface layers become lighter in density, which slightly strengthens the forces that cause stratification. The CE-QUAL-ICM 3D eutrophication model was subjected to Lake Washington [130]. The modelling application sediment diagenesis model, in which 18 state variables are activated in the water column, is connected and sediment–water flows of dissolved oxygen, methane, ammonium, nitrate and phosphate are taken into account. Thanks to the model, it has been revealed that nutrient loads reaching the lake, especially phosphorus, accumulate in the sediment.

4.3 Delft3D-WAQ

The Delft3D-WAQ model, which includes both the water column and the sediment, is a well-established water quality model that is often combined with the hydrodynamic model (Delft3D-FLOW) in the Delft3D modelling system [131, 132]. Delft3D-FLOW can simulate the transport of both matter and heat. Fundamental processes including inorganic carbon balance, nutrient cycling, phytoplankton dynamics, and ecological processes like species competition, nutrient limitations, light, and temperature are all taken into consideration by the Delft3D-WAQ model [133]. Therefore, Delft3D-FLOW and Delft3D-WAQ should be used together to simulate more complex processes such as eutrophication, biological and/or chemical processes [134]. Surface CO_2 emissions and eutrophication in the drinking water reservoir, which change due to climate change, were simultaneously simulated with the Delft3D-WAQ model. The model demonstrates the effects of inorganic carbon on phytoplankton growth [135]. The development of hypoxia was evaluated by using the Delft3D-WAQ model to develop adaptation suggestions against extreme flood events due to climate change in Japan. It has been determined that flood events can significantly affect the development of hypoxia in the substrate [136]. The meteorological effects on the advection and mixing process of nutrients from the side discharge within the lake body were examined using the Delft3D-WAQ model in Taihu Lake to offer quantitative comparison of nutrient loads from different regions of the basin river networks [137]. As a result of the study, it was determined that the decrease in wind speed caused by climate change has negative effects on both internal and external nutrient source release and causes deterioration in water quality.

4.4 EFDC

EFDC can simulate water hydrodynamics, carbon, phosphorus, nitrogen and silica cycles, phytoplankton, macrophytes, dissolved oxygen, faecal coliform bacteria, temperature dynamics [121, 138]. The kinetic processes involved in the EFDC water quality model, including sediment diagenesis, are derived from the CE-QUAL-ICM water quality model [105]. In addition, the model can generate hydrodynamic transport files formatted for WASP and CE-QUAL-ICM [128]. Using the EFDC model, a three-dimensional, hydrodynamic-water quality-sediment model for Taihu Lake was developed and the lake's water quality responses to climate change scenarios were evaluated [139]. As a result of the study, it was determined that the temperature stratification in the lake tends to be early and long-term, the dissolved oxygen decreases, the sediment flux increases, and finally the initiation times, durations and areas of algal blooms increase. Yindong et al. [140] estimated the lake water quality responses to nutrient loads in Lake Chaohu under climate change scenarios with the EFDC model. The results showed that internal nutrient loading is responsible for the annual variation of nutrient concentrations in the lake and that internal loadings fluctuate much more than external nutrient inputs between different seasons. And they speculated that this could create favourable conditions for longer bursts of cyanobacteria throughout the year. As a result of the study, they emphasized that water managers should take into account internal nutrient loads in the restoration of eutrophic lakes under the effects of climate change.

5 Conclusion

The main goal of the Green Deal, the most current agenda on climate change, is to make Europe climate neutral by the year 2050 through switching to a circular economy, tackling biodiversity loss, and reducing pollution. A key target area of the Green Deal [141] and the Circular Economy Action Plan [142] is acxhustainable use of water [143]. EU policy provides incentives to take wastewater treatment beyond the traditional approach with the aim of converting wastewater into a merchantable resource [144]. Reusing wastewater will be a substantial alternative to traditional water resources when faced with diminishing water resources and prolonged droughts. It is considered that reusing properly treated wastewater has a lower environmental impact than other alternative water supply methods such as water transfers or desalination and would contribute to the reduction of greenhouse gas emissions [145]. The use of treated wastewater would be to use it in agriculture or to recharge groundwater reservoirs. Here are modelling approaches to be able to model large-scale approaches und developing tailored solution for recharching groundwater reservoirs. Due to climate change, the hydrodynamic and water quality conditions of water bodies have become more complex and made the management of water resources more difficult. Hydrodynamic properties such as flow velocity/direction, circulation, dispersion, stratification, etc. are critical driving forces for water quality.

Studies on determining the effects of climate change on water resources by integrating hydrodynamic and water quality models are not sufficient [146, 147]. Limited use of integrated hydrodynamic and water quality models in the field of the impacts of climate change are due to (i) models need high resolution data both temporally and spatially (ii) need to analyse detailed forms of water quality parameters (iii) these models must use comprehensive, high-resolution bathymetry conditions for water bodies, (iv) insufficient datasets of climate projections, roughly gridded with daily timescales, for those models that require high resolution data temporally and spatially. Therefore, the development of monitoring networks around the world, especially in developing countries, is indispensable for effective management of water resources.

References

1. IPCC (Intergovernmental Panel on Climate Change). Climate change 2014: synthesis report. In: Core Writing Team, Pachauri RK, Meyer LA, editors. Contribution of working groups I, II and III to the fifth assessment report of the intergovernmental panel on climate change. Geneva, Switzerland: IPCC; 2014. p. 151.
2. WMO (World Meteorological Organization). State of the global climate 2021: WMO Provisional report; 2021.
3. Meinshausen M, Smith SJ, Calvin K, Daniel JS, Kainuma ML, Lamarque JF, Matsumoto K, Montzka SA, Raper SCB, Riahi K, Thomson A, Velders GJM, Van Vuuren DPP. The RCP greenhouse gas concentrations and their extensions from 1765 to 2300. Clim Change. 2011;109(1):213–41.
4. Thirkell TJ, Pastok D, Field KJ. Carbon for nutrient exchange between arbuscular mycorrhizal fungi and wheat varies according to cultivar and changes in atmospheric carbon dioxide concentration. Glob Change Biol. 2020;26:1725–38.
5. Fouladkhah AC, Thompson B, Camp JS. The threat of antibiotic resistance in changing climate. Microorganisms. 2020;8:748.
6. WHO (World Health Organization), UNICEF (United Nations Children's Fund). Progress on drinking water, sanitation and hygiene: 2017 update and SDG baselines. Geneva; 2017.
7. Weerasooriya RR, Liyanage LPK, Rathnappriya RHK, Bandara WBMAC, Perera TANT, Gunarathna MHJP, Jayasinghe GY. Industrial water conservation by water footprint and sustainable development goals: a review. Environ Dev Sustain. 2021;23(9):12661–709.
8. UNESCO (United Nations Educational, Scientific and Cultural Organization), UN-Water. United nations world water development report 2020: water and climate change. Paris: UNESCO; 2020.
9. Jeong G, Kang D. Hydro-economic water allocation model for water supply risk analysis: a case study of Namhan River Basin, South Korea. Sustainability. 2021;13(11):6005.
10. Irannezhad M, Ahmadi B, Liu J, Chen D, Matthews JH. Global water security: a shining star in the dark sky of achieving the sustainable development goals. Sustain Horiz. 2022;1:100005.
11. Stringer LC, Mirzabaev A, Benjaminsen TA, Harris RMB, Jafari M, Lissner TK, Stevens N, Tirado-von der Pahlen C. Climate change impacts on water security in global drylands. One Earth. 2021;4 (6):851–64.
12. Liu L. Application of a hydrodynamic and water quality model for inland surface water systems. Appl Water Syst Manage Modeling. 2018;87–109.
13. Ejigu MT. Overview of water quality modeling. Cogent Eng. 2021;8(1):1891711.
14. Chaturvedi A, Pandey B, Yadav AK, Saroj S. An overview of the potential impacts of global climate change on water resources. Water Conserv Era Global Clim Change. 2021;99–120.
15. Middelkoop H, Daamen K, Gellens D, Grabs W, Kwadijk JC, Lang H, Parmet BWAH, Schädler B, Schulla J, Wilke K. Impact of climate change on hydrological regimes and water resources

management in the Rhine basin. Clim Change. 2001;49(1):105–28.
16. Jahandideh-Tehrani M, Zhang H, Helfer F, Yu Y. Review of climate change impacts on predicted river streamflow in tropical rivers. Environ Monit Assess. 2019;191(12):1–23.
17. Jahandideh-Tehrani M, Helfer F, Jenkins G. Impacts of climate change and sea level rise on catchment management: a multi-model ensemble analysis of the Nerang River catchment, Australia. Sci Total Environ. 2021;777: 146223.
18. Donat MG, Lowry AL, Alexander LV, O'Gorman PA, Maher N. More extreme precipitation in the world's dry and wet regions. Nat Clim Chang. 2016;6(5):508–13.
19. Donat MG, Angélil O, Ukkola AM. Intensification of precipitation extremes in the world's humid and water-limited regions. Environ Res Lett. 2019;14 (6): 065003.
20. Yao J, Chen Y, Chen J, Zhao Y, Tuoliewubieke D, Li J, Yang L, Mao W. Intensification of extreme precipitation in arid Central Asia. J Hydrol. 2021;598: 125760.
21. Arraiza MP, Santamarta JC, Ioras F, García-Rodríguez JL, Abrudan IV, Korjus H, Borála G, editors. Climate change and restoration of degraded land. Madrid: Colegio de Ingenieros de Montes; 2014.
22. Omar MEDM, Moussa AMA, Hinkelmann R. Impacts of climate change on water quantity, water salinity, food security, and socioeconomy in Egypt. Water Sci Eng. 2021;14(1):17–27.
23. Chim K, Tunnicliffe J, Shamseldin AY, Bun H. Assessment of land use and climate change effects on hydrology in the upper Siem Reap River and Angkor Temple Complex, Cambodia. Environ Develop. 2021;39: 100615.
24. Oeurng C, Cochrane TA, Chung S, Kondolf MG, Piman T, Arias ME. Assessing climate change impacts on river flows in the Tonle Sap Lake Basin, Cambodia. Water. 2019;11(3):618.
25. Schmid HP, Grimmond CSB, Cropley F, Offerle B, Su HB. Measurements of CO_2 and energy fluxes over a mixed hardwood forest in the mid-western United States. Agric For Meteorol. 2000;103 (4):357–74.
26. Hu Q, Willson GD, Chen X, Akyuz A. Effects of climate and landcover change on stream discharge in the Ozark Highlands, USA. Environ Model Assess. 2005;10(1):9–19.
27. Dumanski S, Pomeroy JW, Westbrook CJ. Hydrological regime changes in a Canadian Prairie basin. Hydrol Process. 2015;29(18):3893–904.
28. Zaremehrjardy M, Victor J, Park S, Smerdon B, Alessi DS, Faramarzi M. Assessment of snowmelt and groundwater-surface water dynamics in mountainous, foothill, and plain regions in northern latitudes. J Hydrol. 2022;127449.
29. MacDonald RJ, Byrne JM, Boon S, Kienzle SW. Modelling the potential impacts of climate change on snowpack in the North Saskatchewan River Watershed, Alberta. Water Resourc Manage. 2012;26(11):3053–76.
30. Byrne MP, O'Gorman PA. The response of precipitation minus evapotranspiration to climate warming: why the "wet-get-wetter, dry-get-drier" scaling does not hold over land. J Clim. 2015;28(20):8078–92.
31. Yang H, Piao S, Huntingford C, Ciais P, Li Y, Wang T, Peng S, Yang Y, Yang D, Chang J. Changing the retention properties of catchments and their influence on runoff under climate change. Environ Res Lett. 2018;13(9): 094019.
32. Xue J, Gavin K. Effect of rainfall intensity on infiltration into partly saturated slopes. Geotech Geol Eng. 2008;26(2):199–209.
33. Kundzewicz ZW, Mata LJ, Arnell NW, Doll P, Kabat P, Jimenez B, Miller K, Oki T, Zekai S, Shiklomanov I. Freshwater resources and their management. 2007.
34. Pekel JF, Cottam A, Gorelick N, Belward AS. High-resolution mapping of global surface water and its long-term changes. Nature. 2016;540 (7633):418–22.
35. USGCRP (U.S. Global Change Research Program). Impacts, risks, and adaptation in the United States: fourth national climate assessment, vol. II. U.S. Washington, DC.: Global Change Research Program; 2018.
36. Ryberg KR, Chanat JG. Climate extremes as drivers of surface-water-quality trends in the United States. Sci Total Environ. 2022;809:152165.
37. Senhorst HAJ, Zwolsman JJG. Climate change and effects on water quality: a first impression. Water Sci Technol. 2005;51(5):53–9.
38. Vincent WF. Effects of climate change on lakes. In: Likens GE, editor. Encyclopedia of inland waters, vol. 3. Oxford: Elsevier; 2009. p. 55–60.
39. Mujere N, Moyce W. Climate change impacts on surface water quality. In: Ganpat W, Isaac W, editors. Environmental sustainability and climate change adaptation strategies. Hershey: IGI Global; 2017. p. 322–40.
40. Benateau S, Gaudard A, Stamm C, Altermatt F. Climate change and freshwater ecosystems: impacts on water quality and ecological status. Bern, Switzerland: Hydro-Ch2018 Project, Federal Office for the Environment (FOEN); 2019. 110p.
41. Brosse M, Benateau S, Gaudard A, Stamm C, Altermatt F. The importance of indirect effects of climate change adaptations on alpine and pre-alpine freshwater systems. Ecol Solut Evid. 2022;3(1): e12127.
42. Solheim AL, Austnes K, Eriksen TE, Seifert I, Holen S. Climate change impacts on water quality and biodiversity. Background report for EEA

European environment state and outlook report; 2010.
43. Mosley LM. Drought impacts on the water quality of freshwater systems; review and integration. Earth Sci Rev. 2015;140:203–14.
44. Murdoch PS, Baron JS, Miller TL. Potential effects of climate change on surface-water quality in North America 1. JAWRA J Am Water Resourc Assoc. 2000;36(2):347–66.
45. Rodgers EM. Adding climate change to the mix: responses of aquatic ectotherms to the combined effects of eutrophication and warming. Biol Let. 2021;17(10):20210442.
46. Moss B. Cogs in the endless machine: lakes, climate change and nutrient cycles: a review. Sci Total Environ. 2012;434:130–42.
47. Glibert PM. Harmful algae at the complex nexus of eutrophication and climate change. Harmful Algae. 2020;91: 101583.
48. Schindler DW, Smol JP. Cumulative effects of climate warming and other human activities on freshwaters of Arctic and subarctic North America. AMBIO: J Human Environ. 2006; 35(4):160–168.
49. Morales-Marin LA, Carr M, Sadeghian A, Lindenschmidt KE. Climate change effects on the thermal stratification of Lake Diefenbaker, a large multipurpose reservoir. Can Water Resourc J/Revue canadienne des ressources hydriques. 2021;46(1–2):1–16.
50. Woolway RI, Merchant CJ, Van Den Hoek J, Azorin-Molina C, Nõges P, Laas A, Mackay EB, Jones ID. Northern Hemisphere atmospheric stilling accelerates lake thermal responses to a warming world. Geophys Res Lett. 2019;46(21):11983–92.
51. Gray E, Elliott JA, Mackay EB, Folkard AM, Keenan PO, Jones ID. Modelling lake cyanobacterial blooms: disentangling the climate-driven impacts of changing mixed depth and water temperature. Freshw Biol. 2019;64(12):2141–55.
52. Hickey CW, Gibbs MM. Lake sediment phosphorus release management-decision support and risk assessment framework. NZ J Mar Freshwat Res. 2009;43(3):819–56.
53. Xia R, Zhang Y, Critto A, Wu J, Fan J, Zheng Z, Zhang Y. The potential impacts of climate change factors on freshwater eutrophication: implications for research and countermeasures of water management in China. Sustainability. 2016;8(3):229.
54. Ihnken S, Eggert A, Beardall J. Exposure times in rapid light curves affect photosynthetic parameters in algae. Aquat Bot. 2010;93(3):185–94.
55. Nazari-Sharabian M, Ahmad S, Karakouzian M. Climate change and eutrophication: a short review. Eng Technol Appl Sci Res. 2018;8(6):3668.
56. Castenholz RW, Garcia-Pichel F, Whitton BA. Cyanobacterial responses to UV radiation. Ecology of cyanobacteria II. Dordrecht, Netherlands: Springer; 2012. p. 481–499.
57. Nesa B, Baird AH, Harii S, Yakovleva I, Hidaka M. Algal symbionts increase DNA damage in coral planulae exposed to sunlight. Zool Stud. 2012;51:12–7.
58. Zhang Y, Loiselle S, Shi K, Han T, Zhang M, Hu M, Jing Y, Lai L, Zhan P. Wind effects for floating algae dynamics in eutrophic lakes. Remote Sensing. 2021;13(4):800.
59. Pryor SC, Barthelmie RJ, Clausen NE, Drews M, MacKellar N, Kjellström E. Analyses of possible changes in intense and extreme wind speeds over northern Europe under climate change scenarios. Clim Dyn. 2012;38(1):189–208.
60. Hdidouan D, Staffell I. The impact of climate change on the levelised cost of wind energy. Renew Energy. 2017;101:575–92.
61. Andrew JT, Norman DY, Keller B, Girard R, Heneberry J, Gunn JM, Hamilton DP, Taylor PA. Cooling lakes while the world warms: effects of forest regrowth and increased dissolved organic matter on the thermal regime of a temperate, urban lake. Limnol Oceanogr. 2008;53(1):404–10.
62. Butcher JB, Nover D, Johnson TE, Clark CM. Sensitivity of lake thermal and mixing dynamics to climate change. Clim Change. 2015;129(1):295–305.
63. Meerhoff M, Audet J, Davidson TA, De Meester L, Hilt S, Kosten S, Liu Z, Mazzeo N, Paerl H, Scheffer M, Jeppesen E. Feedback between climate change and eutrophication: revisiting the allied attack concept and how to strike back. Inland Waters. 2022;1–18.
64. George G, Hurley M, Hewitt D. The impact of climate change on the physical characteristics of the larger lakes in the English Lake District. Freshw Biol. 2007;52(9):1647–66.
65. Whitehead PG, Wilby RL, Battarbee RW, Kernan M, Wade AJ. A review of the potential impacts of climate change on surface water quality. Hydrol Sci J. 2009;54(1):101–23.
66. IPCC (Intergovernmental Panel on Climate Change). Summary for Policymakers. In: Masson-Delmotte VP, Zhai P, Pirani A, Connors SL, Péan C, et al., editors. Climate change 2021: the physical science basis. Contribution of working group I to the sixth assessment report of the intergovernmental panel on climate change. Cambridge (UK): Cambridge University Press; 2021.
67. Glibert PM, Icarus Allen J, Artioli Y, Beusen A, Bouwman L, Harle J, Holmes R, Holt J. Vulnerability of coastal ecosystems to changes in harmful algal bloom distribution in response to climate change: projections based on model analysis. Glob Change Biol. 2014;20(12):3845–58.
68. Wåhlström I, Höglund A, Almroth-Rosell E, MacKenzie BR, Gröger M, Eilola K, Plikshs M, Andersson HC. Combined climate change and nutrient load impacts on future habitats and eutrophication indicators in a eutrophic coastal sea. Limnol Oceanogr. 2020;65(9):2170–87.
69. Garnier J, Billen G, Lassaletta L, Vigiak O, Nikolaidis NP, Grizzetti B. Hydromorphology of coastal

zone and structure of watershed agro-food system are main determinants of coastal eutrophication. Environ Res Lett. 2021;16(2): 023005.
70. Özen A, Karapınar B, Kucuk I, Jeppesen E, Beklioglu M. Drought-induced changes in nutrient concentrations and retention in two shallow Mediterranean lakes subjected to different degrees of management. Hydrobiologia. 2010;646(1):61–72.
71. Prein AF, Rasmussen RM, Ikeda K, Liu C, Clark MP, Holland GJ. The future intensification of hourly precipitation extremes. Nat Clim Chang. 2017;7(1):48–52.
72. Coffey R, Paul MJ, Stamp J, Hamilton A, Johnson T. A review of water quality responses to air temperature and precipitation changes 2: nutrients, algal blooms, sediment, pathogens. JAWRA J Am Water Resourc Assoc. 2019;55(4):844–68.
73. Moss B, Kosten S, Meerhoff M, Battarbee RW, Jeppesen E, Mazzeo N, Havens K, Lacerot G, Liu Z, Meester LD, Paerl H, Scheffer M. Allied attack: climate change and eutrophication. Inland Waters. 2011;1:101–5.
74. Bergstrom S. The HBV model-its structure and applications. SMHI Hydrology, RH No. 4, Sweden, Norrkoping; 1992.
75. Bergstrom S. The HBV model. In: Singh VP, editor. Computer models of watershed hydrology. Highlands Ranch: Water Resources Publications; 1995. p. 443–76.
76. Moore RJ. The probability-distributed principle and runoff production at point and basin scales. Hydrol Sci J. 1985;30(2):273–97.
77. Graham DN, Butts MB. Flexible, integrated watershed modelling with MIKE-SHE. In: Singh VP, Frevert DK, editors. Watershed models; 2006. p. 245–72.
78. Arnold JG, Srinivasan R, Muttiah RS, Williams JR. Large area hydrologic modeling and assessment. Part I: model development. J Am Water Resourc Assoc. 1998;34(1):73–89.
79. Neitsch SL, Arnold JG, Kiniry JR, Williams JR. Soil and water assessment tool (SWAT) user's manual version 2000. Grassland Soil and Water Research Laboratory, ARS, Temple, TX; 2001.
80. Kuchar L, Iwański S, Jelonek L, Szalińska W. A modeling framework to assess the impact of climate change on river runoff. Meteorol Hydrol Water Manage Res Oper Appl. 2014;2(2).
81. Visser PM, Verspagen JMH, Sandrini G, Stal LJ, Matthijs HCP, Davis TW, Paerl HW, Huisman J. How rising CO_2 and global warming may stimulate harmful cyanobacterial blooms. Harmful Algae. 2016;54:145–59.
82. Janssen ABG, Janse JH, Beusen AHW, Chang M, Harrison JA, Huttunen I, Kong X, Rost J, Teurlincx S, Troost TA, van Wijk D, Mooij WM. How to model algal blooms in any lake on earth. Curr Opin Environ Sustain. 2019;36:1–10.
83. Sawyer CN. Basic concepts of eutrophication. J (Water Pollut Control Federation). 1966;737–44.
84. Mulholland PJ, Best GR, Coutant CC, Hornberger GM, Meyer JL, Robinson PJ, Stenberg JR, Turner RE, Vera- Herrera F, Wetzel RG. Effects of climate change on freshwater ecosystems of the south-eastern United States and the Gulf Coast of Mexico. Hydrol Process. 1997;11(8):949–70.
85. Huang M, Tian Y. An integrated graphic modeling system for three-dimensional hydrodynamic and water quality simulation in lakes. Int J Geo-Inform. 2019;8(1):18.
86. Ji ZG. Hydrodynamics and water quality: modeling rivers, lakes and estuaries. Hoboken, New Jersey: Wiley-Interscience, John Wiley & Sons Inc.; 2008.
87. Kuo JT, Lung WS, Yang CP, Liu WC, Yang MD, Tang TS. Eutrophication modelling of reservoirs in Taiwan. Environ Model Softw. 2006;21(6):829–44.
88. Seiffert R, Hesser F, Schulte-Rentrop A, Seiß G. Potential effects of climate change on the brackish water zone in German estuaries. In: Proceedings of HIC 2012–10th international conference on hydroinformatics, Hamburg; 2012.
89. Dai H, Mao J, Jiang D, Wang L. Longitudinal hydrodynamic characteristics in reservoir tributary embayments and effects on algal blooms. PLoS ONE. 2013;8(7): e68186.
90. Ji D, Wells SA, Yang Z, Liu D, Huang Y, Ma J, Berger CJ. Impacts of water level rise on algal bloom prevention in the tributary of Three Gorges Reservoir, China. Ecol Eng. 2017;98:70–81.
91. Xu H, Yan M, Long L, Ma J, Ji D, Liu D, Yang Z. Modeling the effects of hydrodynamics on thermal stratification and algal blooms in the Xiangxi Bay of Three Gorges Reservoir. Front Ecol Evol. 2021;453.
92. Hua R, Zhang Y. Assessment of water quality improvements using the hydrodynamic simulation approach in regulated cascade reservoirs: a case study of drinking water sources of Shenzhen, China. Water. 2017;9(11):825.
93. Gao X, Zhang S, Sun B, Li N, Liu Y, Wang Y. Assessing the effects of restoration measures on water quality in a large shallow reservoir. Sustainability. 2019;11(19):5347.
94. Ziemińska-Stolarska A, Kempa M. Modeling and monitoring of hydrodynamics and surface water quality in the sulejów dam reservoir. Poland Water. 2021;13(3):296.
95. Sharip Z, Shah SA, Jamin A, Jusoh J. Assessing the hydrodynamic pattern in different lakes of Malaysia, vol. 5. Croatia: IntechOpen; 2018. p. 71–85.
96. Yang Z, Wang T, Voisin N, Copping A. Estuarine response to river flow and sea-level rise under future climate change and human development. Estuar Coast Shelf Sci. 2015;156:19–30.
97. Azadi F, Ashofteh PS, Shokri A, Loáiciga HA. Simulation-optimization of reservoir water quality under climate change. J Water Resour Plan Manag. 2021;147(9):04021054.

98. Iglesias I, Bio A, Melo W, Avilez-Valente P, Pinho J, Cruz M, Gomes A, Vieira J, Bastos L, Veloso-Gomes F. Hydrodynamic model ensembles for climate change projections in estuarine regions. Water. 2022;14(12):1966.
99. Ishikawa M, Gonzalez W, Golyjeswski O, Sales G, Rigotti JA, Bleninger T, Mannich M, Lorke A. Effects of dimensionality on the performance of hydrodynamic models for stratified lakes and reservoirs. Geosci Model Develop. 2022;15(5):2197–220.
100. Chuo M, Ma J, Liu D, Yang Z. Effects of the impounding process during the flood season on algal blooms in Xiangxi Bay in the Three Gorges Reservoir, China. Ecol Model. 2019;392:236–49.
101. Bailey MC, Hamilton DP. Wind induced sediment resuspension: a lake-wide model. Ecol Model. 1997;99(2–3):217–28.
102. Friedrichs MA, Hood RR, Wiggert JD. Ecosystem model complexity versus physical forcing: quantification of their relative impact with assimilated Arabian Sea data. Deep Sea Res Part II. 2006;53(5–7):576–600.
103. Allen JI, Holt JT, Blackford J, Proctor R. Error quantification of a high-resolution coupled hydrodynamic-ecosystem coastal-ocean model: Part 2. Chlorophyll-a, nutrients and SPM. J Marine Syst. 2007;68(3–4):381–404.
104. Ganju NK, Brush MJ, Rashleigh B, Aretxabaleta AL, Del Barrio P, Grear JS, Harris LA, Lake SJ, McCardell G, O'Donnell J, Ralston DK, Signell RP, Testa JM, Vaudrey JM. Progress and challenges in coupled hydrodynamic-ecological estuarine modeling. Estuaries Coasts. 2016;39 (2):311–32.
105. Gupta A, Farjad B, Wang G, Eum H, Dubé M. Integrated environmental modelling framework for cumulative effects assessment. Calgary, AB, Canada: University of Calgary Press; 2021.
106. ASCE (American Society of Civil Engineers). Total maximum daily load analysis and modelling assessment of the practice. Reston, VA: American Society of Civil Engineers; 2017.
107. Wang L, Xu B, Zhang C, Fu G, Chen X, Zheng Y, Zhang J. Surface water temperature prediction in large-deep reservoirs using a long short-term memory model. Ecol Ind. 2022;134:108491.
108. Shourian M, Moridi A, Kaveh M. Modeling of eutrophication and strategies for improvement of water quality in reservoirs. Water Sci Technol. 2016;74(6):1376–85.
109. Xu C, Zhang J, Bi X, Xu Z, He Y, Gin KY. Developing an integrated 3D-hydrodynamic and emerging contaminant model for assessing water quality in a Yangtze Estuary Reservoir. Chemosphere. 2017;188:218–30.
110. Wool T, Ambrose RB, Martin JL, Comer A. WASP 8: the next generation in the 50-year evolution of USEPA's water quality model. Water. 2020;12 (5):1398.
111. Cerco CF, Cole T. User's guide to the CE-QUAL-ICM: three-dimensional eutrophication model. Technical Report EL-95el 5. Vicksburg: U.S. Army Corps of Engineers; 1995.
112. Deltares. D-water quality, versatile water quality modelling in 1D, 2D or 3D systems including physical, (bio)chemical and biological processes User Manual, Version: 5.06; 2020.
113. Hamrick JM. A three-dimensional environmental fluid dynamics computer code: theoretical and computational aspects. The College of William and Mary, Virginia Institute of Marine Science Special Report 317; 1992. 63 pp.
114. DHI (Danish Hydraulic Institute). Coastal and Inland waters in 3D; 2005.
115. Hydroqual. User's guide for RCA, Release 3.0. New Jersey: Hydroqual, Inc.; 2004.
116. Chao X, Jia Y, Shields FD Jr, Wang SSY, Cooper CM. Three-dimensional numerical simulation of water quality and sediment-associated processes with application to a Mississippi Delta lake. J Environ Manage. 2010;91(7):1456–66.
117. Zhang W, Rao YR. Application of a eutrophication model for assessing water quality in Lake Winnipeg. J Great Lakes Res. 2012;38(3):158–73.
118. Bai J, Zhao J, Zhang Z, Tian Z. Assessment and a review of research on surface water quality modeling. Ecol Model. 2022;466: 109888.
119. Zhang L, Lu J, Chen X, Liang D, Fu X, Sauvage S, Sanchez Perez JM. Stream flow simulation and verification in ungauged zones by coupling hydrological and hydrodynamic models: a case study of the Poyang Lake ungauged zone. Hydrol Earth Syst Sci. 2017;21(11):5847–61.
120. Wang Y, Hu W, Peng Z, Zeng Y, Rinke K. Predicting Lake Eutrophication responses to multiple scenarios of lake restoration: a three-dimensional modeling approach. Water. 2018;10 (8):994.
121. Anagnostou E, Gianni A, Zacharias I. Ecological modeling and eutrophication-a review. Nat Resour Model. 2017;30(3):1–27.
122. Obin N, Tao H, Ge F, Liu X. Research on water quality simulation and water environmental capacity in Lushui River based on WASP model. Water. 2021;13(20):2819.
123. Peng S, Fu GYZ, Zhao XH. Integration of USEPA WASP model in a GIS platform. J Zhejiang Univ-Sci A. 2010;11(12):1015–24.
124. Wool TA. WASP8 stream transport-model theory and user's guide; 2017.
125. Ertürk A. Modelling the progress and effects of eutrophication in inland and coastal waters. Aquatic Res. 2019;2(2):92–133.
126. Alam A, Badruzzaman ABM, Ali MA. Assessing effect of climate change on the water quality of the Sitalakhya River using WASP model. J Civ Eng. 2013;41:21–30.
127. Park JY, Park GA, Kim SJ. Assessment of future climate change impact on water quality of Chungju

Lake, South Korea, using WASP coupled with SWAT. JAWRA J Am Water Resourc Assoc. 2013;49(6):1225–38.
128. Camacho RA, Zhang Z, Chao X. Receiving water quality models for TMDL development and implementation. J Hydrol Eng. 2019;24(2):04018063.
129. Wang P, Linker L, Wang H, Bhatt G, Yactayo G, Hinson K, Tian R. Assessing water quality of the Chesapeake Bay by the impact of sea level rise and warming. IOP Conf Ser: Earch Environ Sci. 2017;82: 012001.
130. Cerco CF, Noel MR, Kim SC. Three-dimensional management model for Lake Washington, part II: eutrophication modeling and skill assessment. Lake Reservoir Manage. 2006;22(2):115–31.
131. Wang H, Chen Q, Hu K, La Peyre MK. A modeling study of the impacts of Mississippi River diversion and sea-level rise on water quality of a deltaic estuary. Estuaries Coasts. 2017;40(4):1028–54.
132. Guénand Y, Perga ME, Chanudet V, Bouffard D. Hydropower operations modulate sensitivity to meteorological forcing in a high altitude reservoir. Aquat Sci. 2020;82(3):1–13.
133. Silva FLD, Fushita ÂT, Cunha-Santino MBD, Bianchini Júnior I. Advantages, disadvantages and methods of applying mathematical models to evaluate water quality in reservoirs: a systematic review. Revista Ambiente & Água. 2022;17.
134. Deltares. Delft3D-FLOW, Simulation of multi-dimensional hydrodynamic flows and transport phenomena, including sediments. User manual, Version: 3.15.; 2020.
135. Chen Z, Huang P, Zhang Z. Interaction between carbon dioxide emissions and eutrophication in a drinking water reservoir: a three-dimensional ecological modeling approach. Sci Total Environ. 2019;663:369–79.
136. Lin H, Yano S, Sato Y, Haraguchi N, Maruya Y. Assessment of effects of effluent patterns on the hypoxia development due to climate change in the Ariake Sea, Japan. In: 18th annual meeting of the Asia Oceania geosciences society: proceedings of the 18th annual meeting of the Asia Oceania geosciences society (AOGS 2021); 2022. p. 97–9.
137. Liu S, Ye Q, Wu S, Stive MJ. Wind effects on the water age in a large shallow lake. Water. 2020;12 (5):1246.
138. Zhao L, Li Y, Zou R, He B, Zhu X, Liu Y, Wang J, Zhu Y. A three-dimensional water quality modeling approach for exploring the eutrophication responses to load reduction scenarios in Lake Yilong (China). Environ Pollut. 2013;177:13–21.
139. Tang C, Li Y, Jiang P, Yu Z, Acharya K. A coupled modeling approach to predict water quality in Lake Taihu, China: linkage to climate change projections. J Freshw Ecol. 2015;30(1):59–73.
140. Yindong T, Xiwen X, Miao Q, Jingjing S, Yiyan Z, Wei Z, Mengzhu W, Xuejun W, Yang Z. Lake warming intensifies the seasonal pattern of internal nutrient cycling in the eutrophic lake and potential impacts on algal blooms. Water Res. 2021;188: 116570.
141. EC (European Commission). The European green deal, 11 December 2019, COM (2019) 640 Final; 2019.
142. EC (European Commission). A new circular economy action plan for a cleaner and more competitive Europe, 11 March 2020 COM (2020) 98 Final; 2020.
143. Mannina G, Badalucco L, Barbara L, Cosenza A, Di Trapani D, Gallo G, Laudicina VA, Marino G, Muscarella SM, Presti D, Helness H. Enhancing a transition to a circular economy in the water sector: the eu project wider uptake. Water. 2021;13(7):946.
144. Bakan B, Bernet N, Bouchez T, Boutrou R, Choubert JM, Dabert P, Duquennoi C, Ferraro V, García-Bernet D, Gillot S, Mery J, Rémond C, Steyer JP, Trably E, Trémier A. Circular economy applied to organic residues and wastewater: research challenges. Waste Biomass Valorization. 2022;13 (2):1267–76.
145. EU (European Union). Regulation (EU) 2020/741 of the European Parliament and of the Council of 25 May 2020 on minimum requirements for water reuse (Text with EEA relevance), vol. 177; 2020.
146. Gelda RK, Mukundan R, Owens EM, Abatzoglou JT. A practical approach to developing climate change scenarios for water quality models. J Hydrometeorol. 2019;20(6):1197–211.
147. Liu B, Xia J, Zhu F, Quan J, Wang H. Response of hydrodynamics and water-quality conditions to climate change in a Shallow Lake. Water Resour Manage. 2021;35(14):4961–76.

Determination of Risky Areas for Diffuse Pollution Transport Using GIS: Porsuk Dam Basin Example

Firdes Yenilmez and Aysegul Aksoy

Abstract

In this study, high-risk areas for the transport of diffuse pollution to Porsuk Dam Reservoir (PDR) are identified using geographic information systems (GIS) tools. PDR is located on the Porsuk River at 25 km west of Eskişehir. Majority of the basin of PDR is sited within the borders of Kütahya and Eskişehir provinces. The water quality of PDR is extremely important because it is a drinking water supply for Eskişehir. Yet, the water quality of Porsuk River and PDR have been deteriorated and the trophic state of the reservoir has been declared as hypereutrophic or eutrophic. Previous studies have shown the importance of non-point pollution sources as well as point sources on the pollution, and therefore the need for their control. In this regard, this study aims to identify the high-risk areas for the transport of pollutants through surface drainage. Slope, erosion, hydrogeological permeability, land use, and great soil groups are used to determine the high-risk areas for the transport of pollution loads from agricultural and mining activities within the watershed of PDR. Large parts of the high-risk areas are located at the north of PDR and at the northern part of the basin. In that area, land uses constitute of agricultural areas, magnesite, manganese, and chromium mines, rural areas, and forests. In order to control the water quality in PDR, activities at identified high-risk areas should be managed and necessary precautions should be taken to minimize pollution transport such as applying filtration strips.

Keywords

Diffuse pollution · Geographic information systems · Porsuk Dam Basin · Risky areas · Water pollution

1 Introduction

There is a significant pollution pressure on water resources as a result of population growth, economic development, intensive agricultural activities, urbanization rate, and growing tourism sector [1]. Moreover, the pressure surges with the impacts of climate change as well. According to the IPCC Climate Change Report published in 2022 [2], risks associated with physical water availability and water-related hazards will continue to increase in mid- to long-term in all

F. Yenilmez (✉)
Department of Environmental Engineering, Akdeniz University, 07058 Antalya, Türkiye
e-mail: firdesyenilmez@akdeniz.edu.tr

A. Aksoy
Department of Environmental Engineering, Middle East Technical University, 06800 Ankara, Türkiye
e-mail: aaksoy@metu.edu.tr

A. M. Bahadir et al. (eds.), *Hydrology and Urban Water Supply*, Water and Wastewater Management,
https://doi.org/10.1007/978-3-031-72589-0_2

assessed regions, with greater risks at higher global warming levels. Challenges for water management will be worsened in the near-, mid-, and long-terms, depending on the magnitude, rate, and regional characteristics of future climate change [2]. Situation will be particularly challenging for regions with constrained resources for water management (at high confidence level) [2]. Considering these difficulties and projections, sustainable management of existing water resources will become more critical to cope with the expected water scarcity in the coming years. European Union Water Framework Directive (numbered 2000/60/EC) and the surface water quality-related regulations include measures to achieve good water status in water resources. These regulations set criteria for determining the quality and classification of surface waters in line with sustainable development goals that take the balance of protection and different uses into account [3–5]. In terms of water quality, environmental goal is to reach the status of "good water quality" in terms of biological, hydromorphological, general chemical, and physico-chemical quality parameters, as well as priority substances and certain pollutants at the highest level [1].

In order to achieve environmental targets, it is necessary to determine the pressures from point and diffuse sources by first performing basin characterization. Majority of the pressure on the quality of both surface and ground water may originate from diffuse pollution sources. Diffuse pollution sources can be caused by a variety of land use activities like agriculture, mining, forestry, etc. It is generally easier to manage point pollution sources because they can be pinpointed and controlled through regulations and by efforts of regulators. However, management of diffuse (non-point source) pollution sources is relatively challenging. There may be difficulties to identify and regulate the origin of non-point sources [6, 7]. Yet, transport of diffuse pollution to surface waters via runoff can be reduced by taking some measures. For example, control options can be identified after evaluating sensitive areas for diffuse pollution transport. GIS can be used to identify such areas and non-point pollution sources effectively [8–12]. This would help in assessment and better management of water quality.

One of the basins in which water quality of surface waters has been under stress due to diffuse pollution transport is the Porsuk Dam Basin. Previous studies conducted related to Porsuk Dam Basin have revealed the necessity of focusing on diffuse pollution sources to protect and improve the surface water quality in the basin. In this regard, Koçal (2006) determined the loads originating from point and non-point sources. In the study, the eutrophication problem in PDR was evaluated via GROWA model [13]. Then, the maps of total annual evapotranspiration, groundwater recharge, and runoff were constituted using the model. 50–200 **mm/a** runoff was estimated from agricultural areas. It was mentioned that the major pollution sources for the reservoir were agricultural activities and fertilizer usage in the area. Therefore, it was suggested to minimize the fertilizer usage in the basin to reduce the transport of excess nutrients to Porsuk River and the reservoir.

Küçük (2013) used ArcSWAT to display the comparative effects of different agricultural practices in the basin of PDR. Vulnerable areas for phosphorus transport were identified by applying different scenarios. In the scenarios, different fertilizer and irrigation rates, land use, and point source phosphorus load reductions were considered. Küçük (2013) stated that on top of 50% reduction in fertilizer usage, 20% reduction in irrigation should be applied to obtain 40% decrease in the overall soluble phosphorus load to the PDR [14].

Bayazıt et al. (2019) investigated the trends in the minimum, average, and maximum flows in 19 flow gauging stations in Porsuk Basin by Mann–Kendall test. The data between 1961 and 2013 was evaluated. Decreasing trends were observed in average and maximum flows in most of the basin. Attention was drawn to the expected water scarcity-related problems in the basin in the future [15].

Previous studies have shown that diffuse pollution is essential for the protection and improvement of the water quality in PDR. In this regard, the determination of high-risk areas for diffuse pollution transport would be beneficial to aid in efforts to improve the water quality in PDR. This study implements a GIS-based methodology to determine the high-risk areas for the transport of diffuse pollution causing nutrient loss to surface waters located in Porsuk Dam Basin through surface runoff. The results of this study can be beneficial to decision-makers for better management of water quality in the basin through identifying the areas that may require special attention.

2 Study Area

GIS-based determination of risky areas for diffuse pollution transport was applied for PDR. Porsuk Dam is located on the Porsuk River, at 25 km west of Eskişehir (Fig. 1). The stage-wise construction of the dam was completed in 1972. The dam was designed to control floods, irrigate the Eskişehir and Alpu Plains, and provide drinking water to Eskişehir, the 12th most populated city in Türkiye with the 2021 population of 898,369 [16, 17]. The volume of the dam is 525 million m^3 for the average water depth. PDR

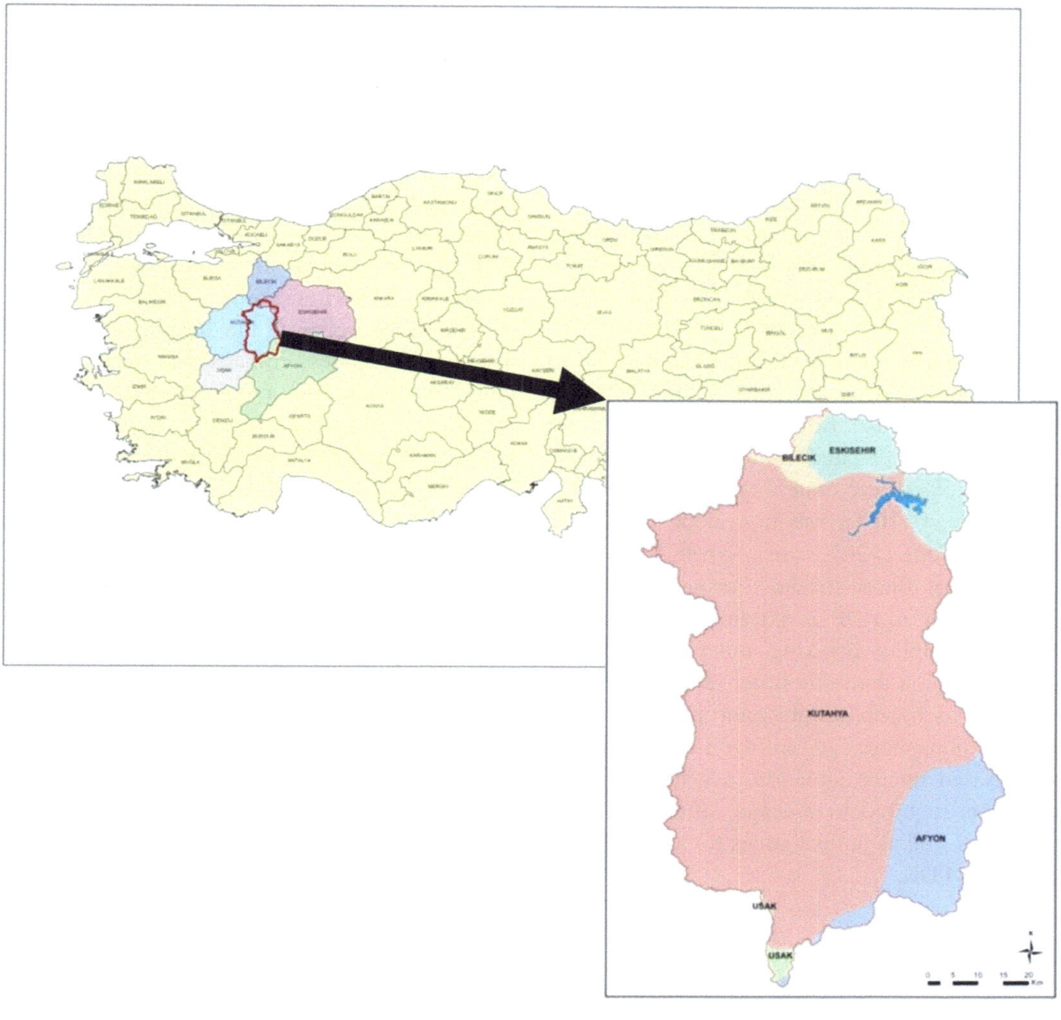

Fig. 1 The location and the watershed of Porsuk Dam Reservoir (PDR)

supplies irrigation water for 16,600 ha of agricultural land [18]. The drainage and surface areas of PDR are about 5104 km^2 and 24.32 km^2, respectively. The depth of water in the reservoir ranges between 4.5 and 20 m [13].

PDR is mainly fed by the Porsuk River, Kunduzlu Stream, and the water input from the Kargın Regulator operated by the State Hydraulic Works [19]. The only outlet of the dam is the Porsuk River. The average hydraulic retention time in the reservoir is 5.3 years [20]. A major portion of the PDR watershed is within the borders of Kütahya province. Relatively smaller portions are within the provincial areas of other cities including Eskişehir, Afyon, Bilecik, and Uşak. Therefore, the water quality in the reservoir is affected by point and diffuse pollution sources arising from several provinces. Land uses, topography, and slopes within the PDR watershed are important especially in evaluating the water quality and impacts of diffuse sources. Forests and agricultural areas constitute the majority of land uses in the PDR Watershed. While forests occupy 38% of the total watershed area, agricultural lands cover 42%. The rest of the total area is composed of settlements, industrial areas, and mining sites. About 61% of the watershed has an elevation within the range of 1000–1200 m. Lower elevation regions (<1000 m) are located around the PDR and covers 11.5% of the total area of the watershed. Low slope (0–10%), medium slope (10–20%), and high slope (>20%) areas cover 52%, 23%, and 25% of the watershed, respectively [21]. Water quality in PDR is extremely important as the dam supplies drinking water to Eskişehir. Despite being a domestic water resource, PDR is classified as hypereutrophic/eutrophic with high aqueous phosphorus and nitrogen concentrations [20]. Based on the situation, a watershed-scale control is required to manage diffuse pollution sources as well as point sources for a better water quality in PDR.

3 Methodology

In order to determine the risky areas that contribute to diffuse pollution, first the watershed was divided into sub-basins to estimate the potential of such pollution in different parts of the watershed. For this purpose, ArcSWAT [22] was used as a tool. ArcSWAT is a GIS-integrated version of SWAT, a river basin-scale model that can predict nutrient and sediment loads arising from point and diffuse pollution sources. SWAT has been widely used in the world to simulate surface runoff, the effect of different land use management practices on water quality, and the effect of climate on the hydrology of watersheds [23–27]. However, in this study, ArcSWAT was used only to determine the sub-basins within the watershed of PDR. Then, sub-basins were evaluated for their potential contribution to diffuse pollution transport based on selected parameters and their quantities/states using ArcGIS, as will be provided. Finally, the most important areas within the watershed were determined based on their higher potential for contribution to diffuse pollution and its transport.

ArcSWAT requires the designation of land use, soil properties, climate data (daily precipitation, maximum/minimum air temperature, solar radiation, wind speed, relative humidity, etc.), groundwater, water use, land use management information, soil chemistry, and water quality data to make predictions. A single watershed or a system of multiple hydrologically connected watersheds can be simulated by ArcSWAT. Each watershed can be divided into a number of sub-basins. Then sub-basins are converted into one or more hydrologic response units (HRUs) based on land use, soil, and slope characteristics within the sub-basins. HRUs are lumped land areas within a sub-basin that are comprised of similar land cover, soil, and management combinations [23, 28]. In order to determine the sub-basins in the PDR watershed, a Digital Elevation Model

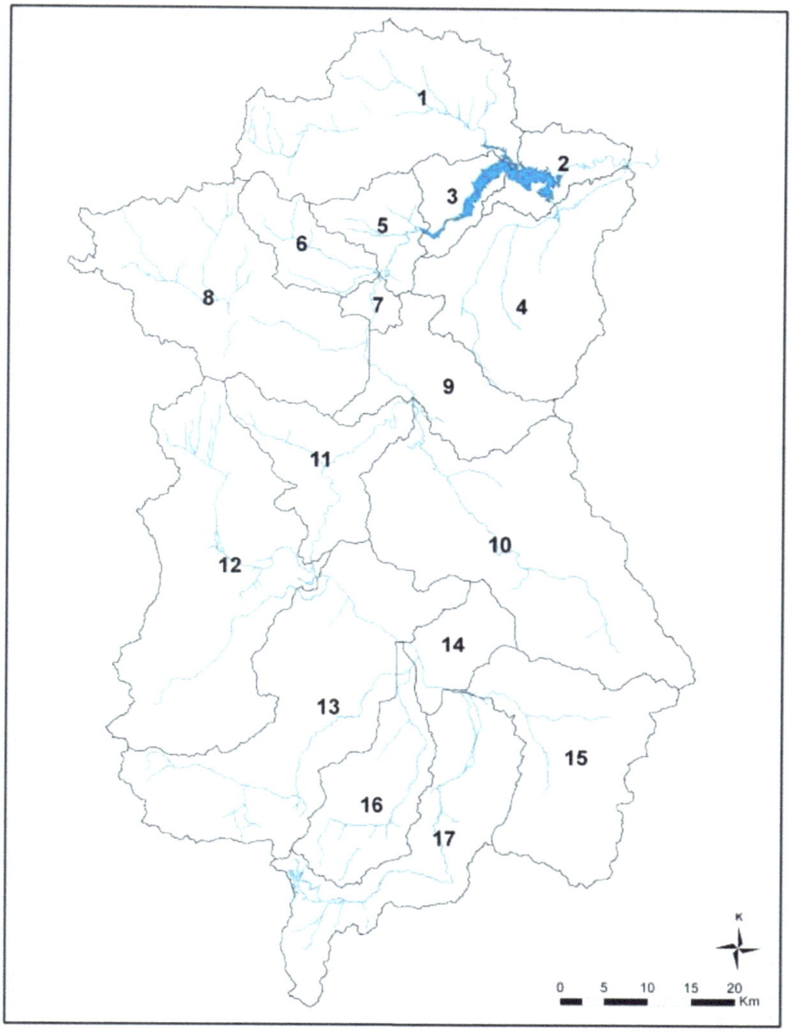

Fig. 2 Sub-basins constituted for PDR watershed

(DEM) of 20 × 20 m resolution was used. Using the DEM, the digital stream network of PDR was constructed. The inlets and outlets in the PDR Watershed were defined. Overall, the watershed was divided into 17 sub-basins as depicted in Fig. 2.

Following the division of the watershed into sub-basins, levels were assigned for slope, erosion intensity, hydrogeological permeability, land use, and great soil groups. These levels defined the degree of vulnerability of an HRU and a sub-basin for diffuse pollution transport. Levels were assigned for each parameter or data group depending on their impact on permeability, drainage, and surface runoff (Table 1). Levels were assigned in a descending order such that a high value indicated higher contribution to pollution and transport. For example, bare rock and rubble exhibited lower infiltration and higher surface runoff potential compared to other great soil groups. Therefore, it had the highest level within its data group. In determining the levels

Table 1 Levels assigned for each data group used in high-risk area determination for diffuse pollution transport

Evaluation parameters	Value or state	Levels
Land slope (%)	0–10	1
	10–20	2
	20–30	3
Erosion intensity	1st degree	1
	2nd degree	2
	3rd degree	3
	4th degree	4
Hydrogeological permeability	Pervious	1
	Semi-pervious	2
	Impervious	3
Land use	Surface water, marsh, reeds	1
	Forest, lemur, heath, brushwood, wooded	2
	Agriculture, pasture, grassland	3
	Residential, industry, military etc	4
	Rocky, stony, openness, mining	5
Great soil groups	Hydromorphic, alluvial	1
	Delime brown forest	2
	Colluvial	3
	Brown forest, auburn, reddish auburn, high mountain-meadow	4
	Rocky, stony, openness, mining	5
	Delime brown, brown	6
	Bare rock and rubble	7

for different parameter groups, the study of Küçükönder and Karabulut (2007) was used as well [29]. The compositions (i.e., percentages of clay, sand, silt) of given soil types were taken into account to establish the levels for great soil groups.

In order to determine the areas that have higher potential of contribution to diffuse pollution transport, all data layers were overlapped using ArcGIS. Following the layer overlapping, areas highlighted were evaluated according to their parameter levels. Areas containing aggregated levels of 2 or 3 for slope, 2, 3, or 4 for erosion intensity, 2 or 3 for hydrogeological permeability, 3, 4, or 5 for land use, 4, 5, 6, or 7 for great soil groups contributed to high-risk areas for surface drainage and enhanced diffuse pollution.

4 Results and Discussion

Based on the methodology and assumptions used, the high-risk areas for diffuse pollution transport in the basin of the PDR are depicted in Fig. 3. Grey areas in the figure show the highly risky regions in terms of transport of diffuse pollution to surface waters through surface runoff. Contamination of runoff is mainly through agricultural and mining activities. In these regions all parameters considered provide favourable conditions to enhance diffuse pollution transport. As can be seen, highly risky areas are generally located in the north of PDR. Therefore, this area requires special attention as diffuse pollutants have higher potential to reach the dam.

Fig. 3 Risky Areas for diffuse pollution transport within the watershed of PDR

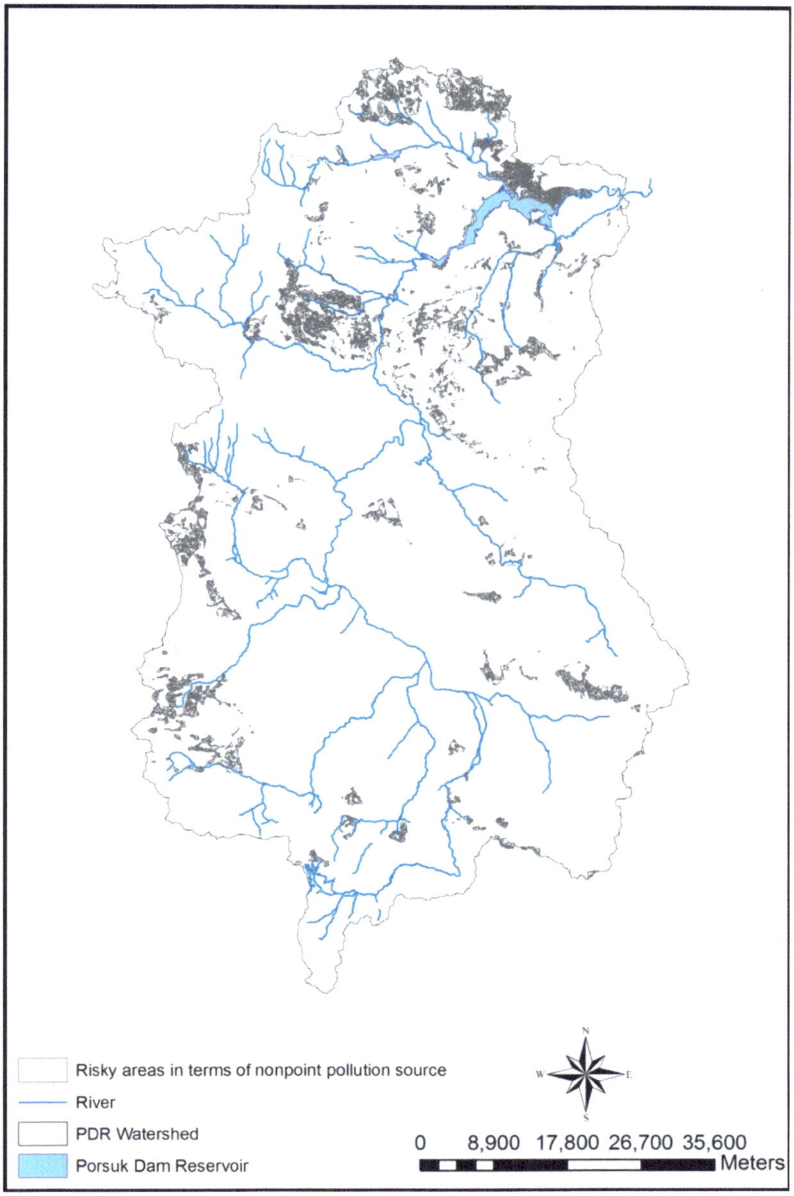

In order to have a whole picture, a map of potential pollution sources in PDR was generated by taking all point and diffuse pollution sources into account. Agricultural areas, mining sites, and solid waste disposal sites were integrated into the pollution map as diffuse pollution sources, whereas industrial facilities, wastewater treatment plants, and a slaughterhouse in the area were marked as point pollution sources. High-risk areas for surface drainage for the whole watershed and potential point and diffuse pollution sources can be seen in Fig. 4.

As depicted in Fig. 4, large parts of the high-risk areas for diffuse pollution transport are located at the north of PDR (close to shores) and at the north border of the watershed. In this area, land uses are mainly comprised of agricultural areas, magnesite, manganese, and chromium

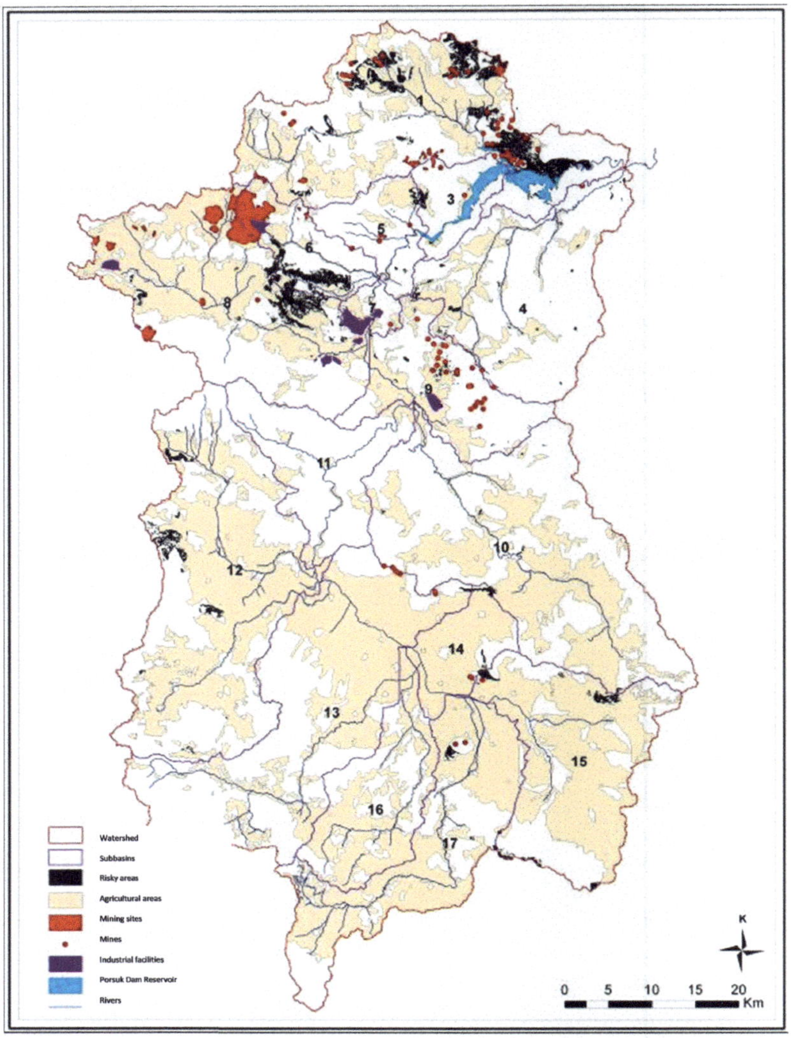

Fig. 4 Risky areas for surface drainage, sub-basins, and potential pollution sources

mining areas, rural areas, and forests. It is possible that solid waste created as a result of mining activities in this region constitutes an issue, as this area has high potential for enhanced diffuse pollution transport. Therefore, special care should be given in the closure plans of mining sites in those areas.

In Fig. 4, high-risk areas around flowing water bodies can be seen as well. In these locations, there is a need to establish storm water management facilities such as establishment of filtration strips. Such facilities should be constructed/applied around Kunduzlu Stream and PDR. In addition, agricultural activities at risky areas for diffuse pollution transport should be handled with caution. As PDR suffers from high nutrient pollution, this may be important. In risky areas for diffuse pollution transport, plants that would require less fertilizer usage and pesticides may be preferred. Moreover, best agricultural management techniques can be applied in agricultural activities to minimize transport through surface drainage.

5 Conclusion

In this study, high-risk areas that may impact the pollution level in PDR through surface drainage were identified using GIS-based tools and relevant data. Various locations within the basin of PDR were found as likely to contribute to the pollution of the reservoir originating from agricultural areas, mining sites, rural areas, and forests. Especially, areas close to the northern shores of the dam were critical.

Advances in data evaluation techniques such as GIS provide valuable resources for better diffuse pollution control and water resources management. With such tools, risky areas that have relatively high potential to contribute to pollutant transport through surface runoff can be found. This information is valuable for decision makers to designate land uses in risky areas to minimize pollution and take precautions to decrease pollution transport. Management alternatives can be better determined in this regard.

References

1. Ministry of Agriculture and Forestry, General Directorate of Water Management. Guidance document for determination of environmental targets to be achieved in order to succeed good status of surface water bodies (Yerüstü Su Kütlelerinin İyi Duruma Ulaşabilmesi İçin Sağlanması Gereken Çevresel Hedeflerin Belirlenmesine Yönelik Rehber Doküman). https://www.tarimorman.gov.tr/SYGM/Belgeler/kamag/Rehber_Doküman_Çevresel_Hedefler.pdf. Last access date: 01 July 2022.
2. IPCC. Summary for policymakers, climate change 2022: impacts, adaptation, and vulnerability. In: Pörtner H-O, Roberts DC, Tignor M, Poloczanska ES, Mintenbeck K, Alegría A, Craig M, Langsdorf S, Löschke S, Möller V, Okem A, Rama B, editors. Contribution of working group II to the sixth assessment report of the intergovernmental panel on climate change. Cambridge University Press; 2022.
3. Water Framework Directive, European Union (WFD E.U.). Establishing a framework for community action in the field of water policy; Directive 2000/60/EC of the European Parliament and of the Council of 23 October 2000; EU: Brussels, Belgium; 2000.
4. Surface Water Quality Regulation in Türkiye. Official Gazette, No. 29327, 30 April Wednesday, Ankara, Türkiye; 2015.
5. Yenilmez F. Statistical evaluation of water quality and meteorological conditions in Karacaören-II Dam Lake. Kocaeli J Sci Eng. 2022;5(1):97–104. https://doi.org/10.34088/kojose.1079170.
6. Campbell N, D'Arcy B, Frost A, Novotny V, Sansom A. Diffuse pollution, an introduction to the problems and solutions. IWA Publishing; 2005. e-ISBN: 9781780402956.
7. Reaney SM, Mackay EB, Haygarth PM, Fisher M, Molineux A, Potts M, Benskin C, McW H. Identifying critical source areas using multiple methods for effective diffuse pollution mitigation. J Environ Manage. 2019;250: 109366. https://doi.org/10.1016/j.jenvman.2019.109366.
8. Tim US, Mostaghimi S, Shanholtz VO. Identification of critical nonpoint pollution source areas using geographic information systems and water quality modeling. JAWRA J Am Water Resourc Assoc. 1992;28:877–87. https://doi.org/10.1111/j.1752-1688.1992.tb03189.x.
9. Basnyat P, Teeter LD, Lockaby BG, Flynn KM. The use of remote sensing and GIS in watershed level analyses of non-point source pollution problems. For Ecol Manage. 2000;128(1–2):65–73. https://doi.org/10.1016/S0378-1127(99)00273-X.
10. Sivertun A, Prange L. Non-point source critical area analysis in the Gisselö watershed using GIS. Environ Modell Softw. 2003;18(10):887–898. https://doi.org/10.1016/S1364-8152(03)00107-5
11. Grayson R, Kay P, Foulger M. The use of GIS and multi-criteria evaluation (MCE) to identify agricultural land management practices which cause surface water pollution in drinking water supply catchments. Water Sci Technol. 2008;58(9):1797–802. https://doi.org/10.2166/wst.2008.569.
12. Orlikowski D, Bugey A, Périllon C, Julich S, Guégain C, Soyeux E, Matzinger A. Development of a GIS method to localize critical source areas of diffuse nitrate pollution. Water Sci Technol. 2011;64(4):892–8. https://doi.org/10.2166/wst.2011.672.
13. Koçal M. Investigation of water quality in Porsuk Dam lake using a mathematical model. M.S. Thesis, Gebze Institute of Technology, The Department of Environmental Engineering; 2006.
14. Küçük, E. Investigation of non-point source pollution potential in the watershed of Porsuk Dam reservoir. MSc. Thesis, Middle EastTechnical University, Department of Environmental Engineering; 2013.
15. Bayazıt Y, Bakış R, Koç C. Mapping of stream flow trends in Porsuk basin using GIS environment. J Geosci Environ Protect. 2019;7:58–66. https://doi.org/10.4236/gep.2019.79005.
16. TÜİK. Address based population registration system results; 2022. https://www.tuik.gov.tr/indir/duyuru/favori_raporlar.xlsx. Last access date: 01 July 2022.
17. Ministry of Environment and Forestry. Environmental status report of Eskişehir; 2006. p. 26–7.
18. Eskişehir Governorship. Environmental status report, Eskişehir provincial environmental and urban administration; 2020. p. 167. https://webdosya.csb.gov.tr/

db/ced/icerikler/esk-seh-r_-cdr2019-20200814100218.pdf.
19. Muhammetoğlu A, Muhammetoğlu H, Oktaş S, Özgökçen L, Soyupak S. Impact Assessment of different management scenarios on water quality of Porsuk River and Dam system-Turkey. Water Resour Manage. 2005;19:199–210. https://doi.org/10.1007/s11269-005-3473-z.
20. AKS Consulting. Final report of the protection of the Porsuk Dam as a source of drinking and utility water and determination of special provisions (İçme ve kullanma suyu kaynağı olarak Porsuk Barajı'nın korunması ve özel hüküm belirlenmesi çalışması nihai raporu). Ministry of Environment and Urbanization, Ankara; 2012.
21. Yenilmez F, Düzgün Ş, Aksoy A. Evaluation of 3-D spatial distribution of dissolved oxygen concentrations in a eutrophic lake. Environ Sci Pollut Res. 2023. https://doi.org/10.1007/s11356-023-26143-w.
22. SWAT. The soil & water assessment tool; 2022. https://swat.tamu.edu/.
23. Bulut E. Assessment of management policies for lake Uluabat basin using AVSWAT. M.S. Thesis, Middle East Technical University, The Department of Environmental Engineering; 2005.
24. Özcan Z, Kentel E, Alp E. Evaluation of the best management practices in a semi-arid region with high agricultural activity. Agric Water Manag. 2017;194:160–71. https://doi.org/10.1016/j.agwat.2017.09.007.
25. Li Y, Wang H, Deng Y, Liang D, Li Y, Shen Z. How climate change and land-use evolution relates to the non-point source pollution in a typical watershed of China. Sci Total Environ. 2022;839: 156375. https://doi.org/10.1016/j.scitotenv.2022.156375.
26. Song Y, Wu Y, Sun C, Zhao F, Hu J, Chen J, Qiu L, Lian Y. Spatiotemporal features of pollutant loads in the Yan River Basin, a typical loess hilly and gully watershed in the Chinese Loess Plateau. Geosci Lett. 2022;9(1). https://doi.org/10.1186/s40562-022-00220-3.
27. Uy HC, Quebada AEE, Amante MLP. Hydropower generation potential of Samar river system based on GIS and SWAT model. GMSARN Int J. 2023;17(1):33–9.
28. Neitsch SL, Arnold JG, Kiniry JR, Srinivasan R, Williams JR. Soil and water assessment tool input/output file documentation version 2009. Texas Water Resources Institute Technical Report No. 365, Texas A&M University System; 2010.
29. Küçükönder M, Karabulut M. Landfill site selection in Kahramanmaraş using multiple criteria analyses. Turk J Geograph Sci. 2007;5(2):1–23.

Artificial Intelligence-Based Sea Level Prediction: A Review of Recent Studies from a Coastal Engineering Perspective

Erkin Tas, Rifat Tur, and Ali Danandeh Mehr

Abstract

Sea level changes is one of the most noteworthy events that affected the World population living in coastal regions. Therefore, sea level change prediction on coastal regions are extremely important challenges for management of coastal zones and mitigating the adverse impact of its variations. As one of the major indicators of global climate change, several sea level prediction studies have been published by the researchers using tide gauge observations. This chapter presents a review of the recent studies on coastal sea level change across different coastal zones of the World. Based on the review, most advanced and recent methodologies for sea level prediction were presented. Also, the results of the studies were interpreted, and main outcomes have been summarized in this chapter. Our review showed that utilizing AI-based sea level prediction modelling require to combine various data combinations with advance statistical analysis techniques. Artificial Neural Networks (ANN)-based models was found as the most commonly used artificial intelligence method in sea level prediction studies. In addition, meteorological factors such as temperature and air pressure have been reported by the authors as the increase prediction accuracy. Lastly, future directions of the sea level prediction studies were interpreted.

Keywords

Sea level · Prediction · Artificial intelligence · Machine learning · Meteorological factors

1 Introduction

Climate and its dynamics has never been stationary ever since. Annual or seasonal movements of the Earth, change in axial tilt, change in rate of gases in the atmosphere, and change on the earth surface such as land uplift phenomenon or tectonic movements have been determinants of short and long-term changes in climate. All these changes are part of the natural processes that have been going on for millions of years on the Earth. On the other hand, the negative pressure of anthropogenic activities on the atmosphere is quite high in the last 100–150 years with the industrial revolution. In addition, world population has increased day by day and has reached 7 billion in late October 2011. We are currently added 80 million people per year to the

E. Tas · R. Tur (✉)
Department of Civil Engineering, Faculty of Engineering, Akdeniz University, Antalya, Turkey
e-mail: rifattur@akdeniz.edu.tr

A. D. Mehr
Department of Civil Engineering, Antalya Bilim University, Antalya, Turkey

A. M. Bahadir et al. (eds.), *Hydrology and Urban Water Supply*, Water and Wastewater Management,
https://doi.org/10.1007/978-3-031-72589-0_3

planet [1]. Increasing population and anthropogenic activities have led to an increase in greenhouse gases (GHG) released into the atmosphere. Global warming is the name given to the increase in average temperatures measured throughout the year on earth because of the GHG emissions. Due to increasing GHG emissions, climate change has become one of the biggest threats facing humankind during twenty-first century [2, 3]. The ever-escalating emissions of GHG can be traced to all parts of the planet and to all sections of society and the economy [1]. Climate change effects the natural environment of Earth resulting in atmospheric temperature rise, water temperature rise, sea level rise, tropical cyclone intensification, precipitation change, ocean current change and causes serious socio-economic impacts on landscape, ecosystem, recreation, agriculture, water sources, aquaculture, and infrastructure [4]. Recently, catastrophic disasters such as devastating flooding in South America, and Southeast Asia, wildfires in Australia, devastating hurricanes in Western United States, and devastating cyclones in Africa continent, South Asia, and Pacific region clearly show us the effects of climate change in our planet [5].

To emphasize climate change and climate related threats scientist from 50 nations met at the First World Climate Conference in Geneva 40 years ago. Then, similar meetings have been made through the 1992 Rio Summit, the 1997 Kyoto Protocol, and 2015 Paris Agreement to draw attention of policy makers on climate change. Focusing on academic community, more than 11,000 scientists from 153 countries around the world have issued a warning declaring a "climate emergency" [5]. Ripple et al. [6] mentioned that more than 2,800 scientist, 1,990 jurisdictions, and 34 countries declared or recognized a climate emergency. Besides all this, climate activists around the world covered extensively in the media raising societal awareness. For this reason, climate change has become a phenomenon that concerns all society and institutions today.

Sea level change (SLC) is well-known and one of the most significant effects of climate change. According to a special report published by International Panel of Climate Change (IPCC) [7], the global sea level has increased 1.4 mm/year, 2.1 mm/year, 3.2 mm/year, and 3.6 mm/year over the period of 1901–1990, 1970–2015, 1993–2015, and 2006–2015, respectively. Also, Hay et al. [8] reported that global sea level rise reached an upward trend rate of 3.0 ± 0.7 mm/year during the period of 1993–2010. A recent review [9] showed that most of the researchers verified that global sea level rise is continuing and accelerating day by day. Global SLC are influenced by thermal expansion, melting of glaciers, land water storage changes [10, 11], whereas local SLC are influenced by land subsidence, ocean dynamics, and climate change [12]. No doubt, academic community verify that SLC is real and effected by climate change.

Considering that most of the World's population has gathered in coastal areas throughout the history, coastal area and coastal structures are directly influenced by the SLC [13]. Focusing on the impacts of SLC on densely populated coastal zones, more frequent storm surge [14], higher wave [15], beach erosion and inundation of low-lying regions [16, 17] are experienced. In addition, SLC has an adverse effect on port operations, and stability of coastal structure in changing water depths. Therefore, analysis of the SLC due to Climate change is a vital coastal engineering challenge for the long-term performance of coastal structures. On the other hand, prediction of SLC is another significant issue for coastal engineering due to three major reasons as (1) developing early warning system for storm surge and coastal flooding, (2) operation, planning, or maintenance of coastal structures, and (3) renewable energy such as tidal dams or turbines. Storm surges is the rising of sea level due to meteorological changes such as high winds and low atmospheric pressures [18]. Mohanty et al. [19] reported that there is an upward trend in the frequency of severe cyclones. Vitousek et al. [20] emphasized that numerous low-lying coastal regions are vulnerable to storm surge and flooding due to SLC depending on atmospheric forces. Also, coastal structures such as seaports are designed to be resilient to various stresses along their life span, but they can be vulnerable to extreme meteorological events that may cause disruption to operations [21]. SLC depend on atmospheric forces are the major affect the reliability of ports and logistic chain. Therefore, operation of port can be interrupted by these natural disasters. Lastly, the slope of the shoreline is critical to determining the degree to which SLC will affect it. As the sea level gradually rises, the shoreline retreats, and structures that were once thought to be safe from wave attack become vulnerable [1]. Afore-mentioned reasons reveals that accurate future sea level prediction is needed to mitigation of SLC impacts on coastal structures.

Wherefore that sea level height in a specific location is the sum of mean sea level (MSL), effect of meteorological factors (MF), and effect of tidal characteristics [22], SLC includes many uncertainties and nonlinearity. MF also referred as to be non-tidal components of sea level height. Therefore, it can be concluded that MSL in a region is the combination of tidal and non-tidal components. The generating mechanism of SLC is highly complex and depending on several variables. Since there is no empirical or deterministic way of evaluating future sea level heights, artificial intelligence (AI)-based models (e.g., Artificial Neural Networks (ANN), Adaptive Neuro Fuzzy Inference System (ANFIS), Genetic Programming (GP), Support Vector Machine (SVM)) are frequently used to predict future values. In the recent years, AI methods in forecasting coastal engineering related issues have become viable such as wave prediction [23–32], and tidal level prediction [33–39]. These two concepts can be confused with each other as the difference between sea level prediction and tide prediction is often not fully understood. Sea level includes contribution of static water mass, and effect of dynamic process such as atmospheric circulations, MF, tides due to gravitation forces, ocean/sea currents, land uplift phenomenon etc. However, tides are the one of the generating factors of SLC. Therefore, these two concepts are very similar but different mechanism. This difference between the SLC prediction and tide prediction should not be forgotten while reading this study.

In this study, most up-to-date AI-based modelling approaches are reviewed considering the importance of SLC prediction for coastal engineering challenges. The literature review on SLC prediction is summarized in Sect. 2. The general characteristics of these studies and the applied models are examined in Sect. 3. Concluding remarks are presented in Sect. 4.

2 Summary of Reviewed Studies

In this part AI-based SLC prediction studies reviewed, and each paper presented with a summary. All studies summarized and evaluated to answer the following questions:

- Where is the study area?
- What is the period of data used?
- What are the input parameters?
- Is there any pre-processing or feature selection method used?
- Which AI-modelling technique(s) were used for the prediction?
- How was model performance evaluated?
- What were the study's findings?

The studies reviewed in this chapter and presented by dividing into two groups. First group (Group-1 Studies) includes studies that use only lagged sea level values to predict future SLC. Second group (Group-2 Studies) includes studies that use lagged MF values and lagged sea level values to predict future SLC. The list of the selected papers which are indexed by web of science (during 2011–2020) can be seen in Table 1 with their basic features.

Tide gauge is the major technique for sea level observations near coast. They can measure sea level heights periodically as well as sea climate related MF such as wind, temperature, pressure, and humidity. In this study, the articles that predict SLC using tide gauge observations were reviewed.

Table 1 The list of reviewed papers indexed by Web of Science during (2011–2022)

Studies	Author(s) (year)	Tide gauge location	Obtained data period or duration	Used data type
Group-1 studies	Shiri et al. (2011) [40]	Australia coast	1991–2002	Hourly
	Pashova and Popova (2011) [41]	Bulgaria coast	1942–1944 and 1960–2008	Daily
	Karimi et al. (2013) [42]	Australia coast	App. 2-years data (16,001 sample)	Hourly
	Cong and Meesad (2014) [51]	Vietnam coast	2006–2007	Hourly
	Kisi et al. (2014) [43]	Japan coast	2004–2006	Hourly
	Kurniawan et al. (2014) [44]	Singapore coast	2004	Hourly
	Minero-Re et al. (2014) [45]	Mexico coast	More than a-year (480 sample)	Daily
	Kaloop et al. (2016) [52]	Egypt coast	2008–2011	Hourly
	Ertuğrul et al. (2017) [46]	Antarctica, Iceland, Italy, Japan, Mauritius, Portugal, and USA coast	Range between 1951 to 2013	Monthly
	Kaloop et al. (2017) [53]	Canada Coast	Different years (1997, 1999, and 2001) short-term observations	Hourly
	Amuah and Boye (2018) [47]	Ghana coast	1965–1984	Monthly
	Imani et al. (2018) [48]	Taiwan coast	2004–2011	Daily
	Altunkaynak and Kartal (2019) [54]	Turkey coast	2004–2006	Daily
	Khaledian (2020) [49]	Iran coast	App. 6-years (73-month sample)	Hourly
	Accarino et al. (2021) [50]	Italy coast	2013–2016	Daily
	Altunkaynak and Kartal (2021) [55]	Turkey coast	Different years (2003, and 2005)	Daily
	Filippo et al. (2012) [63]	Brazil coast	1982–1983	Hourly

(continued)

Table 1 (continued)

Studies	Author(s) (year)	Tide gauge location	Obtained data period or duration	Used data type
Group-2 studies	Gohernejad et al. (2013) [69]	Iran coast	Range between 1979 to 2008	Monthly
	Nitsure et al. (2014) [70]	USA Coast	2 years data	Hourly
	El-Diasty and Al-Harbi (2015) [64]	Saudi Arabia, and Canada Coast	One-month and a week data	Hourly
	El-Diasty et al. (2017) [71]	Saudi Arabia, and Canada Coast	2-month data	Hourly
	Moghadam (2017) [65]	Iran coast	1990–2013	Monthly
	Hieronymus et al. (2019) [72]	Sweden coast	1961–2005	Hourly
	Lai et al. (2019) [73]	Malaysia coast	2007–2017	Monthly
	Roshni et al. (2019) [74]	India coast	1985–2012	Monthly
	Bruneau et al. (2020) [75]	Global	1980–2015	Hourly
	Ishida et al. (2020) [66]	Japan coast	1980–2018	Hourly
	Muslim et al. (2020) [76]	Malaysia coast	2007–2016	Monthly
	Wang et al. (2020) [77]	USA Coast	2015–2018	Hourly
	Zubier and Eyouni (2020) [67]	Saudi Arabia	2011	Daily
	Ishid et al. (2020) [66]	Japan coast	1980–2018	Hourly
	Karsavran and Erdik (2021) [78]	Turkey coast	2004–2006	Hourly
	Raj and Gharineiat (2021) [79]	Australia coast	1993–2019	Hourly
	Tur et al. (2021) [32]	Turkey coast	2014–2015	Hourly

2.1 Group-1 Studies

As early mentioned, AI-based SLC prediction studies are performed by using different inputs. The lagged values of sea level measurement are one of the most frequently used ways of prediction. Several researches have been published to evaluate model performance of different AI-based techniques. For example, in 2011, Shiri et al. [40] compared several AI-based sea level prediction models using the hourly sea level collected from a tide gauge located in Hilarys Boat Harbor. Authors indicated that 12, and 24-h moving average method used to smoothing out high frequency noise in the collected data. Then, Autocorrelation Function (ACF), and Partial

Autocorrelation Function (PACF) analysis were conducted to investigate interrelation between the lagged times of sea level measurements. The model performance of ANFIS, ANN, linear regression (LR), and autoregressive (AR) was evaluated in terms of Correlation Coefficient (R), Root Mean Square Error (RMSE), and Scatter Index (SI) indicators. The results showed that ANFIS, and ANN models outperform to LR, and AR models for sea level prediction. A detailed comparison of the sea level prediction capabilities of different type of ANN models were examined by Pashova and Popova [41]. To this end, daily sea level measurements were obtained from a tide gauge in Burgas Harbor located in Bulgarian coast. Different ANN models were developed based on five different architectures such as Feed-Forward (FF), Cascade-Feed-Forward (CFF), Feed-Forward Time-Delay (FFTD), Radial Basis Function (RBF), and Generalized Regression (GR). ACF analysis was used to select input combination of these models. In addition, multiple linear regression (MLR) model was developed to compare with ANN models. The model assessment was performed by using R, RMSE and SI indicators. Authors indicated that all ANN models produce suitable results. However, the best one can be chosen as FFBP-based ANN models based on the results of the study. Focusing on the performance of MLR model, it was observed that MLR give more accurate results than RBF, and GRNN-based ANN models. Karimi et al. [42] developed Autoregressive Moving Average (ARMA), ANN, and ANFIS models to predict 1, 24, 48, and 72-h ahead sea level height in Darwin Harbor, Australia. To this end, hourly sea level observations were collected from a tide gauge located in Darwin Harbor. The input combinations of lagged sea level values were selected by using MLR. The almost 60% of obtained sea level were used to train developed models, while 40% of sea level values were used to test and validate developed models. The performance assessment of each model was evaluated using R2, RMSE and variance indicators. The results showed that all models accurately predict 1, and 24-h ahead sea level heights. However, it was observed that the ANN and ANFIS models are superior to ARMA models for long-term sea level height prediction. Kisi et al. [43] compared the sea level prediction performance of Gene Expression Programing (GEP), ANN, and ANFIS models. The hourly sea level measurement, and hourly MF measurements data such as rainfall, air temperature, wind speed, and wind direction were derived from Mukho Station located in Sea of Japan. ACF, PACF, and Cross-correlation function (CCF) analysis were performed to select best input combinations for 1, 24, 48, 72, 96, 120-h ahead sea level prediction. R, RMSE, and SI were used to evaluate model performance. It was observed that a quintuple-input model, which uses current and four lagged sea water heights was selected as best input combination. Whereas all models gave accurate results for 1-ahead prediction in terms of statistical indicators, increasing prediction horizon decreases the model accuracy. Kurniawan et al. [44] investigated the prediction performance of GP approach for sea level prediction using observed hourly sea level data in Singapore coast. Average mutual information (AMI) and correlation analysis were used to select optimum input combinations to predict 1, 2, 4, 6, 12, and 24 h ahead sea levels. The performance evaluations were performed in terms of RMSE and Mean absolute error (MAE). Authors indicated that combination of AMI with GP is increase the prediction accuracy. Minero-Re et al. [45] developed non-linear autoregressive neural network (NAR-ANN) and harmonic analysis (HA) models to compare their prediction performance. Firstly, hourly sea level measurements were collected from southern coast of Gulf of California, Mexico. The harmonic components were determined to develop HA models. Also, several NAR-ANN models with different network configurations were developed. The performance of each model was evaluated in terms of R. The results showed that NAR-ANN models gave better prediction results for short-term prediction (<10 days). For the first time Generalized behavioral learning method (GBLM) models were developed by Ertuğrul et al. [46] to predict long-term sea level fluctuations. To this end monthly sea level observations were collected from seven tide gauges which are in different coasts of Earth. The lagged values of sea level observations were used as input combinations. More than one-month ahead predictions of sea level was performed. Also, authors developed ANN models for the same coasts to test the accuracy of GBLM model. The validation of each model was done in terms of RMSE and Mean Absolute Percentage Error (MAPE) statistics. It was observed that GBLM model can accurately predict sea level fluctuations. Amuah and Boye [47] presented Multivariate Adaptive Regression Spline (MARS), and ANN models to predict for one-month ahead sea level. To this end, monthly tidal observations during the period 1965–1984 were collected from Takoradi Harbor tide gauge station, Ghana. 70% of the collected data was used for training stage, while the rest were used for testing stage. R, and MAE statistics were used to compare accuracy of the developed models. The results showed that ANN model produce more accurate sea level prediction than MARS model. Imani et al. [48] developed SVM, RBF, Extreme Learning Machine (ELM), and Relevance Vector Machine (RVM) models to predict 1, 3, and 7-days ahead sea level in Dongshi tide gauge, Taiwan. The daily mean, minimum, and maximum sea level height measurement were collected during the period 2004–2011. The lagged sea level measurements were used as inputs and optimum input combinations were selected using ACF analysis. The performance evaluation was conducted by using RMSE, MAE, and R indicators. It was observed that developed ELM, and RVM models are outperformed the SVM, and RBF models. Also, authors indicated that RVM is a promising alternative method for sea level prediction. Khaledian [49] presented SVM, and ANN models to predict sea level in Caspian Sea coast. The hourly sea level observations were derived from Anzali station. Gamma Test was conducted to observed sea level data for selection of best input combinations. Also, authors indicated that month number and monthly mean sea level values were involved as input parameters to increase model performance. The models were compared using R, RMSE, Nash–Sutcliffe Efficiency (NSE). The results showed that SVM models is superior to ANN models. Also, it was observed that adding auxiliary parameters such as month number and monthly mean sea level height increase the model performance. Accarino et al. [50] developed several Long Short-Term Memory (LSTM) neural network models to compare its prediction capabilities with Southern Adriatic Northern Ionian Forecasting System (SAN-IFS). Sea level measurements were collected from five tide gauges located in Italy Coast. Lagged values of sea level measurements were adopted to the model as input parameters. The results of models were evaluated in terms of R2, and RMSE indicators. The future sea levels up to 3 days were predicted. The results showed that developed LSTM network models produce more accurate prediction results than SAN_IFS forecasts.

On the other hand, different approaches have been used for SLC prediction in recent years. Hybrid models that combine the two or more models together are another use of prediction. Focusing on hybrid AI-based prediction models, our reviewed showed that a few applications exist in the relevant literature. For example, Cong and Meesad [51] presented interval type-2 TSK-FL models using fuzzy c-means clustering algorithms and hybrid chaos firefly algorithm with GP (CFGA) to predict 1, and 12-h ahead sea level heights in Nhatrang Sea, Vietnam. The hourly sea level measurements were collected for the period of 2006–2007. The input combinations of lagged sea level values were used to feed developed models (75% for train, 25% for test). Performance evaluations were done using RMSE, and SI indicators. The results showed that developed model is superior to standalone fuzzy logic (FL), and GP models for prediction of future sea levels. Kaloop et al. [52] compared the prediction performance of neural network auto regressive moving average (NNARMA), and ANFIS models. The hourly sea level measurements were collected from Alexandria tide gauge located in Egypt coast. The collected sea level time-series (during 2008–2011) smoothed via moving average filtering, which is a simple low-pass finite-duration impulse response

(FIR) method. Then lagged sea level values were used as input combinations of prediction models. Authors indicated that using FIR filtering the tidal components of the sea level variations were subtracted. The performance evaluation of the developed models was tested using R, RMSE, and ACF. The results showed that NNARMA model is superior to ANFIS model for sea level prediction. Kaloop et al. [53] compared the sea level prediction accuracy of ANFIS, and Wavelet Neural Network (WNN) models. The required hourly sea level time-series were derived from three tide gauges in Canadian coast. Then several input combinations were selected trial and error process. The lagged values of hourly sea level measurements were used as inputs of the developed models. The performance assessment was performed by using RMSE, MAE, Maximum Absolute Error (XAE), Model Correlation (MC), and Percentage of RMSE (PE) indicators. The results showed that ANFIS model is superior to WNN model for short-term sea level predictions. Hybrid AI-based models such as DWT-FL, and CWT-FL were presented by Altunkaynak and Kartal [54] to predict sea level heights up to 7 days in Bosporus Strait of Marmara Sea, Turkey. The daily sea level data were used which are measured by Japanese Taisei Corporation during Marmaray Project during 2004–2006 for 15 months. Discrete Wavelet Transform (DWT), and Continuous Wavelet Transform (CWT) methods were used to achieve spectral band components of obtained sea level data. Then, lagged sea level values, and decomposed band values were used as the inputs of developed FL models. The performance of each model was tested using RMSE, and NSE indicators. CWT-FL model was found as more accurate than DWT-FL model, and standalone FL model in terms of performance measurement for sea level prediction. Altunkaynak and Kartal [55] developed several AI models to predict future sea level up to 7 days. Daily sea level measurements were obtained from three different station located in Bosporus Strait, Turkey. Firstly, obtained sea level data were separated into spectral bands by using DWT. Then, several MLR models were established to identify input combinations. Three input combination alternatives including lagged sea level heights, and decomposed bands of sea level were used to developed prediction models. DWT was combined with SVM, k-Nearest Neighbors (KNN), Decision Tree (DT) to achieve advance AI-based prediction models. The prediction results of each model were compared using RMSE, and NSE indicators. Authors indicated that SVM, and KNN models gave similar results, whereas DT model results far behind the others.

2.2 Group-2 Studies

Focusing on the MF that influence the SLC, several studies have been published by researchers in literature [56–62]. These studies investigated the statistical relationship between MF and SLC. The findings showed that MF such as wind speed, wind direction, air temperature, and air pressure directly influence non-tidal SLC. The influence of MF and tidal components on SLC has been also investigated by using AI-based models to capture their prediction capabilities on SLC. For example, Filippo et al. [63] developed ANN model to predict real time sea level in Brazilian coast. The hourly sea level and MF data (including wind and atmospheric pressure) were derived from four different tide gauges located in Brazilian coast. Firstly, gaps in the obtained time-series were filled via fast Fourier transform (FFT). Then, tidal harmonic constants (astronomical tide) were determined. The sea level, atmospheric pressure, wind, and astronomical tide were used as the input of the developed ANN model. Authors indicated that calculated harmonic constants are the most important input variable in the sea level prediction in the region. El-Diasty and Al-Harbi [64] compared the sea level prediction performance of traditional harmonic analysis, ANN, and Wavelet Network (WN) models. In the first part of the study, short-term (1 month and a week) sea level measurement from different tide gauges in Canadian and Saudi coast were used to compare the performance of the three different methods. The results revealed that the Wavelet Network method is superior to others. Second part of the study includes short-term (1 month and a week) sea level and MF measurements from three different tide gauges in Saudi Arabia. The lagged values of sea level measurement, air pressure, air temperature, relative humidity, and wind speed and wind direction were used to feed wavelet network as predictor. Authors indicated that the meteorological factors improve the accuracy of the Wavelet Network method by about 20%. Moghadam [65] focused on the identification of MF affecting sea level changes on the northern coast of Persian Gulf. To this end, eight MF were examined. 24-year period hourly water level records of fourteen tide gauge station were used. Concept of monthly anomaly is adopted the study to remove vertical land movement. Then, monthly mean water anomaly is calculated. Principal Component Analysis (PCA) was performed to investigate the relative relation of MF to sea level data. An accurate ANN based algorithm under statistic hypothesis framework was developed to detect of significant predictors. The algorithm revealed nine MF, which can predict up 56% of total variance in the mean water anomaly. Also, it is found that more than half of the predicted variability is manifested by zonal wind, sea surface temperature and sea level pressure patterns.

Ishida et al. [66] developed LSTM network model to predict coastal sea level. Hourly sea level measurements were gathered from Osaka tide gauge located in Japan coast, whereas MF data were gathered from a reanalysis dataset of ERA5. The current and lagged values of wind speed, sea level pressure, surface temperature, azimuth and altitude of sun and moon were used as inputs of the developed model. The performance assessment was done using RMSE, and NSE. Authors indicated that developed model can reflect the effects of gravitational attractions of sun and moon on coastal sea levels. The results showed that global mean air temperature increase the model accuracy. Zubier and Eyouni [67] investigated the prediction capabilities of MF in the Eastern Central Red Sea. The daily sea level measurements were derived from Jeddah tide gauge during 2013. Also, daily MF measurements from a weather station in the Jeddah were obtained for same period. A Nonlinear Auto-Regressive Network (NARX) type ANN model was developed. The combination of lagged values of sea level and MF were selected as predictors and 1-day ahead sea level was selected as predictand of the model. The model performance was assessed using R, and MSE indicators. The results showed that the NARX-ANN model with MF predictands accurately simulate the sea level changes. Tur et al. [68] compared MLR, and ANFIS models to predict 1, 12, and 24-h ahead sea level. To this end, hourly sea level, and MF measurements such as air temperature, and air pressure were obtained from Antalya tide gauge near Southern coast of Turkey during 2014–2015. ACF, PACF, and CCF analysis were performed to select best input combinations. Based on the analysis two scenarios were constructed for predictive modelling. The model performance was evaluated by using NSE, RMSE, MAE, and SI indicators. The results showed that ANFIS is superior to MLR for sea level prediction. Also, authors indicated that adding MF as input parameters increases the performance accuracy of MLR models up to 33%.

On the other hand, several AI-based modelling techniques have been compared by the researchers to reveal their performance capabilities on sea level prediction using MF as predictor. For example, Gohernejad et al. [69] developed hybrid DWT-ANN, and DWT-ANFIS models to simulate future sea level under climate change impacts. The outputs of CGCM3 (3rd version of the coupled Canadian global climate model) under two climate change scenarios A1b and A2 were used to evaluate climate change impact on sea level. To this end, a large set of data such as sea level, geopotential height, zonal wind, relative humidity, sea surface temperature, sea level pressure, and sea precipitable water were derived from four tide gauge points near the coastline of southern Iran. Then, stepwise regression analysis was performed to select optimum input parameters. Geopotential height, sea surface temperature, sea level pressure, and sea precipitable water were selected as input parameters of developed models. DWT was used to decomposes the input and output time-series into smoother time series to be used in simulation models. The performance assessment was done using R, Mean Square Error (MSE), RMSE, and NSE indicators. The results showed that DWT-ANFIS produce better prediction results than DWT-ANN. Also, authors indicated that average sea level rise at the southern coast of Iran is approximately 33 cm, and 59 cm under A1b and A2 scenarios, respectively. Nitsure et al. [70] developed ANN, and GP models to predict sea level anomalies in eastern USA coast. To this end, hourly sea level, harmonically predicted tidal levels, and MF data such as wind speed and wind direction were collected from four tide gauges near the USA coastline. Optimum input selection was done using serial correlations, saliency analysis, and CCF analysis. In addition, authors indicated that trial and error method was used for input selection due to its simplicity. The model performance was evaluated using R, MAE, Mean Squared Relative Error (MSRE), SI, and NSE indicators. Authors concluded that developed models predicts sea levels indirectly by predicting sea level anomalies. It was observed that GP produce more accurate results than ANN based on performance evaluation indicators. El-Diasty et al. [71] compared prediction accuracy of the formerly presented WN model [64] and, traditional HA model, and new hybrid HA-WN model for sea level prediction. The sea level observations were collected from four tide gauges located in Saudi Arabia, and USA coast. Tidal harmonic components were determined using HA. Then, new hybrid model was developed using tidal components, and sea level values. Authors indicated that hybrid HA-WN model is superior to others. Also, it was found that HA-WN model produce more accurate prediction results than standalone HA, and standalone WN with a percentage of 85%, and 20%, respectively. Hieronymus et al. [72] compared the prediction performance of MLR and different ANN models such as TD-ANN, NAR-ANN, and NARX-ANN using both tide gauge station data and remote sensing data. Hourly sea level measurements were derived from nine tide gauges near the Sweden coast during 1961–2005. Also, winds, and sea level pressure data were obtained from ERA-40 dataset. Both sea level and MF were used as predictor of sea level. The results showed that the ANN is superior to MLR model when local sea level measurements and remote dataset were all used, while performance of these two techniques much similar in the other cases. The performance of SVM and GP models were compared by Lai et al. [73] for different long-term prediction horizons of sea level. Monthly mean sea level observations were taken from three tide gauges during 2007–2012 which are located near Malaysia coast. Monthly rainfall, mean cloud cover, and sea surface temperature observations were also adopted to the study as MF. Several SVM models were developed with different functions such as Normalized Polynomial Kernel (NP), RBF, and Pearson Universal Kernel (PUK) to compare ramped half-half (RHH)-based GP models. The model performance was evaluated via R, RMSE, SI, MAE, and MAPE indicators. The results showed that SVM outperforms GP using all MF and sea level measurements as input parameters. Authors indicated that PUK-SVM, and RHH-GP models can be used to predict future sea level. Roshni et al. [74] compared the prediction performance of three AI techniques such as RVM, ELM, and Gaussian Process Regression (GPR) using MF as input parameters. The monthly sea level measurements were derived from a tide gauge in Haldia Port near the Indian coast. The MF variables such as air temperature, sea surface pressure, humidity, precipitation, and geopotential heights were taken from NCEP/NCAR reanalysis dataset. Correlation analysis, and Wavelet Coherence Analysis were performed to select optimum input combinations of MF variables to predict monthly sea level measurements in the location. The model performance evaluated via NSE, R, R2, RMSE, Root Square Error (RSR), Bias, and index of agreement indicators. The results showed that all three AI-based modelling techniques is accurate the monthly sea level prediction. Authors indicated that the prediction performance of GPR model is slightly better than others. Bruneau et al. [75] investigated the global skill of ANN to predict non-tidal SLC (total sea water minus harmonic analysis based on 37 tidal constituents). Hourly sea level data collected from over 600 tide gauges around the world during 1980–2015 were adopted to the study. Also, MF were derived from high-resolution atmospheric reanalysis ERA5 of ECMWF. Each gauge was modelled independently using ANN, and continuous rank probability score (CRPS) was used to assess the performance of models. The randomly selected MF variables (50%) were used to feed the developed model. Authors indicated that the ANN model produce significant results in reconstruction extreme surges. However, it was mentioned that ANN has still lack accuracy to capture strongest surge events. The results showed that developed model is a simple and consistent to assess the uncertainty through a probabilistic prediction of SLC. Muslim et al. [76] investigated the effects of MF to predict future sea level trends for different time horizons such as 2019, 2023, 2028, 2048, and 2068. To this end, sea level measurements were collected from three tide gauges located in Malaysia coast. Also, four MF such as wind speed, wind direction, precipitation, and mean cloud cover were obtained from Malaysia Meteorology Department for the tide gauge locations. Two different scenarios were selected via trial-and-error method. These were grouped as SC1, and SC2. SC1 was constructed using wind speed, and wind direction values, while SC2 was constructed precipitation, and mean cloud cover values. Then, MLP-ANN, ANFIS models based on the mentioned scenarios were developed to predict sea level. The model performance was evaluated in terms of R, RMSE, and SI. The results showed that ANFIS model is superior to ANN model. Afterwards, the sea level trends were analysed using collected sea level data for 2019, 2023, 2028, 2048, and 2068. These trends were compared with the trends obtained based on the results of developed prediction models. For each year trends, comparison was established considering upper limits, lower limits, mean values, and average changes. Authors observed that the predicted SLC for different time periods is higher for SC1 in two tide gauges. Wang et al. [77] presented hybrid wavelet-ANN, and wavelet-ANFIS models to predict sea level. To this end, sea level, tidal level, wind speed, and wind direction measurements during 2015–2019 were collected from two tide gauges near the New Jersey, and Los Angeles coast of USA. Sea level anomalies time-series were constructed by subtracting tidal level time-series from sea level time-series. Then several standalone ANN, and ANFIS models were developed using lagged values of sea-level, sea-level anomalies, and MF values as input parameters. Hybrid wavelet models were developed by considering different types of decomposition of input parameters. The model performance of developed models was tested by NSE, MAE, RMSE, and MSRE indicators. The results showed that hybrid wavelet-ANFIS models improve the prediction accuracy of sea level. Also, authors indicated that the inputs of sea level anomalies, and wind speed provide superior prediction capability compared the traditional lagged sea level-only models. Karsavran and Erdik [78] developed ANN, SVM, and MLR models to predict sea level. The hourly sea level observations were collected from four tide gauges located in Bosporus Strait, Turkey. In addition, measured water surface salinity, temperature, wind speed, and atmospheric pressure were adopted to study as inputs of developed models. Pareto chart was employed to decide optimum input combinations for modelling. The model performance evaluation was done using R, and RMSE indicators. The results showed that ANN model is superior to other models. Also, authors investigated influence of the river discharge on sea level in location. To this end, Danube River discharge measurements were obtained to test its effect on SLC. Authors indicated that the discharge of Danube River by the lag time of 70 days produce the highest performance of ANN with increasing statistical performance. Raj and Gharineiat [79] compared the performance of ANN, and MARS models for sea level trend prediction. 26-years of hourly sea level measurements were derived from four tide gauges near the Australian coast. The mean sea level value, and inverse barometric effect was subtracted from the original sea level measurements to decrease the noise. ACF and PACF analysis were conducted to select optimum lagged sea level values as input parameters. Also, CCF analysis was performed to investigate relationship between sea level measurements and MF measurements such as water temperature, air temperature, and atmospheric pressure. Lagged MF values were selected based on CCF analysis as inputs of the developed models. The prediction results of the models were compared in terms of several statistical indicators. The results showed that ANN model produce slightly better performance than MARS model. Authors applied ANN, and MARS techniques to modelling storm surges associated with cyclones. It was observed that ANN is superior to MARS to predict peak storm surges.

3 Overview of Reviewed Studies

SLC prediction concept used in reviewed studies consist of five major stages (see Fig. 1) as:

- Obtaining sea level and MF from tide gauge stations
- Data pre-processing
- Feature selection
- AI-based modelling
- Performance evaluation

Several sea level and MF measurements were obtained for predictive modelling of SLC from tide gauges across the World. Based on Table 1, it was observed that most of the coastal areas of the World have been studied by the researchers including Canada, Brazil, Mexico, and USA coast in America continent; Bulgaria, Italy, Portugal, Turkey, and Sweden coast in Europa continent; Iran, Japan, Ghana, Mauritius, Malaysia, Saudi Arabia, Singapore, Taiwan, and Malaysia coast in Asia-Africa continents; as well as Antarctica, and Australia continents. Also, Bruneau et al. (2020) [75] studied the global SLC using more than 600 tide gauge stations across the World. However, most of these studies considered the open basin sea/ocean. Although several studies have been published for close basin sea such as Mediterranean Sea [46, 50, 68], Caspian Sea [49], Red Sea [67], they are less in number compared to the studies published on the open basin sea/ocean. Tur et al. [68] stated that it is important to distinguish between open basin sea, and closed basin sea, since SLC in open basin sea is may less affected by oceanographic and meteorological effects. Focusing on the duration of obtained sea level measurements, it was observed that even one or two-month data [64, 71] were used for predictive modelling. However, most of the data were obtained for more than a-year (see Table 1). Several researchers indicated that the long-term sea level observations may not be available due to the high cost of monitoring [64, 68]. Therefore, researchers used AI-based models as they require fewer pieces of data. On the other hand, the dominant type of sea level and MF data were found as hourly data for both Group-1 and Group-2 studies. However, some researchers that investigate the influence of MF on SLC found appropriate the use of monthly data.

AI-based predictive modelling is required some data engineering techniques to prepare data for prediction such as data pre-processing, and

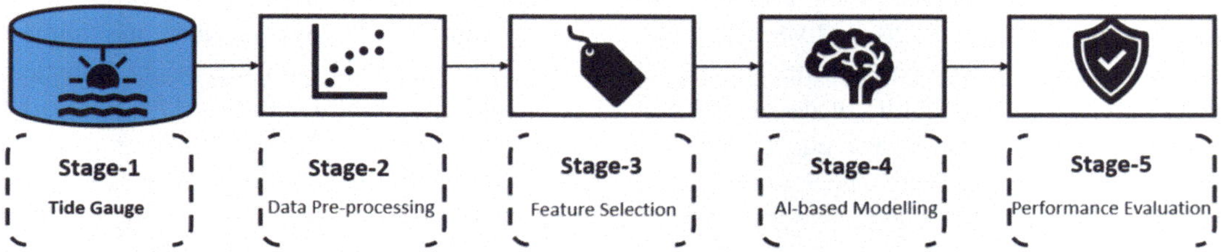

Fig. 1 Major stages of SLC prediction based on literature

feature selection. A flowchart was presented in Fig. 1 to demonstrate the basic stages of SLC prediction based on summarized studies. The tide gauge observations contain missing values in it. To overcome this situation, researchers are mostly used interpolation techniques. Lagged values of observed sea level, and MF were selected as inputs of SLC predictions. The results of El-Diasty and Al-Harbi [64], Ishida et al. [66], Karsarvan [78], Tur et al. [68] showed that using combination of sea level and MF inputs increase the SLC prediction accuracy.

Feature selection or input selection is performed via statistical techniques. It was observed that mostly correlation analysis such as ACF, PACF, and CCF were used to investigate relationship between lagged time of input parameters and output parameters. Although some researchers used basic trial and error method to select optimum input parameters, it was observed that advanced mathematical techniques such as AMI, Gamma Test, FFT, and PCA-based investigations are also adopted to the studies. Kurniawan [44] indicated that AMI-based input combination is produce more accurate prediction results than correlation-based input combinations for GP modelling.

The SLC prediction based on ANN were frequently proposed in literature. Focusing on Group-1 studies it was observed that ANN models are superior to traditional linear time-series analysis models such as MLR, AR, ARMA. Another research presented by Minero-Re [45] showed that ANN is superior to HA which is the traditional sea and tidal level prediction model. Pashova and Popova [41] compared the five ANN models with different architectures to reveal best ANN-type. Authors indicated that FFBP-based ANN is found as most accurate as others. However, Khaledian [49] proposed that SVM give more accurate results than ANN. It is not possible to say one model is the best for SLC prediction due to SLC depends on site-specific observations. However, it can be concluded as ANN is a common and strong AI-based technique for predictive modelling of SLC. On the other hand, most of the developed ANFIS models gave more accurate prediction results than ANN models for Group-2 studies. Another study that published by Nitsure et al. [70] showed that GP is superior to ANN for prediction of SLC. However, Bruneau et al. [75] indicated that ANN can be used for the re-construction of extreme storm surge events excluding SLC prediction. Also, NARX type ANN [67] was found as accurate for daily prediction. Lai et al. [73] compared the several SVM, and GP models using different input combinations. Authors indicated that SVM is superior to GP. Considering the findings of [49, 73], SVM is an accurate AI-based modelling technique for SLC prediction. Another important point for both Group-1, and Group-2 studies is that the hybrid models for example DWT-ANFIS, DWT-ANN, NNARMA or interval Type 2 TSK-FL increase the prediction performance comparing with standalone models. Therefore, hybrid models were suggested by the researchers as an advance modelling technique in SLC prediction. Lastly, the model performance evaluation criteria were overviewed in this study. Although the R, R2 and RMSE indicators were found as most frequently used statistical indicators in the studies, NSE, MAE, SI, and more were adopted to the studies. It was observed that more recent studies tend to include NSE indicators.

4 Main Conclusion and Directions for Future Studies

Considering climate change impact on coastal zones and coastal structures, many researchers have been developed models to predict SLC. In this study, 34 SLC prediction studies were reviewed, and their main findings were summarized. Our review shows that:

- ANN was the most frequently used AI-based modelling technique on SLC prediction studies.
- Due to a strong correlation between SLC and MF, adopting MF to SLC prediction models can improve the model accuracy.
- Hourly observations are used for prediction purpose, whereas monthly observations are

used to investigate the relationship between the MF and SLC.

– It was observed that most of the studies consider short-term prediction intervals. The studies which are consider long-term (more than 1 day) prediction intervals showed that increasing prediction intervals, decrease the model accuracy. Therefore, future studies need to pay more attention to long-term prediction of SLC.

Although AI-based models showed reliable prediction results, combining outputs of different models by ensemble modelling can improve prediction accuracy. In future studies, implementing AI-based ensemble modelling using various data combinations would be instrumental in increasing SLC prediction accuracy.

References

1. Tarbuck EJ, Lutgens FK, Tasa D. Earth science global edition. Pearson; 2015.
2. Kont A, Jaagus J, Aunap R. Climate change scenarios and the effect of sea-level rise for Estonia. Global Planet Change. 2003;36(1–2):1–15.
3. Watson PJ. Status of mean sea level rise around the USA (2020). GeoHazards. 2021;2(2):80–100.
4. Isobe M. Impact of lobal warming on coastal structures in shallow water. Ocean Eng. 2013;71:51–7.
5. Ripple WJ. World scientists' warning of a climate emergency 2021. Bioscience. 2021;71(19):894–8.
6. Ripple WJ, Wolf C, Newsome TM, Barnard P, Moomaw WR. World scientists' warning of a climate emergency. Bioscience. 2020;70(1):8–12.
7. Oppenheimer M, Glavovic B, Hinkel J, van de Wal R, Magnan A, Abd-Elgawad A, Cai R, Cifuentes-Jara M, DeConto R, Ghosh T, Hay J, Isla F, Marzeion BB, Meyssignac B, Sebesvari Z. Sea level rise and implications for low-lying islands, coasts and communities. In: Pörtner H-O, Roberts DC, Masson-Delmotte V, Zhai P, Tignor M, Poloczanska E, Mintenbeck K, Alegría A, Nicolai M, Okem A, Petzold J, Rama B, Weyer NM, editors. IPCC special report on the ocean and cryosphere in a changing climate. In press; 2019.
8. Hay CC, Morrow E, Kopp RE, Mitrovica JX. Probabilistic reanalysis of twentieth-century sea-level rise. Nature. 2015;517:481–4.
9. Adebisi N, Balogun AL, Min TH, Tella A. Advances in estimating sea level rise: a review of tide gauge, satellite altimetry and spatial science approaches. Ocean Coast Manag. 2021;208: 105632.
10. Cazenave A, Nerem RS. Present-day sea level changes: Observations and causes. Rev Geophys. 2004;42(3):RG3001.
11. Oppenheimer M, Glavovic BC, Hinkel J, van de Wal R, Magnan AK, Abd-Elgawad A, Cai R, Cifuentes-Jara M, DeConto RM, Ghosh T, Hay J, Isla F, Marzeion B, Meyssignac B, Sebesvari Z. Sea level rise and implications for low-lying islands, coasts and communities. In: Pörtner H-O, Roberts DC, MassonDelmotte V, Zhai P, Tignor M, Poloczanska E, Mintenbeck K, Alegría A, Nicolai M, Okem A, Petzold J, Rama B, Weyer NM, editors. IPCC special report on the ocean and cryosphere in a changing climate. In press; 2019.
12. Dawod G, Haggag GG. Variations of sea level and atmospheric parameters along the Egyptian coast over 2008–2020. J Sci Eng Res. 2022;9(5):85–100.
13. Cheon SH, Suh KD. Effect of sea level rise on nearshore significant waves and coastal structures; 2016.
14. Samaras AG, Karambas TV. Modelling the impact of climate change on coastal flooding: implications for coastal structures design. J Marine Sci Eng. 2021;9:1008.
15. Lobeto H, Menendez M, Losada IJ. Future behavior of wind wave extremes due to climate change. Sci Rep. 2021;11:7869.
16. O'Grady JG, Hemer MA, McInnes KL, Trenham CE, Stephenson AG. Projected incremental changes to extreme winddriven wave heights for the twenty-first century. Sci Rep. 2021;11:8826.
17. Kirezci E, Young IR, Ranasinghe R, Muis S, Nicholls RJ, Lincke D, Hinkel J. Projections of global-scale extreme sea levels and resulting episodic coastal flooding over the 21st century. Sci Rep. 2020;10:11629.
18. Kaveh NA, Ghaheri A, Chegini V, Nazarali M. Prediction of nontidal sea level variations in the Persian Gulf using data assimilation techniques. Coast Eng J. 2018;60(3):340–55.
19. Mohanty UC, Mohapatra M, Singh OP, Bandyopadhyay BK, Rathore LS, editors. Monitoring and prediction of tropical cyclones in the Indian Ocean and climate change. Springer Science & Business Media; 2013.
20. Vitousek S, Barnard PL, Fletcher CH, Frazer N, Eriksom L, Storlazzi CD. Doubling of coastal flooding frequency within decades due to sea-level rise. Sci Rep. 2017;7(1399):1–9.
21. Christodoulou A, Christidis P, Demirel H. Sea-level rise in ports: a wider focus on impacts. Marit Econ Logistics. 2019;21:482–96.
22. Pugh DT. Tides, surges and mean sea level. John Wiley and Sons Ltd. Wiley, Chichester; 1987. 472 pp.
23. Makarynskyy O. Improving wave predictions with artificial neural networks. Ocean Eng. 2004;31:709–24.

24. Fernandez JC, Salcedo-Sanz S, Gutierrez PA, Alexandre E, Hervas-Martinez C. Significant wave height and energy flux range forecast with machine learning classifiers. Eng Appl Artif Intell. 2015;43:44–53.
25. James SC, Zhang Y, O'Donncha F. A machine learning framework to forecast wave conditions. Coast Eng. 2017;137:1–10.
26. Shahabi S, Khanjani MJ, Kermani MRH. Significant wave height modelling using a hybrid wavelet-genetic programming approach. Coast Harbor Eng. 2017;21:1–10.
27. Ali M, Prasad R. Significant wave height forecasting via an extreme learning machine model integrated with improved complete ensemble empirical mode decomposition. Renew Sustain Energy Rev. 2019;104:281–95.
28. Ali M, Prasad R, Xiang Y, Deo CD. Near real-time significant wave height forecasting with hybridized multiple linear regression algorithms. Renew Sustain Energy Rev. 2020;132:110003.
29. Shamshirband S, Mosavi A, Rabczuk T, Nabipour N, Chau K. Prediction of significant wave height; comparison between nested grid numerical model, and machine learning models of artificial neural networks, extreme learning, and support vector machines. Eng Appl Comput Fluid Dyn. 2020;14(1):805–17.
30. Chen D, Liu F, Zhang Z, Lu X, Li Z. Significant wave height prediction based on wavelet graph neural network. In: IEEE 4th international conference on big data and artificial intelligence (BDAI), Qingdao, China; 2021.
31. Park SB, Shin SY, Jung KH, Lee BG. Prediction of significant wave height in Korea Strait using machine learning. J Ocean Eng Technol. 2021;35(3):336–46.
32. Tur R, Yöntem S. A comparison of soft computing methods for the prediction of wave height parameters. Knowl-Based Eng Sci. 2021;2(1):31–46.
33. Lee W, Resdi TABT. Neural network approach to coastal high and low water level prediction. In: Proceedings of international civil and infrastructure engineering conference (In CIEC); 2014. p. 273–86.
34. Meena BL, Agrawal JD. Tidal level forecasting using ANN. Procedia Eng. 2015;116(1):607–14.
35. Muhammad A et al. Weekly prediction of tides using neural networks. Procedia Eng. 2015;116(Apac):678–82.
36. Salim AM, Dwarakish GS, Liju KV, Thomas J, Gayathri D, Rajeesh R. Procedia Eng. 2015; 116:678–82.
37. Abubakar AG, Mahmud MR, Tang KKW, Husaaini A. Presentation of a wavelet-based harmonic model for tidal level forecasting at Sabah and Sarawak. Geomatics Environ Eng. 2020;14(4):5–23.
38. Elbisy MS, Alijahdali AH, Natto AH, Bakhsh AA, Almaliki AF, Alharthi MA, Hassan AO. Prediction of daily tidal levels along the central coast of eastern Red Sea using artificial neural network. Int J Geomate. 2020;19(76):54–61.
39. Granata F, Di Nunno F. Artificial intelligence models for prediction of the tide level in Venice. Stoch Env Res Risk Assess. 2021;35:2537–48.
40. Shiri J, Makarynskyy O, Kisi O, Dierickx W, Fard AF. Prediction of short-term water levels using an adaptive neuro-fuzzy inference system. J Waterway Port Coast Ocean Eng (ASCE). 2011;137(6):344–54.
41. Pashova L, Popova S. Daily sea level forecast at tide gauge Burgas, Bulgria using artificial neural networks. J Sea Res. 2011;66(2):154–61.
42. Karimi S, Kisi O, Shiri J, Makarynskyy O. Neuro-fuzzy and neural network techniques for forecasting sea level in Darwin Harbor, Australia. Comput Geosci. 2013;52:50–9.
43. Kisi O, Karimi S, Shiri J, Makarynskyy O, Yoon H. Forecasting sea water levels at Mukho station, South Korea using soft computing techniques. Int J Ocean Clim Syst. 2014;5(4):175–188.
44. Kurniawan A, Ooi SK, Babovic V. Improved sea level anomaly prediction through combination of the data relationship and genetic programming in Singapore regional waters. Comput Geosci. 2014;72:94–104.
45. Minero-Re EM, Mohedano JGC, Fernandez ACR, Cabeza JAS. Comparison of artificial neural networks and harmonic analysis for sea level forecasting. Ciencias Marinas. 2014;40(4):251–61.
46. Ertuğrul ÖF, Tağluk ME. Forecsting local mean sea level by generalized behavioural learning method. Arab. J. Sci. Eng. 2017;42:3289–98.
47. Amuah VI, Boye CB. Performance evaluation for mean sea level prediction using multivariate adaptive regression spline and artificial neural network; 2018.
48. Imani M, Kao HC, Lan WH, Kuo CY. Daily sea level prediction at Chiayi coast, Taiwan using extreme learning machine and relevance vector machine. Global and Planetary Change, 161, 211–221. Ghana Mining J. 2018;18(1):1–8.
49. Khaledian MR, Isazadeh M, Biazar SM, Pham QB. Simulating Caspian Sea surface water level by artificial neural network and support vector machine models. Acta Geophys. 2020;68:553–63.
50. Accarino G, Chiarelli M, Fiore S, Federico I, Causio S, Coppini G, Aloisio G. A multi-model architecture based on long short-term memory neural networks for multi-step sea level forecasting. Futur Gener Comput Syst. 2021;124:1–9.
51. Long NC, Meesad P. An optimal design for type-2 fuzzy logica system using hybrid of chaos firefly algorithm and genetic algorithm and its application to sea level prediction. J Intell Fuzzy Syst. 2014;27:1335–46.
52. Kaloop MR, Rabah M, Elnabwy M. Sea level change analysis and models identification based on short tidal gauge measurements in Alexandria, Egypt. Mar Geodesy. 2016;39(1):1–20.
53. Kaloop MR, El-Diasty M, Hu JW. Real-time prediction of water level change using adaptive

neuro-fuzzy inference system. Geomat Nat Haz Risk. 2017;8(2):1320–32.
54. Altunkaynak A, Kartal E. Performance comparison of continous wavelet-fuzzy and discrete wavelet-fuzzy models for water level predcitions at the northern and southern boundary of Bosphorus. Ocean Eng. 2019;186: 106097.
55. Altunkaynak A, Kartal E. Transfer Sea level learning in the Bosphorus Strait by wavelet-based machine learning methods. Ocean Eng. 2021;233: 109116.
56. Cui M, von Storch H, Zorita E. Coastal sea level and the large-scale climate state: a downscaling exercise for the Japanese Islands. Tellus A: A Dyn Meteorol Oceanogr. 1995;47(1):132–44.
57. Wroblewski A. The effect of the North Sea on oscillations of the mean monthly sea levels in Baltic Sea. Cont Shelf Res. 1998;18(5):501–14.
58. Andersson HC. Influence of long-term regional and large-scale atmospheric circulation on the Baltic Sea level. Tellus A: Dyn Meteorol Oceanogr. 2002;54 (1):76–88.
59. Omstedt A, Pettersen C, Rodhe J, Winsor P. Baltic Sea climate: 200 yrs. of data on air temperature, sea level variation, ice cover, and atmospheric circulation. Clim Res. 2004;25(3):205–16.
60. Chen D, Omstedt A. Climate-induced variability of sea level in Stockholm: influence of air temperature and atmospheric circulation. Adv Atmos Sci. 2005;22(5):655–64.
61. Hünicke B, Zorita E. Influence of temperature and precipitation on decadal Baltic Sea level variations in the 20th Century. Tellus A: Dyn Meteorol Oceanogra. 2006;58(1):141–53.
62. Johansson MM, Kahma KK. On the statistical relationship between the geostrophic wind and sea level variations in the Baltic Sea. Boreal Environ Res. 2016;21:25–43.
63. Filippo A, Torres AR Jr, Kjervfe B, Monat A. Application of artificial neural network (ANN) to improve forecasting of sea level. Ocean Coast Manag. 2012;55:101–10.
64. El-Diasty M, Al-Harbi S. Development of wavelet network model for accurate water levels prediction with meteorological factors. Appl Ocean Res. 2015;53:228–35.
65. Moghadam FM. Neural network-based approach for identification of meteorological factors affecting regional sea-level anomalies. J Hydrol Eng (ASCE). 2017;22(3):04016058.
66. Ishida K, Tsujimoto G, Ercan A, Tu T, Kiyama M, Amagasaki M. Hourly-scale coastal sea level modelling in a changing climate using long short-term memory neural network. Sci Total Environ. 2020;720:137613.
67. Zubier KM, Eyouni LS. Investigating the role of atmospheric variables on sea level variations in the eastern central Red Sea using an artificial neural network approach. Oceanalogia. 2020;62(3):267–90.
68. Tur R, Tas E, Haghighi AT, Mehr AD. Sea level prediction using machine learning. Water. 2021;13:3566.
69. Gohernejad H, Shamsai A, Hosseini SA. Vulnerability assessment of southern coastal areas of Iran to sea level rise: evaluation of climate change impact. Oceanalogia. 2013;55(3):611–37.
70. Nitsure SP, Londhe SN, Khare KC. Prediction of sea water levels using wind information and soft computing techniques. Appl Ocean Res. 2014;47:344–51.
71. El-Diasty M, Al-Harbi S, Pagiatakis S. Hybrid harmonic analysis and wavelet network model for sea level prediction. Appl Ocean Res. 2018;70:14–21.
72. Hieronymus M, Hieronymus J. On the application of machine learning techniques to regression problems in sea level studies. J Atmos Oceanic Technol. 2019;36(9):1889–902.
73. Lai V, Ahmed AN, Malek MA, Afan HA, Ibrahim RK, El-Shafie A, El-Shafie A. Modeling the nonlinearity of sea level oscillations in the Malaysian coastal areas using machine learning algortihms. Sustainability. 2019;11(17):4643.
74. Roshni T, Samui P, Drisya J. Operational use of machine learning for sea-level modelling. Indian J Geo Marine Sci. 2019;48(09):1427–34.
75. Bruneau N, Polton J, Williams J, Holt J. Estimation og global coastal sea level extremes using neural networks. Environ Res Lett. 2020;15: 074030.
76. Muslim TO, Ahmed AN, Malek MA, Afan HA, Ibrahim RK, El-Shafie A, Sapitang M, Sherif M, Sefelnasr A, El-Shafie A. Investigating the influence of meteorological parameters on the accuracy of sea-level prediction models in Sabah, Malaysia. Sustainability. 2020;12(3):1–18.
77. Wang B, Wang B, Wu W, Xi C, Wang J. Sea-water-level prediction via combined wavelet decomposition, neuro-fuzzy and neural networks using SLA and wind information. Acta Oceanol Sin. 2020;39 (5):157–67.
78. Karsavran Y, Erdik T. Artificial intelligence-based prediction of seawater level: a case study for Bosphorus strait. Int J Math Eng Manage Sci. 2021;6(5):1242–54.
79. Raj N, Gharineiat Z. Evaluation of multivariate adaptive regression splines and artificial neural network for prediction of mean sea level trend around the northern Australian coastlines. Mathematics. 2021;9:2696.

Modelling Climate Change Impacts on Watershed Hydrology Using SWAT

Filiz Dadaser-Celik

Abstract

Modelling climate change impacts is important for sustainable water management under changing climatic conditions. Soil and Water Assessment Tool (SWAT) is among the most used watershed-scale models for modelling climate change impacts. In this study, a review of the literature about the use of SWAT in climate change impact assessment is provided. The review was conducted on the articles published between the years 1998 and 2021 in journals listed in the Web of Science Core Collection. It classified the studies based on the geographical location, hydrologic variable investigated, climate model used, and emission scenario selected. The review also examined the challenges encountered during model development and applications. This analysis showed that the majority of the studies in the literature focused on simulation of streamflow for evaluation of changes in watershed hydrology. A few studies examined other water balance components, such as groundwater recharge, evapotranspiration, soil moisture. SWAT models were run in various watersheds over different continents by using a variety of global circulation models. SRES A1, A2 scenarios and RCP4.5 and RCP8.5 scenarios were the emission scenarios most studied. SWAT is advantageous over the other models, as it has robust theoretical background, straightforward application with user-friendly interface, and applicability under data-scarce conditions. Uncertainties in the hydrologic models and climate models and scenarios pose challenges for proper determination of climate change impacts at the watershed scale with SWAT.

Keywords

Climate change · Watershed hydrology · Hydrologic modelling · SWAT

1 Introduction

Climate change caused by global warming has become one of the major environmental problems. Climate change is expected to cause higher temperatures and alter the magnitude, timing and spatial distribution of precipitation [1]. The characteristics and intensity of the expected changes can be very variable over the World. Some regions can become drier while some others are coping with floods.

The changes in the climatic conditions can have profound impacts on the hydrologic system

F. Dadaser-Celik (✉)
Environmental Engineering Department, Erciyes University, Kayseri, Turkey
e-mail: fdadaser@erciyes.edu.tr

A. M. Bahadir et al. (eds.), *Hydrology and Urban Water Supply*, Water and Wastewater Management,
https://doi.org/10.1007/978-3-031-72589-0_4

[2]. Water availability and supply [3–5], water quality [6, 7], and aquatic ecosystems [8] can be affected. Extreme events, such as droughts and floods, can become more frequent and their magnitudes can be stronger. The climatic changes and their impacts on the hydrology and water resources have received significant attention from the scientific community in recent decades [9, 10]. The trends in streamflow and other hydrologic variables, the changes in intensity and frequency of extreme events have been studied by using historical hydrologic and climatic data. These studies have identified the changes already occurred in the hydrologic system and examined the drivers for these changes. Another group of studies have focused on the future and seek to make projections about future conditions based on watershed models. Projections about future hydrologic conditions form the basis for developing strategies for efficient planning and management of water resources.

Modelling techniques provide valuable tools for understanding and quantifying the linkage between the hydrologic and climatic systems. Hydrological models present simplified representations of the physical relationships between different components of the hydrologic cycle. These models can help characterize hydrologic processes in watersheds, determine the sensitivity of the system to various human and climate induced changes, and make predictions about future conditions. Understanding the climate change impacts on the watershed hydrology is critical for understanding the sensitivity of the hydrologic system to expected climatic variations.

Various hydrologic models have been used for modelling hydrologic processes at the watershed scale. The first mathematical models developed for modelling of water systems appeared in the 1950s. With the Stanford Watershed Model (SWM), which was developed by Crawford and Linsley [11] in 1966, single-component hydrological cycle models were replaced by multi-component models, in which watershed characteristics were considered. The SWM model has led to the development of process-based models that allow further research and development for watershed models. Among the most popular models used today, Soil and Water Integrated Model (SWIM) [12], Soil and Water Assessment Tool (SWAT) [13], Dynamic Watershed Simulation Model (DWSM) [14], Hydrological Simulation Program Fortran (HSPF) [15] can be listed.

In this study, the objective is to examine the use of a watershed-scale hydrologic model, SWAT, in climate change impact assessment. SWAT is a semi-distributed and physically-based hydrologic model, which can be used to simulate water, sediment, and nutrient transport at the watershed scale over long time periods [16, 17]. It has been used for modelling watershed hydrology since 1998 and have found applications in different regions worldwide. SWAT has been used for simulating water balance components and streamflow and water quality at the watershed scale. The effects of land use changes [e.g., 18–20] and climatic changes [e.g., 21–23] have also been simulated with SWAT. In this study, a review of studies conducted with SWAT to analyze the hydrologic impacts of climate change is provided. A discussion about the challenges encountered during model applications is also provided.

2 Methods

The studies included in this review were selected from the Web of Science database. Four main criteria were used for selection of the articles. First, to ensure that the articles selected specifically focus on the topic, the articles, whose title contain the words "climate change" and "SWAT" was selected. Second, only peer-reviewed journal articles that were included in Web of Science Core Collection were included. This means that conference proceedings or articles published in journals not included in the Web of Science Core Collection were not included. Third, the analyses cover the period, 1998–2021. The first paper on the theoretical aspects of the SWAT was published in 1998 [13], therefore we aim to reach the literature since this date. Finally, the articles that focus on the climate change projections with SWAT were included, in other words, the studies that focus on climate sensitivity analyses were excluded. Of course, there is an extensive literature on the use of SWAT in hydrologic modelling studies. The studies about the climate change impacts are also numerous. However, with these four criteria, the general characteristics of the studies conducted with SWAT were captured. The selected studies were classified according to the geographical location, hydrologic variable investigated, climate model used, and emission scenario used.

3 SWAT Model Description

SWAT is a physically-based model, where physical relationships between the hydrologic components are represented in semi-distributed way. SWAT can simulate different components of the hydrologic cycle (i.e., precipitation, evapotranspiration, infiltration, percolation, surface runoff, lateral flow, and return flow) together with sediment and nutrient transport at the watershed scale [16, 17]. The physical processes associated with water, sediment, and nutrient transport, nutrient cycling, plant growth can be simulated with SWAT. The inputs required by SWAT include data on weather, soils, and topography, vegetation, and land management practices.

A schematic representation of the SWAT algorithm is provided in Fig. 1. The first step in creating a SWAT model is the delineation of watershed and sub-basin boundaries. Digital elevation model (DEM) data is used for this purpose. DEM is analysed by filling in sinks, creating flow direction, and accumulation grids, and drawing the stream network. After delineation of watershed and sub basin boundaries, reservoir locations can be added, if necessary. Sub-basin maps are combined with land use, soil

Fig. 1 SWAT Algorithm

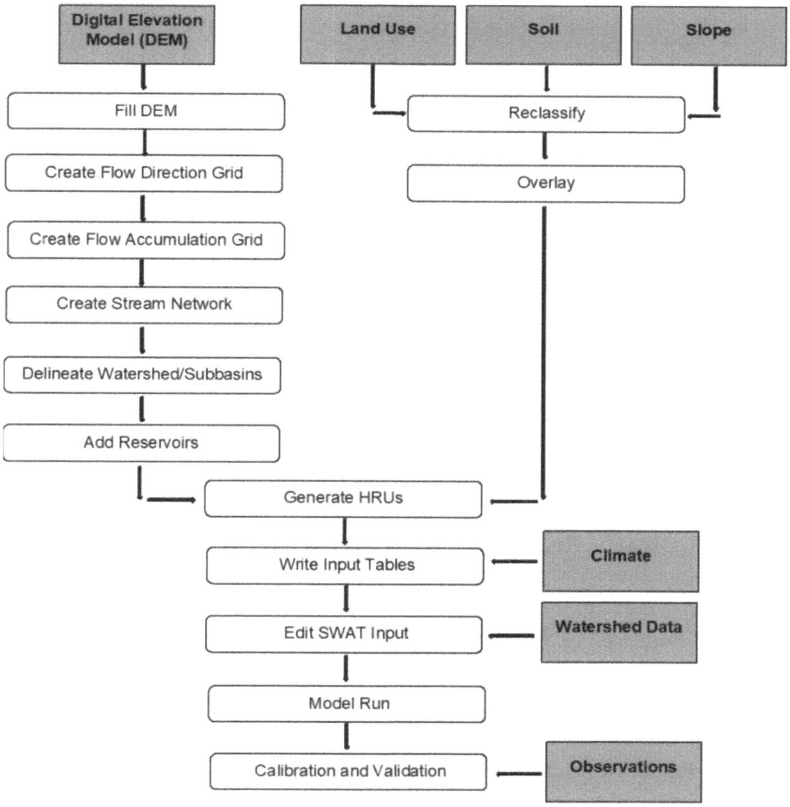

and slope layers to create hydrologic response units (HRUs). HRUs are the basic units in a SWAT model and water, sediment, and nutrient flows are calculated for each HRU. After addition of climate data (e.g., precipitation, air temperature, relative humidity, solar radiation, and wind speed), SWAT input tables can be created. These tables can be edited with relevant information about land management practices from the watershed. At this step, the model can be run, but calibration and validation are necessary in order to get the best fit between model results and observations.

In SWAT, the basic calculation units are the HRUs. The water balance equation provided in Eq. 1 is solved for each HRU [24]. In Eq. 1, SWt is the soil water content at time step t, SWo is the initial soil water content, Rday is daily precipitation, Qsurf is surface runoff, Ea is evapotranspiration, Wseep is percolation, and Qgw is groundwater flow. At each time step, the model calculates the change in soil moisture based on inflows and outflows.

$$SW_t = SW_0 + SW_t = \sum_{i=1}^{n}(R_{day} - Q_{surf} - E_a - W_{seep} - Q_{gw}) \quad (1)$$

4 Geographical Region

SWAT was used to evaluate the climate change impacts in different regions of the world distributed over different continents.

Starting from 1990s, SWAT was used for modelling hydrology, water quality, and climate change impacts in the North and South America. Upper Mississippi River Basin, Maumee River Basin, Clackamas River Watershed, some other watersheds in Rhode Island and Texas were modelled with SWAT in the USA [7, 25–29] and Athabasca River Basin and Little Smoky River Basin in Canada [30, 31]. Although the number of studies conducted in the South America were lower compared to that of North America, some studies were identified from Mexico and Uruguay. Deng et al. [22] studied climate change impacts in a river basin in Mexico and Aznarez et al. [32] examined the response of Laguna del Sauce in Uruguay to climatic changes.

Different river basins have been studied in Asia. Majority of the studies were conducted in China [33–37] or South Korea [6, 21, 38–43]. In China, the climate change impacts were investigated with SWAT in Luohe River Basin, Lanjiang Catchment, Liao River Watershed, and Ru River Basin. In South Korea, Seolma-Cheon Watershed, Chungju Dam Watershed, and Hongshui River Basin were studied. Alejo et al. [44] studied Maasin River Watershed in Philippines and Tamor River Basin and Budhigandaki River Basin were studied in Nepal [45, 46]. Several studies were conducted in India, in Baitarani River Basin, Upper Sind River Basin, Bhavani River Basin [47–50]. Shrestha et al. [51] studied Swat River Basin and Awan et al. [52] studied Indus Basin in Pakistan. Chueh et al. [53] conducted a study in the Erren River Basin in Taiwan and Li and Fang [23] in the Mun River in the Mekong Basin in Thailand. [54]. In the Middle Eastern part of Asia, many studies were conducted in various river basins in Iran [55, 56] including the Gharesou River Basin and Karkheh River Basin.

In Europe, SWAT was applied in the Eo River Basin and Segura River Basin in Spain [57, 58], Uggerby River Catchment in Denmark [59], and Nemunas River Basin in Belarus [60]. Peker and Sorman [61] applied SWAT in the Euphrates-Tigris Basin in Türkiye.

In Africa, climate change impacts were evaluated in Nzoia Catchment in Kenya [62], Volta River Basin in West Africa [63], and Lake Victoria Basin [64]. Tanteliniaina et al. [65] simulated climate change impacts at the Mangoky River Basin in Madagascar and Negewo et al. [66] simulated them at the Genale Dawa River Basin in Ethiopia.

Previous literature show that SWAT has been applied in several watersheds in different continents. Some of the studies were conducted by using high spatial and temporal resolution data. In some other regions, data availability was the major challenge encountered, however, using data from global datasets provided an alternative solution. Overall, these applications show that SWAT can be applied successfully in different types of watersheds and can be used for climate change impact assessment. SWAT has strong theoretical background and user-friendly application. It also provides satisfactory model performance under data-scarce conditions. Costs for the waste collection, minimize the viability of the unit.

5 Hydrological Variables

SWAT is capable of simulating a number of hydrologic variables, such as streamflow, groundwater recharge, evapotranspiration, reservoir volumes, etc.

A majority of the studies in the literature focused on streamflow simulation for understanding the climate change impacts [6, 7, 21, 23, 25, 33, 35, 36, 38, 63, 67–72]. Streamflow is a key parameter in watershed hydrology and it is directly related to climatic conditions as well as watershed characteristics. Streamflow is also a parameter, which combines base flow, subsurface flow, and surface runoff components, therefore behaves as an indicator for obtaining a combined response of a basin. The response of streamflow to climatic variations was variable. For example, Ficklin et al. [73] estimated increases in streamflow as a result of increased precipitation in the Lake Victoria Basin in Kenya, whereas Fereidoon and Koch [56] predicted an increase in streamflow during the dry season and a decrease during the wet season in the Karkheh River Basin in Iran. Wang et al. [35] estimated a decrease in streamflow due to increase in evapotranspiration in the northeastern China. Although many studies preferred to focus on streamflow alone, some other studies focused on different components such as surface runoff, lateral flow, and base flow [31, 43, 62]. Water yield from the basin [43, 59, 63, 66] and reservoir inflows and outflows [54, 64] were also estimated with SWAT under climate scenarios.

SWAT provides several techniques for calculating potential and actual evapotranspiration, which are critical parameters, particularly in agricultural basins. Evapotranspiration constitutes the major water loss item in the basin water budget, particularly in arid and semi-arid regions. A few studies focused on how evapotranspiration rates could change with expected climatic changes [31, 41, 43, 49, 50, 63, 72]. For example, Park et al. [6] estimated an increase in evapotranspiration by 23.1% under A1B scenario in a mountainous dam watershed. Possible changes in crop yields [29, 56] or crop productivity [47], irrigation and irrigated area [29, 56, 59] were also examined with SWAT. Chen et al. [29] estimated that irrigation of corn land use could decrease by 15–42% under climatic changes.

Snowmelt runoff [21], snow water equivalent and snow-covered days [61], soil moisture [38, 63], groundwater recharge [6], and consumptive water use [52] were the other variables simulated with SWAT, however the number of these studies were limited. Kim et al. [21] estimated increases in contribution of snowmelt runoff to streamflow due to changes in precipitation and temperature patterns. Peker and Sorman [61] estimated that snow water equivalent would go down by 30–39% and snow-covered days would get shorter by 37–43 days in the Euphrates-Tigris Basin in Türkiye. Joh et al. [38] estimated a downward trend in soil moisture due upward trends in air temperature, precipitation, and evapotranspiration in the Seolma-Cheon watershed in South Korea. Awan et al. [52] estimated a 7% and 11% increase in consumptive water use in the Indus River Basin in Pakistan. Some other studies examined water quality issues based on variables such as total nitrogen, total phosphorus, nitrates, etc. together with hydrologic variables due to interdependence between them [6, 7, 21, 35, 36, 39, 42, 53]. Majority of these studies showed that pollutant transport is directly related to streamflow and a decrease in pollutant loads would be detected when streamflow gets lower and an increase in pollutant loads would be detected when streamflow gets higher.

6 Global Circulation Models (GCM)

Global circulation models (GCMs) are widely used for evaluating climate change impacts on precipitation and air temperatures. GCMs simulate the processes in the atmosphere, ocean, and land surface with mathematical equations [74]. These models run either at the global or continental scales. Although GCMs provide valuable data on expected climatic changes, the configurations of GCMs are variable and the data from GCMs can include significant amount of uncertainty. It is challenging to represent real-world conditions due to large spatial coverage. Therefore, data from several GCMs are used in a single study to identify the range of uncertainty in predictions. Another approach for reducing uncertainties is preparing multi-model ensembles, which use data from numerous GCMs to produce a single composite dataset [75]. Complementary to GCMs, regional climate models (RCMs) are also needed. RCMs provide more reliable data for simulation of regional and local climatic conditions [76].

Several studies run simulations with data from different GCMs and compared the results with each other [21, 25, 28, 30, 35, 58, 62, 72], but some other studies focused on outputs from single GCMs [39, 40, 48, 54, 56]. Earlier studies used projections created by the CISRO-RegCM2, CCC, CCSR, CISRO-Mk2, GFDL, HadCM3, CGCM2, ECHAM, MIROC3.2hires, EH5OM, PRECIS, ECHAM4, COSMO-CLM, CGCM3, NorESM models [6, 7, 21, 25, 33, 38, 40, 47, 49, 52, 54, 55, 62, 63, 68, 70]. Chen et al. [29] used data from 19 CGMs (ACCESS1-0, BCC-CSM1-1, CANESM2, CCSM4, CESM1-BGC, CNRM-CM5, CSIRO-MK3-6-0, GFDL-ESM2G, GFDL-ESM2M, INMCM4, IPSL-CM5A-LR, IPSL-CM5A-MR, MIROC5, MIROC-ESM, MIROC-ESM-CHEM, MPI-ESM-LR, MPI-ESM-MR, MRI-CGCM3, NORESM1-M). Chun et al. [30] used the outputs of CanESM2, CCSM4, CNRM-CM5, CSIRO-MK3, MIROC5 models. Du et al. [31] used ACCESS 1-0, CNRM-CM5, CanESM2 model outputs. Deng et al. [72] used the outputs of the BCC-CSM 1.1, CanESM2, CCSM4, CSIRO-MK 3.6.0, CNRM-CM5, GFDL-CM3, and MRI-CGCM 3 models. HadGEM2-ES, MPI-ESM-MR, and GFDL-ESM2M were used by Peker and Sorman [61]. Touseef et al. [37] used projections created by the GFDL-ESM2 M, HadGEM2-ES, IPSL-CM5A-LR, MIROC-ESM-CHEM, and NorESM1-M models. GFDL, MIROC5, and HADGEM outputs were used by Cheng and Chang [28].

Some studies used ensemble projections by using data from several GCMs. Verma et al. [7] obtained projection data from CGCM3, GFDL-CM2, and HadCM3 models and created an ensemble dataset. Senent-Aparicio et al. [57] used different combinations of GCMs, CNRM-CM5, MPI-ESM-LR, EC-EARTH, IPSL-CM5A-MR together with RCMs CCLM4-8-17, HIRHAM5, REMO2009, RACMO22E, RCA4, WRF331F to identify climate change impacts. Output from five GCMs (GFDL-ESM2M, HadGEM2-ES, IPSL-CM5A-LR, MIROC, and NoerESM1-M) were combined to model climate change impacts in a rice paddy catchment [77]. Tanteliniaina et al. [65] used an ensemble of data from the NorESM and CanESM models. Li et al. [23] produced climate projection data by using data from 34 GCMs. Jeon et al. [43] used five RCMs (HadGEM3-RA, RegCM4, MM5, WRF, and GRIMs) employed in the Coordinated Regional Climate Downscaling Experiment. Marahatta et al. [46] used an advanced envelop-based climate selection method to identify GCMs from a large pool. They selected HadGEM2, GFDL, CanESM2, and MPI-ESM-LR for RCP4.5 scenarios and HadGEM2, GFDL, CanESM2, and MPI-ESM-LR, and MIROC-ESM for RCP 8.5 scenarios.

7 Emission Scenarios

Complex relationships among human activities, CO2 emissions and concentrations and temperature variations led to the development of climate projections, concentration pathways [78]. Since 1990, four generations of emission scenarios were created: Scientific Assessment 1990 (SA90)[79],

Table 1 Emission scenarios

SRES scenarios	A1	Very rapid economic growth; global population peaks in mid-century and declines thereafter; rapid introduction of new and more efficient technologies
	A2	Heterogeneous World, The underlying theme is self-reliance and preservation of local identities, continuously increasing population, regionally oriented economic development
	B1	A convergent World, population peaks in midcentury and declines thereafter, rapid change in economic structures toward a service and information economy, with reductions in material intensity and the introduction of clean and resource-efficient technologies
	B2	Local solutions to economic, social and environmental sustainability, continuously increasing global population, at a rate lower than A2, intermediate levels of economic development, less rapid and more diverse technological change than in the B1 and A1
RCPs	RCP1.9	Limits global warming to below 1.5 C
	RCP2.6	Carbon dioxide (CO_2) emissions start declining by 2020 and go to zero by 2100
	RCP4.5	Considerable removal of greenhouse gases from the atmosphere; between RCP2.6 and RCP4.5
	RCP8.5	Emissions peak around 2040, then decline
	RCP6	Emissions peak around 2080, then decline
	RCP7	Baseline outcome
	RCP8.5	Emissions continue to rise throughout the twenty-first century

1992 IPCC Scenarios (IS92)[80], Special Report on Emission scenarios (SRES) [81], Representative Concentration Pathways (RCPs) [34] and Shared Socioeconomic Pathways (SSPs) [82]. SA90 scenarios were defined in the first IPCC Report [79]. SA90 emission scenarios were revised and four new scenarios were created as IS92 emission scenarios in 1992. Special Report on Emissions Scenarios defined SRES scenarios as A1, A2, B1, B2 (Table 1). Finally, in 2010, scenarios based on Representative Concentration Pathways (RCPs) were created (Table 1).

Earlier studies focused on SRES scenarios [6, 7, 21, 33, 38, 40, 47, 49, 55, 62, 63, 68]. Among SRES scenarios, A1B [6, 7, 21, 38, 40, 47, 49, 63, 68], A2 [6, 21, 33, 49, 55, 62, 68], B1 [6, 21, 38, 40], and B2 [33, 34, 49, 62] scenarios were widely studies. The majority of the studies in the literature conducted in recent years considered RCP scenarios. Among RCP scenarios, RCP4.5 [21, 23, 31, 32, 36, 37, 41, 43, 46, 51, 53, 56, 58, 60, 61, 66, 72, 75, 77] and RCP8.5 [21, 23, 28, 30–32, 36, 37, 41–43, 46, 51–53, 56–61, 66, 72, 75, 77] scenarios were the most popular. Only a few studies evaluated the effects of RCP2.6 [23, 32, 37, 43, 59, 77] and RCP6.0 [53, 77] scenarios. Kim et al. [21] evaluated the impacts under both SRES and RCP scenarios together. Tanteliniaina et al. [65] investigated climate change impacts under two shared socioeconomic pathways (SSP) scenarios, SSP 3–7.0 and SSP 5–8.5.

8 Challenges of SWAT Model Application

Two important challenges encountered in application of SWAT in climate change studies originate from SWAT model's structure and model development process or uncertainties in climate change projections [83]. In some studies, it was argued that the uncertainty coming from GCMs can be much higher than those due to hydrologic model structure [84]. However, some other studies such as Zhang et al. [68] argued that hydrologic model uncertainties can be comparable with uncertainties due to GCMs.

Several challenges were reported in modelling the hydrologic system with SWAT. These challenges included problems with data availability, data quality, and limitations of model structure. Jha et al. [25] reported uncertainty in model results due to the use of low resolution land use/cover data. Sood et al. [63] and Chen and Chang [28] reported that the use of outdated land use/cover, availability of limited information on land management practices in agricultural areas, and misrepresentation of dams, reservoirs, and ponds caused problem in model calibration/validation. Zhang et al. [68] explained that the uncertainties in model parameters can be responsible for comparatively lower model performance during model validation. They also mentioned that the structure of SWAT and calibration method caused lower model performance in estimating extreme events. Uniyal [49] discussed that the SWAT model's estimation of recession caused insufficient estimation of low and high streamflows. Verma et al. [7] discussed that representation of hydrologic processes contain some degree of uncertainty. Lee et al. [41] discussed that SWAT's limitations in simulating groundwater descent characteristics created problem in simulation of groundwater component. Limitations in data on snow water equivalent and snow depth were mentioned for calibration/validation of a SWAT model in a mountainous basin [61]. The hydrologic model structure and misrepresentation of measures for water conservation in the basin were mentioned among the major source of uncertainty [23]. Chen and Chang mentioned that SWAT models can include uncertainty due to the input data used and parameter selection.

GCM related uncertainties are related to GCMs structure, GCM initial conditions, GCM scenarios, and downscaling method used [6, 23, 28, 33, 37, 41, 54, 55]. Jha et al. [25] reported that streamflow projections changed between −6 and 51% when projections from the CISRO-RegCM2, CCC, CCSR, CISRO-Mk2, GFDL, and HadCMS models were used. Ficklin et al. [73] estimated that mean annual streamflow would change by 6% to 115% due to change in rainfall in the range of 2.4%–23.2%. Deng et al. [72] reported very variable hydrologic response based on analysis conducted with seven GCMs. Marahatta et al. [46] used a method to select GCMs for specific conditions and scenarios to cope with uncertainties regarding GCM selection. GCMs produce outputs at very coarse spatial and temporal resolutions. However, hydrologic assessment requires high resolution data. Downscaling can provide the required resolution, however it also brings certain degree of uncertainty [7, 23, 55, 62].

Another challenge with climate change predictions with SWAT is not considering factors other than climate in the analyses [6, 62]. Hydrologic system can be affected by anthropogenic factors such as population growth and land use/cover changes. Also water consumption patterns can change in the future. Kim et al. [21] mentioned that land-use change and the changes in soil environment, leaf canopy inception of snowfall, and snowmelt mechanism and humus soil layer can affect snowmelt mechanisms in the future. Construction and operation of reservoirs, expansion of urban areas can also affect hydrologic processes in the future [23].

9 Conclusion

Modelling tools have been increasingly used to determine watershed-scale impacts of climate change. SWAT is one of the most used models worldwide and is proved to be successful in watershed scale hydrological processes. SWAT model is a continuous model, which provide opportunities for running simulations for long time periods. Climate projections data, therefore, can be used with SWAT to make predictions about future conditions.

In this study, we examined the literature to determine general characteristics of the studies conducted with SWAT for modelling climate change impacts the analyses revealed that data from several different GCMs were used in SWAT simulations in various watersheds worldwide. The majority of the studies focused on prediction of streamflow but some studies also examined other water components. SRES scenarios and

RCP scenarios were the widely applied climate scenarios. The studies with SWAT started in 1998, but the number of studies that focus on climate change impact assessment increased significantly in recent years, which points to the high potential of SWAT for this application.

References

1. IPCC. Climate change 2021: the physical science basis. In: Masson-Delmotte V, Zhai PA, Pirani SL, Connors C, Péan S, Berger N, Caud Y, Chen L, Goldfarb MI, Gomis M, Huang K, Leitzell E, Lonnoy JBR, Matthews TK, Maycock T, Waterfield O, Yelekçi RY, Zhou B, editors. Contribution of working group I to the sixth assessment report of the intergovernmental panel on climate change. Cambridge University Press; 2021.
2. Huntington TG. Evidence for intensification of the global water cycle: review and synthesis. J Hydrol. 2006;319:83–95.
3. O'Hara JK, Georgakakos KP. Quantifying the urban water supply impacts of climate change. Water Resour Manage. 2008;22:1477–97.
4. Wit M, Stankiewicz J (2006) Changes in surface water supply across Africa with predicted climate change. Science. 2006;311:1917–21.
5. Jouma N, Dadaser-Celik F. Assessing hydrologic alterations due to reservoirs and intensified irrigation in a semi-arid agricultural river basin using SWAT*. Irrig Drain. 2022;71:452–71.
6. Park JY, Park MJ, Ahn SR, Park GA, Yi JE, Kim GS, Srinivasan R, Kim SJ. Assessment of future climate change impacts on water quantity and quality for a mountainous dam watershed using SWAT. Trans ASABE. 2011;54:1725–37.
7. Verma S, Bhattarai R, Bosch NS, Cooke RC, Kalita PK, Markus M. Climate change impacts on flow, sediment and nutrient export in a great lakes watershed using SWAT. Clean-Soil Air Water. 2015;43:1464–74.
8. Wrona FJ, Prowse TD, Reist JD, Hobbie JE, Lévesque LMJ, Vincent WF. Climate change effects on aquatic biota, ecosystem structure and function. AMBIO: J Human Environ. 2006;35:359–69.
9. Arnell NW. Climate change and global water resources. Glob Environ Chang. 1999;9:S31–49.
10. Grafton RQ, Pittock J, Davis R, Williams J, Fu G, Warburton M, Udall B, McKenzie R, Yu X, Che N, Connell D, Jiang Q, Kompas T, Lynch A, Norris R, Possingham H, Quiggin J. Global insights into water resources, climate change and governance. Nat Clim Chang. 2013;3:315–21.
11. Crawford NH. Digital simulation in hydrology; Stanford watershed model IV. Department of Civil Engineering, Stanford University, Stanford, Calif.; 1966.
12. Krysanova V, Wechsung F, Arnold J, Srinivasan R, Williams J. SWIM (soil and water integrated model); 2000.
13. Arnold JG, Srinivasan R, Muttiah RS, Williams JR. Large area hydrologic modeling and assessment—part 1: model development. J Am Water Resourc Assoc. 1998;34:73–89.
14. Borah DK, Xia R, Bera M. Chapter 5: DWSM − A dynamic watershed simulation model. In: Mathematical models of small watershed hydrology and applications; 2002.
15. Donigian A, Bicknell B, Imhoff J. Hydrological simulation program—Fortran (HSPF). Comput Models Watershed Hydrol. 1995;12:395–442.
16. Arnold JG, Moriasi DN, Gassman PW, Abbaspour KC, White MJ, Srinivasan R, Santhi C, Harmel RD, Griensven AV, Liew MWV, Kannan N, Jha MK. SWAT: model use, calibration, and validation. Trans ASABE. 2012;55:1491–1508.
17. Neitsch SL, Arnold JG, Kiniry JR, Williams JR, King KW. Soil and water assessment tool theoretical documentation; 2005.
18. Affessa GM, Belew AZ, Tenagashaw DY, Tamirat DM. Land use/cover change impacts on hydrology using SWAT model on Borkena Watershed, Ethiopia. Water Conserv Sci Eng. 2022;7:55–63.
19. Chaemiso SE, Kartha SA, Pingale SM. Effect of land use/land cover changes on surface water availability in the Omo-Gibe basin, Ethiopia. Hydrol Sci J-Journal Des Sciences Hydrologiques. 2021;66:1936–62.
20. dos Santos FM, de Oliveira RP, Di Lollo JA. Effects of land use changes on streamflow and sediment yield in Atibaia River Basin-SP, Brazil. Water. 2020;12
21. Kim SB, Shin HJ, Park M, Kim SJ. Assessment of future climate change impacts on snowmelt and stream water quality for a mountainous high-elevation watershed using SWAT. Paddy Water Environ. 2015;13:557–69.
22. Deng C, Pisani B, Hernandez H, Li YM. Assessing the impact of climate change on water resources in a semi-arid area in central Mexico using a SWAT model. Boletin De La Sociedad Geologica Mexicana. 2020;72.
23. Li CY, Fang HY. Assessment of climate change impacts on the streamflow for the Mun River in the Mekong Basin, Southeast Asia: Using SWAT model. Catena. 2021;201.
24. Arnold JG, Srinivasan R, Muttiah RS, Williams JR. Large area hydrologic modeling and assessment part1: model development. J Am Water Resour Assoc. 1998;34:73–89.
25. Jha M, Arnold JG, Gassman PW, Giorgi F, Gu RR. Climate change sensitivity assessment on Upper Mississippi River Basin streamflows using SWAT. J Am Water Resour Assoc. 2006;42:997–1015.

26. Chambers BM, Pradhanang SM, Gold AJ (2017) Simulating climate change induced thermal stress in coldwater fish habitat using SWAT model. Water. 2017;9
27. Ercan MB, Maghami I, Bowes B, Morsy MM, Goodall JL. Estimating potential climate change effects on the upper Neuse watershed water balance using the SWAT model. J Am Water Resour Assoc. 2020;56:53–67.
28. Chen JJ, Chang H. Relative impacts of climate change and land cover change on streamflow using SWAT in the Clackamas River Watershed, USA. J Water Climate Change. 2021;12:1454–70.
29. Chen Y, Marek GW, Marek TH, Moorhead JE, Heflin KR, Brauer DK, Gowda PH, Srinivasan R. Simulating the impacts of climate change on hydrology and crop production in the Northern High Plains of Texas using an improved SWAT model. Agric Water Manag. 2019;221:13–24.
30. Chunn D, Faramarzi M, Smerdon B, Alessi DS. Application of an integrated SWAT-MODFLOW model to evaluate potential impacts of climate change and water withdrawals on groundwater-surface water interactions in West-Central Alberta. Water. 2019;11.
31. Du XZ, Shrestha NK, Wang JY. Assessing climate change impacts on stream temperature in the Athabasca River Basin using SWAT equilibrium temperature model and its potential impacts on stream ecosystem. Sci Total Environ. 2019;650:1872–81.
32. Aznarez C, Jimeno-Saez P, Lopez-Ballesteros A, Pacheco JP, Senent-Aparicio J. Analysing the impact of climate change on hydrological ecosystem services in Laguna del Sauce (Uruguay) using the SWAT model and remote sensing data. Remote Sens 13.
33. Zhang X, Srinivasan R, Hao F. Predicting hydrologic response to climate change in the Luohe River basin using the SWAT model. Trans ASABE. 2007;50:901–10.
34. Zhang AJ, Zhang C, Fu GB, Wang BD, Bao ZX, Zheng HX. Assessments of impacts of climate change and human activities on runoff with SWAT for the Huifa River Basin, Northeast China. Water Resour Manage. 2012;26:2199–217.
35. Wang Y, Bian JM, Zhao YS, Tang J, Jia Z. Assessment of future climate change impacts on nonpoint source pollution in snowmelt period for a cold area using SWAT. Sci Rep. 2018;8.
36. Yang XY, He RM, Ye JY, Tan ML, Ji XY, Tan L, Wang GQ. Integrating an hourly weather generator with an hourly rainfall SWAT model for climate change impact assessment in the Ru River Basin, China. Atmos Res. 2020;244.
37. Touseef M, Chen LH, Yang WZ. Assessment of surface water availability under climate change using coupled SWAT-WEAP in Hongshui River Basin, China. ISPRS Int J Geo-Inform. 2021;10.
38. Joh HK, Lee JW, Park MJ, Shin HJ, Yi JE, Kim GS, Srinivasan R, Kim SJ. Assessing climate change impact on hydrological components of a small forest watershed through SWAT calibration of evapotranspiration and soil moisture. Trans ASABE. 2011;54:1773–81.
39. Park JY, Park GA, Kim SJ. Assessment of future climate change impact on water quality of Chungju Lake, South Korea, using WASP coupled with SWAT. J Am Water Resour Assoc. 2013;49:1225–38.
40. Park JY, Yu YS, Hwang SJ, Kim C, Kim SJ. SWAT modeling of best management practices for Chungju dam watershed in South Korea under future climate change scenarios. Paddy Water Environ. 2014;12: S65–75.
41. Lee J, Jung C, Kim S, Kim S. Assessment of climate change impact on future groundwater-level behavior using SWAT groundwater-consumption function in Geum River Basin of South Korea. Water. 2019;11.
42. Kim DH, Jang T, Hwang S. Evaluating impacts of climate change on hydrology and total nitrogen loads using coupled APEX-paddy and SWAT models. Paddy Water Environ. 2020;18:515–29.
43. Jeon DJ, Ligaray M, Kim M, Kim G, Lee G, Pachepsky YA, Cha DH, Cho KH. Evaluating the influence of climate change on the fate and transport of fecal coliform bacteria using the modified SWAT model. Sci Total Environ. 2019;658:753–62.
44. Alejo LA, Ella VB. Assessing the impacts of climate change on dependable flow and potential irrigable area using the SWAT model. The case of Maasin River watershed in Laguna, Philippines. J Agricult Eng. 2019;50:88–98.
45. Bhatta B, Shrestha S, Shrestha PK, Talchabhadel R. Evaluation and application of a SWAT model to assess the climate change impact on the hydrology of the Himalayan River Basin. Catena. 2019;181.
46. Marahatta S, Aryal D, Devkota LP, Bhattarai U, Shrestha D. Application of SWAT in hydrological simulation of complex mountainous river basin (Part II: climate change impact assessment). Water. 2021;13.
47. Lakshmanan A, Geethalakshmi V, Rajalakshmi D, Bhuvaneswari K, Srinivasan R, Sridhar G, Sekhar NU, Annamalai H. Climate change adaptation strategies in the Bhavani Basin using the SWAT model. Appl Eng Agric. 2011;27:887–93.
48. Narsimlu B, Gosain AK, Chahar BR. Assessment of future climate change impacts on water resources of upper Sind River Basin, India using SWAT model. Water Resour Manage. 2013;27:3647–62.
49. Uniyal B, Jha M, Verma A. Assessing climate change impact on water balance components of a River Basin using SWAT model. Water Resour Manage. 2015;29:4767–85.
50. Padhiary J, Das DM, Patra KC, Sahoo BC, Singh KK. Prediction of climate change impact on streamflow and evapotranspiration in Baitarani basin using SWAT model. J Agrometeorol. 2018;20:325–8.

51. Shrestha S, Sattar H, Khattak MS, Wang GQ, Babur M. Evaluation of adaptation options for reducing soil erosion due to climate change in the Swat River Basin of Pakistan. Ecol Eng. 2020;158.
52. Awan UK, Liaqat UW, Choi M, Ismaeel A. A SWAT modeling approach to assess the impact of climate change on consumptive water use in Lower Chenab Canal area of Indus basin. Hydrol Res. 2016;47:1025–37.
53. Chueh YY, Fan CH, Huang YZ. Copper concentration simulation in a river by SWAT-WASP integration and its application to assessing the impacts of climate change and various remediation strategies. J Environ Manage. 2021;279.
54. Ahn SR, Jeong JH, Kim SJ. Assessing drought threats to agricultural water supplies under climate change by combining the SWAT and MODSIM models for the Geum River basin, South Korea. Hydrol Sci J-Journal Des Sciences Hydrologiques. 2016;61:2740–53.
55. Zahabiyoun B, Goodarzi MR, Bavani ARM, Azamathulla HM. Assessment of climate change impact on the Gharesou River Basin using SWAT hydrological model, clean—soil air. Water. 2013;41:601–9.
56. Fereidoon M, Koch M. SWAT-MODSIM-PSO optimization of multi-crop planning in the Karkheh River Basin, Iran, under the impacts of climate change. Sci Total Environ. 2018;630:502–16.
57. Senent-Aparicio J, Perez-Sanchez J, Carrillo-Garcia J, Soto J. Using SWAT and fuzzy TOPSIS to assess the impact of climate change in the headwaters of the Segura River Basin (SE Spain). Water. 2017;9.
58. Perez-Sanchez J, Senent-Aparicio J, Santa-Maria CM, Lopez-Ballesteros A. Assessment of ecological and hydro-geomorphological alterations under climate change using SWAT and IAHRIS in the Eo River in Northern Spain. Water. 2020;12.
59. Liu W, Bailey RT, Andersen HE, Jeppesen E, Nielsen A, Peng K, Molina-Navarro E, Park S, Thodsen H, Trolle D. Quantifying the effects of climate change on hydrological regime and stream biota in a groundwater-dominated catchment: a modelling approach combining SWAT-MODFLOW with flow-biota empirical models. Sci Total Environ. 2020;745.
60. Cerkasova N, Umgiesser G, Erturk A. Development of a hydrology and water quality model for a large transboundary river watershed to investigate the impacts of climate change—a SWAT application. Ecol Eng. 2018;124:99–115.
61. Peker IB, Sorman AA. Application of SWAT using snow data and detecting climate change impacts in the mountainous eastern regions of Turkey. Water. 2021;13.
62. Githui F, Gitau W, Mutua F, Bauwens W. Climate change impact on SWAT simulated streamflow in western Kenya. Int J Climatol. 2009;29:1823–34.
63. Sood A, Muthuwatta L, McCartney M. A SWAT evaluation of the effect of climate change on the hydrology of the Volta River basin. Water Int. 2013;38:297–311.
64. Velasco PP, Bauwens W. Climate change analysis of Lake Victoria outflows (Africa) using Soil and Water Assessment Tool (SWAT) and General Circulation Models (GCMs). Asia Life Sci. 2013;22:659–75.
65. Tanteliniaina MFR, Rahaman MH, Zhai J. Assessment of the future impact of climate change on the hydrology of the Mangoky River, Madagascar using ANN and SWAT. Water. 2021;13.
66. Negewo TF, Sarma AK. Estimation of water yield under baseline and future climate change scenarios in Genale Watershed, Genale Dawa River Basin, Ethiopia, using SWAT model. J Hydrol Eng. 2021;26
67. Fathian F, Ahmadzadeh H, Mansouri B, Vaheddoost B. Assessment of water demand reliability using SWAT and RIBASIM models with respect to climate change and operational water projects. Agric Water Manage. 2022;261.
68. Zhang XJ, Xu YP, Fu GT. Uncertainties in SWAT extreme flow simulation under climate change. J Hydrol. 2014;515:205–22.
69. Shimola K, Krishnaveni M. Sensitivity of SWAT simulated reservoir inflow to climate change in a semi-arid basin. Mausam. 2015;66:181–6.
70. Zhou J, He D, Xie YF, Liu Y, Yang YH, Sheng H, Guo HC, Zhao L, Zou R. Integrated SWAT model and statistical downscaling for estimating streamflow response to climate change in the Lake Dianchi watershed, China. Stoch Env Res Risk Assess. 2015;29:1193–210.
71. Zhu Q, Zhang XJ, Ma C, Gao C, Xu YP. Investigating the uncertainty and transferability of parameters in SWAT model under climate change. Hydrol Sci J-Journal Des Sciences Hydrologiques. 2016;61:914–30.
72. Nazari-Sharabian M, Taheriyoun M, Ahmad S, Karakouzian M, Ahmadi A. Water quality modeling of Mahabad Dam watershed-reservoir system under climate change conditions, using SWAT and system dynamics. Water. 2019;11.
73. Ficklin DL, Luo YZ, Luedeling E, Zhang MH. Climate change sensitivity assessment of a highly agricultural watershed using SWAT. J Hydrol. 2009;374:16–29.
74. Shackley S, Young P, Parkinson S, Wynne B. Uncertainty, complexity and concepts of good science in climate change modelling: are GCMs the best tools? Clim Change. 1998;38:159–205.
75. Wang B, Zheng L, Liu DL, Ji F, Clark A, Yu Q. Using multi-model ensembles of CMIP5 global climate models to reproduce observed monthly rainfall and temperature with machine learning methods in Australia. Int J Climatol. 2018;38:4891–902.
76. Rummukainen M. State-of-the-art with regional climate models. WIREs Clim Change. 2010;1:82–96.
77. Wang KY, Onodera SI, Saito M, Okuda N, Okubo T. Estimation of phosphorus transport influenced by climate change in a rice paddy catchment using SWAT. Int J Environ Res. 2021;15:759–72.

78. Hayhoe K, Edmonds J, Kopp RE, LeGrande AN, Sanderson BM, Wehner MF, Wuebbles DJ. Climate models, scenarios, and projections. In: Wuebbles DJ, Fahey DW, Hibbard KA, Dokken DJ, Stewart BC, Maycock TK, editors. Climate science special report: fourth national climate assessment. Washington, DC, USA: U.S. Global Change Research Program; 2017.
79. Santhi C, Arnold JG, Williams JR, Dugas WA, Srinivasan R, Hauck LM. Validation of the SWAT model on a Large River Basin with point and nonpoint sources-1. JAWRA J Am Water Resourc Assoc. 2001;37:1169–88.
80. Leggett JEA. Emissions scenarios for the IPCC: an update. Climate change 1992: the supplementary report to the IPCC scientific assessment; 1992.
81. Nakicenovic N, Swart R. Special report on emissions scenarios. UK: Cambridge University Press; 2000.
82. Riahi K, van Vuuren DP, Kriegler E, Edmonds J, O'Neill BC, Fujimori S, Bauer N, Calvin K, Dellink R, Fricko O, Lutz W, Popp A, Cuaresma JC, Kc S, Leimbach M, Jiang L, Kram T, Rao S, Emmerling J, Ebi K, Hasegawa T, Havlik P, Humpenöder F, Da Silva LA, Smith S, Stehfest E, Bosetti V, Eom J, Gernaat D, Masui T, Rogelj J, Strefler J, Drouet L, Krey V, Luderer G, Harmsen M, Takahashi K, Baumstark L, Doelman JC, Kainuma M, Klimont Z, Marangoni G, Lotze-Campen H, Obersteiner M, Tabeau A, Tavoni M. The Shared Socioeconomic Pathways and their energy, land use, and greenhouse gas emissions implications: an overview. Glob Environ Chang. 2017;42:153–68.
83. Kundzewicz ZW, Krysanova V, Benestad RE, Hov Ø, Piniewski M, Otto IM. Uncertainty in climate change impacts on water resources. Environ Sci Policy. 2018;79:1–8.
84. Kay AL, Davies HN, Bell VA, Jones RG. Comparison of uncertainty sources for climate change impacts: flood frequency in England. Clim Change. 2009;92:41–63.

Challenges and Success Stories for Sustainable Water Management in Mena Region

Ahmed A. Hassan, Arwa Hamaideh, and Peter Riad

Abstract

Water sector in the MENA region faces many challenges that hinder its social and economic development. For this purpose, many workshops were organized and held by EXCEED-SWINDON last 10 years in order to discuss these challenges and to find the best solutions to overcome or at least to minimize their long term negative impacts. This paper is a short review of more than 20 papers, which were published in the EXCEED book series "Water Perspectives in Emerging Countries" from 2015 to 2019. This review presents many challenges and some examples for success stories for good water management in the MENA Region. The challenges could be classified as natural, social, political, technical, financial, governance and institutional, and trans-boundary ones. Natural challenges include frequent droughts and climate change; social challenges include high population growth, social misconceptions that impact water use behavior, and corruption. Technical challenges refer to low water use efficiency, especially in agricultural and domestic sectors. Policy challenges include poor allocation among the different competing demands as most of available water resources are allocated to the agricultural sector, and poor cropping practices in the agricultural sector. Economic challenges include lack of funds and low level of cost recovery in most of the MENA countries. Governance and institutional challenges refer to the lack of effective coordination and unclear lines of responsibilities among the different governmental bodies in charge of water resources management. The success stories for good water management are taken here from Egypt, Tunisia, Turkey and Morocco. Finally, many conclusions and recommendations from this review could be driven.

Keywords

EXCEED-SWINDON · MENA Region · Water challenges · Water management · Water sector

1 Introduction

Middle East and North African (MENA) Region is by far the driest and most water scarce region in the world; and that this is increasingly affecting the economic and social development of most

A. A. Hassan (✉) · P. Riad
Irrigation and Hydraulics Department, Faculty of Engineering, Ain Shams University, Cairo, Egypt
e-mail: ahmad9657@yahoo.co.uk

A. Hamaideh
Water, Energy and Environment Center, University of Jordan, Amman, Jordan

A. M. Bahadir et al. (eds.), *Hydrology and Urban Water Supply*, Water and Wastewater Management,
https://doi.org/10.1007/978-3-031-72589-0_5

countries in the region. MENA has 5% of the world population with less than 1% of the available world's freshwater resources. Despite these characteristics, many countries are still using up considerable amounts of scarce resources in supporting inefficient public sector institutions for both irrigation and water supply. Today, average per capita water availability in the region is about 1,200 cubic meters per year, whereas the world average is close to 7,000. The annual water availability in the region ranges from a high of about 1,800 in Iran to less than 200 m^3 per capita and year in Jordan, West Bank/Gaza, and Yemen. By 2025, the regional average water availability is expected to be just little over 500 m^3 per capita and year.

Irrigated agriculture claims about 90% of the available water resources of the region and contributes to somewhere between 5 and 20% of the GDP, depending on the country. The water resource situation in the region is becoming progressively bleaker due to droughts of higher frequency and longer duration [1]. The interactions between water resources and agricultural production in the MENA countries can be summarized as follows [2]:

Agricultural water footprint (WF) exceeds the total renewable water resources especially in Jordan, Egypt and Syria. In these countries, the internal agricultural WF is higher (to some extend even significantly) than the amount of available water. The existing water resources are not capable to cover the agricultural and domestic water demand, and the existing water dependency on external water resources will continue to increase dramatically in the future.

Water used for agricultural production in Turkey is still much less than the country's total renewable water resources; and it is expected to remain the same for next 30 years in case of keeping the same strategy.

In order to develop sustainable water demand management proposals for the MENA countries, blue and green water footprints (WFs) of agricultural productions were evaluated, where blue WF refers to the volume of surface and groundwater consumed as a result of the production of a good; the green WF refers to the rainwater consumed. For this purpose, ten most produced agricultural goods in each country, representing more than 85% of total production, were evaluated and the percentage of green and blue water used for the production are presented in Fig. 1. The most produced ten agricultural products and their water demands for each respective country are obtained from [3]. The components of WF expressed in green and blue WF are obtained from [4].

The countries especially poor in own water resources are using blue water for the irrigation of these crops. This is again an interesting point indicating poor governance of water resources in water poor countries. In case of Jordan, the most water scarce country in the region, blue water resources, which are stored in surface and ground water resources, are used to produce an agricultural added value of 3.1%, which is almost negligible in the overall economy of the country.

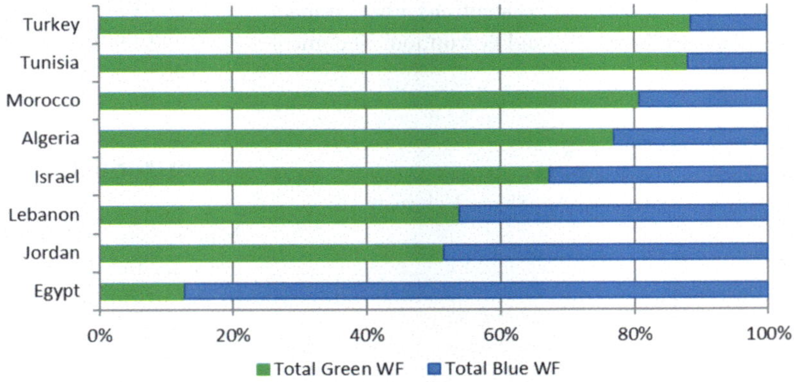

Fig. 1 Share of blue and green WF of ten most produced agricultural goods

Therefore, politicians, decision makers and water professionals should work together to change the plant patterns as well as the irrigation practices so that the scarce resources can be used more economically and environment-friendly.

2 Challenges Facing Water Sector in MENA

Water sector in the MENA Region faces many challenges that hinder the country's social and economic development. These challenges can be classified as natural, social, political, technical, financial, governance and institutional, and trans-boundary ones [5].

2.1 Natural Challenges

Drought is a natural challenge, the impacts of which can be mitigated by sound planning and management of water resources and demands. It is estimated that about 80% of the Arab countries' land is barren, where the average annual precipitation ranges between 0 and 180 mm, while the average annual evapotranspiration is more than 2000 mm [6]. According to the Arab Centre for the Studies of Arid Zones and Dry Lands (ACSAD) [7], the Arab World receives only 2.1% of the world's average annual precipitation, despite the fact that it makes 10% of the world's total area, and 5% of the world's population. In addition, about 2000 billion cubic meters (BCM) of rain fall in the Arab world; however, huge proportion of this rainfall is lost due to high evaporation and evapotranspiration rates [5].

Climate change is another challenge to the water sector in the MENA Region. Based on global and regional models, it is expected that the MENA Region will become drier due to the higher near-surface air temperatures, lower precipitation and higher evaporation rates, which will have threatening consequences on all vital sectors [8]. According to Chenoweth et al. [9], a study of climate change impacts on water resources in MENA predicted a decrease in precipitation of about 4% between 2040 and 2069. However, an increase in precipitation by 30.7% is expected in Kuwait and 27.9% in Qatar, while for Jordan and Lebanon precipitation is expected to decrease by 16.6% and 14.7%, respectively.

2.2 Social Challenges

Population growth rate in MENA countries is among the highest in the world, which is due to social as well as religious factors [10]. The population in the Arab countries has jumped from 128 million in 1970 to 359 million in 2010 [11], which is projected to reach about 692 million by 2050 [11, 12]. This population increase, in addition to the socio economic development, resulted in dramatically enlarging the gap between the limited available resources and the increasing demands [13]. The per capita share of available water resources in several MENA countries has dropped to extremely low levels over the last three decades as a result of high population growth [14]. According to the population projection of the United Nations, the annual per capita water share in Jordan is expected to drop from 110 to 56 m^3 by the year 2050. In Lebanon, the annual per capita water availability is expected to decrease from 900 to 725 m^3 [15].

2.3 Policy Challenges

Despite the fact that most of the MENA countries suffer from severe water shortage, water resources in most of these countries are not managed in an integrated sustainable manner. The mismanagement includes improper allocation among the different sectors, where the agricultural sector consumes the lion's share of the available resources. It is estimated that irrigated agriculture in the Arab world consumes about 87% of the available resources [16], although the contribution of the agricultural sector to the national economy in the Arab World is very low [17]. Public policies of subsidies and underpricing encourage overuse of water in the agricultural sector to meet the high food demand as a result of high population growth rate [18]. It is interesting to know that the benefit from these subsidies went to the rich farmers rather than to the poor.

2.4 Technical Challenges

The extremely low water use efficiency in the agricultural sector in MENA countries is responsible for considerable proportion of the increase in water losses. It is estimated that irrigation water use efficiency in the Arab World is as low as 30–40% [19]. The low water use efficiency in the agricultural sector is mainly attributed to the wide spread traditional inefficient irrigation methods [19–21]. Surface irrigation is practiced in about 80% of irrigated land, sprinkler irrigation is practiced in 23%, and micro-irrigation is practiced in only 2.8% of irrigated areas in the Arab World [22, 23]. For example, about 80% of water resources in North Africa are allocated to agriculture. It is estimated that 52 BCM water can be saved annually by reducing conveyance losses by 50% and raising irrigation efficiency from its current level of 40–50 to 80% [24].

2.5 Poor Governance and Institutional Setup

Improper institutional set up, where no clear boundaries among the responsibilities of the different institutions in charge of the water sector exist, resulted in poor governance in this sector aggravating the impact of water shortage [16]. In MENA countries, in general, irrigation and drinking waters are administered by different governmental bodies without clear guidelines for coordination among them [25]. Institutions responsible for water in MENA take pricing measures to increase revenues without taking into consideration the pricing impact on the behavior of the end-users. In most cases, the poor performance of many water utilities in the MENA region is due to lack of financial resources, mismanagement and the tendency to administer a utility as government department not as a private entity, which creates problems of unclear responsibilities for operation, low tariffs, difficulty retaining qualified personnel, lack of necessary legislation and political interferences [10].

2.6 Lack of Financial Resources

Lack of financial resources to maintain and to upgrade water infrastructures such as water distribution and storm water collection systems in many MENA countries is a major challenge to the sustainability of the water sector. In wet years and extreme rainfall events, the storm water collection system and water harvesting infrastructures are incapable to handle the generated runoff, which results in property damage and some cases of life loss. In the years 1993, 2003 and 2013, the Middle East witnessed intensive rainfalls and snow storms originating from Siberia, causing vast destruction to the fragile low-capacity infrastructures [10].

2.7 Political Instability

The MENA region has been politically instable during the past several decades, as evidenced by several wars and revolutions that took place and are still ongoing. Political instability in the region has resulted in sudden fluxes of refugees to certain neighbouring countries such as Jordan, Lebanon and Turkey, which resulted in huge pressure on all resources, more specifically on water resources. In addition to the instantaneous flux of refugees, political instability in the region has resulted in sustained demographic changes, which also require more efficient use of available natural resources to sustain social and economic development.

2.8 Transboundary Resources

Several countries in MENA Region depend on transboundary water resources to satisfy their demands. Two-thirds of the Arab world's surface water supplies are transboundary [26]. Roughly speaking about 90% of the Euphrates' annual flow, and 50% of the Tigris' water supply originates from Turkey [27]. In addition, more than 50% of Iraq's renewable water resources are trans-boundary. Moreover, about three quarters of Sudan's and Syria's water resources originate outside their borders. Transboundary water resources can lead to competition and conflicts between upstream and downstream countries about water rights. In the absence of agreements among the riparian countries that regulate water rights, transboundary water resources are not a reliable water resource for downstream countries, which is the case in the MENA Region. More sustainable transboundary water resources management practices are needed in the MENA Region in order to protect and to sustain social and economic developments of the downstream countries.

3 Success Stories for Sustainable Water Management in MENA Countries

3.1 Egypt

Can deficit irrigation adapt water scarcity in the MENA Region?

Abu-Hashim and Lilienthal [28] assessed the production of sound yields with limited water usage. They investigated the Faba bean yield that is dependent on irrigation water amount, especially in semi-arid regions like North Sinai, which are irrigated with water supply of mixed waters. Irrigation regimes with 50% of the amount applied by farmers gave high water use efficiency of 2.36 kg/m^3, and soil salinization was relatively decreased by 30%. Soil nutrients decreased with increasing the water stress. Thus deficit irrigation could save water that is extremely important in semi-arid regions, such as North Sinai [28].

Estimation of evaporation losses in High Aswan Dam (HAD) reservoir using remote sensing technique

Egypt is one of the countries facing challenges due to its limited water resources. Lake Nasser is the reservoir of High Aswan Dam (HAD), and its evaporation losses (ET) are an important contributing factor to the lake water budget. Although no universally recognized method is known for evaporation estimation, reliable evaporation data are necessary for efficient management of the reservoir and the scarce water resources. Ahmed Hassan et al. [29] calculated the monthly evaporation rates using Remote Sensing (RS) and GIS techniques. The study applied Surface Energy Balance System (SEBS) method using Terra Moderate Resolution Imaging Spectroradiometer (MODIS) satellite earth observation data for the years 2006 and 2012. Additionally, remote sensing ET product from the MODerate Resolution Imaging Spectrometer (MOD16) at years 2006, 2010 and 2012 were used to determine the monthly spatial distribution evaporation rates at Lake Nasser. The results were processed using ARC-GIS software. The evaporation rates calculated using data of High Aswan Dam Authority (HADA) were compared with SEBS and MOD16 estimation. High correlation between SEBS and MOD16 was found by using IBM SPSS 22.0 software package, which gives a good indication of the use of only one method to estimate the evaporation rates.

3.2 Tunisia

The New Tunisian Strategy from May 2017 highlighted six national priorities [30]: (i) water, energy and food security, (ii) emerging democratic society: education, culture, and youth, (iii) quality healthcare, (iv) digital and industrial transition, (v) governance and decentralization, and (vi) the circular economy.

The currently undertaken Tunisian "Water Code" will be the first Tunisian legal instrument for water resources management. In this water code, attention will be paid to the decentralized and innovative governance of water resources management [31].

The Tunisian water management strategy "WATER 2050" concerns the establishment of national strategies regarding water resources management and environment preservation by 2050 [32].

3.3 Turkey

Actions have been taken in Turkey

The main goal of Turkish National Basin Management Strategy (NBMS) is to provide guidance to medium and long-term decisions and investment programs regarding the protection of Turkey's water basins and their natural resources, development and sustainable use. The strategy targets are strengthening legal and institutional capacity, coordination and cooperation among institutions and stakeholder, sustainable management and use of water resources, raising the awareness of the people living in the aquifer area, and pressure reduction on natural resources [33].

The Turkish Watershed Protection Action Plans—Konya Closed Basin is a national initiative aiming at (i) determining the quantities, characteristics and pollution status of existing surface and subsurface waters in the basin, (ii) determining the pressure and its effects of various activities (urban, industrial, agricultural, economic), (iii) performing a detailed investigation of pollution sources and loads, with creation of water quality maps, and (iv) preparing a Basin Conservation Action Plan with the participation of all the stakeholders [34].

The Turkish National Drought Management Strategy Document and Action Plan aims at developing legislation on drought management, ensuring policy related to drought management, coordinating national and international cooperation, and conducting research, planning and implementation studies in drought management [35].

Environmental and financial evaluation agricultural water resources management

Turkey's 78 million ha of land consists of 28 million ha of agricultural land, which is about one third of the country. In 2011, a total land of 5.61 million ha was irrigated. Approximately 3.32 million ha of irrigation area (83% of the total), which had been developed and opened by DSI (State Hydraulic Works, General Directorate), was irrigated with surface water and 17% from groundwater resources [36]. According to the average of the DSI data for many years, irrigation rate is 65% and irrigation efficiency 45%. Due to surface irrigation methods, irrigation efficiency or the amount of water reaching the product is estimated to be less than 40% worldwide [37].

Since 2003, irrigation systems in Turkey, which are known as open or classical systems, have been abandoned and closed systems (drip irrigation, sprinkler) started to be constructed. From 2003 to 2013, it is stated that more than 1.1 million ha of closed irrigation systems have been applied. In addition, the efficiency of classical irrigation systems, which cause wasting scarce water resources, is around 60%, while the efficiency of modern closed systems is 90%, respectively [38].

Climate change is a critical issue in the management of agricultural water resources. The selection of suitable plant species for complying climate change impacts for countries having water scarcity problems expected to occur in future like in Turkey is of crucial concern, where sustainable water management should be achieved by providing maximum benefits with minimal investment. For example, instead of sugarcane, which consumes high amounts of water, other industrial plants that need less water and provide almost the same economic value can be preferred (e.g., sugar beet) [39, 40]. In horticultural crops, dwarf species should be preferred to provide maximum benefit from the applied water, fertilizer and pesticides. Climate models should be taken into consideration when selecting suitable species for climate change. Thus, the effects of future climate change would be reduced. Erosion prevention techniques are mandatory in agricultural areas in Turkey, typically having steep and multi-directional slopes as they proceed from the West to the East. For this reason, erosion prevention measures can be tightened, and soil, fertilizer and pesticide losses can be reduced.

As a result, if only pressurized irrigation systems are practiced, nitrogen fertilizer losses can be reduced by 50%, whereas phosphorus and potassium losses can be reduced by at least 30%. However, it can be said that the world's total fertilizer consumption will decrease by at least 35%. This can be accounted as US$ 85 billion in financial value for 2020 [41].

Assessment of water management in irrigation applications in Karapinar Basin, Konya

Karapınar basin (Konya, Turkey), also known as the Konya-Çumra-Karapınar basin, has a semi-arid climate with an area of approximately 8,822 km^2 in the Konya Closed Basin. In recent years, decreasing rainfall and global warming have caused water stress in the basin due to the increasing temperatures proving its unsustainability. Having a deficit in water budget, the underground water level drops by 1 m every year, and the ground water reserves are depleting. Subsequently, owing to the falls of water level, a large number of sinkholes are formed in the region threatening the safety of life and property of the people there. In consideration of the constrictions in the agricultural production, life in the region will become dangerous, the existing desertification symptoms will increase, and the basin will become desert eventually [42].

Aydin and Kumcu [43] analyzed and discussed the water potentiality of Konya-Karapınar region, the excess use of available water in agriculture and sustainability of water resources. They stated that irrigation of the agricultural areas has resulted in excess surface and groundwater use in the basin. The main reasons of water scarcity in this area are unplanned increase of the irrigated areas and the current crop patterns. In order to reduce the water demand in the agriculture, they recommended cultivation of less water consuming species rather than high water requiring plants, using renewable solar energy for pumping groundwater.

3.4 Morocco

Dahchour and El Hajjaji [44] presented the actions taken by the Moroccan Government to face the challenge of water management in Moroccan cities. These actions can be briefly presented as follows.

National Strategies to Tackle the Problem of Water Scarcity

Enhancing Irrigation Efficiency

Saving 2 BCM/yr in irrigation by conversion to drip by drip system, saving 0.4 BCM/yr by improvement of the yield in agriculture, saving 0.2 BCM/yr by improvement of the network of drinking water.

Development of Supply

60 new major dams, 1,000 small dams, north–south transfer: 0.8 BCM, desalination of sea water, reuse of treated wastewater, harvesting rainwater.

Protection and preservation of water resources

Reduction in levies and building control, artificial recharge of groundwater, protection of the water quality, Implementation of sanitation programs and wastewater treatment, prevention of and sanctions against Industrial pollution, implementation of a national plan for household and similar wastes.

Reducing vulnerability to climate change

Identifying vulnerable sites to floods (400 sites), strengthening forecasting and warning, combat against the effects of drought (diversification of sources of water supply).

Regulatory reforms and institutional

Review of the law 10–95. It has been reviewed and enriched to adapt to new requirements for the development of water resources.

Infrastructure and capacity building

Development of applied science and research, capacity building and training, development of suitable information systems.

National Water Quality Protection Plan (NWQPP)

Management of water does not rely on the efficiency of its mobilization and usage only. Measures for its protection have to be taken. In this context, the national water quality protection plan (PNAL) was launched in 2005 to cope with the expected shortness in water supply by providing alternative resources and to ensure a better quality of water. Actually, drought in Morocco has become more prevalent in the past decade and is predicted to be more sever due to the expected decrease of 4% over the next 20 years [45].

The current total volume of sewage discharged in Morocco was estimated about 0.6 BCM/yr. 48% of these waters are discharged into rivers or applied to land, the rest is discharged into the sea. This amount is expected to reach 0.9 BCM/yr. The pollution load from wastewater is estimated to 131,715 t of organics, 42,131 t of nitrogen, and 6,230 t of phosphorus annually. On the basis of an average 30 to 60 m^3/yr of wastewater production per capita, a population of 1 million inhabitants and a rate of irrigation of 10,000 m^3/ha, this amount of wastewater would be sufficient to irrigate 3,000–9,000 ha land [46]. In 2016, the rate of treated wastewater in 113 treatment plants has reached 71% (instead of 5% in 1999). The sewer network has attained 10,000 km with a flux of 500,000 m^3/d [47].

Management of Drinking Water

The main producer of drinking water in Morocco is the National Office of Water and Electricity (ONEE) established in 1972. The current water production exceeds 55 m^3/s. ONEE is also in charge for distribution of water in the cities via its regional agencies. Since the mid-1990s, in the big cities such as Casablanca, Rabat, and Tangier, the distribution is delegated through concessions to the private sector (Lydec in 1997, Redal in 1998, and Amendis in 2009). These concessions cover fairly the needs of 40% of the population. At present, 95% of the population is provided with drinking water in urban zones and 91% in rural zones [47].

The control of water distributed is basically under the supervision of ONEE. Its main laboratory was founded in 1968. Private companies involved in the distribution perform some complementary analyses. Some urgent analyses are performed by mobile laboratories at different locations when necessary. The control of drinking water covers more than 4,600 sampling points and 70,000 samples/yr.

Water-Energy-Food strategy for sustainable development in Morocco

National water strategy of Morocco has allowed securing provision and supply of water to the consumers. More than 140 dams with a storage capacity of 17 BCM are destined to irrigate 1.5 million ha farm land. Special authorities and agencies allow a well management of water in terms of providing, supplying and protecting water. Reuse of wastewater is not yet authorized, but tests of acceptability of real benefits are positive so far.

Regarding the energy issues, the shift towards renewable energies will relief the country from the burden associated with the costs of petroleum derivatives and promotes Morocco as a leader in renewable energies. The total current energy capacity installed in Morocco in 2015 is 8,154 MW, with 34% of renewable energies according to the following mix: carbon (32%), natural gas (11%), hydropower (22%), fuel oil and diesel (24%), solar (2%) and wind power (10%). Hydroelectric sources are also improving to achieve a total power capacity of 2,280 MW.

Moroccan economy relies mainly on agriculture and some limited industrial sectors as food and textile industry that represent fairly 15% of GDP. During the last two decades, Morocco has launched a new economic strategy aiming at diversifying and boosting the industrial sector through an industrial emergency plan. Management of water has been dealt with through a national strategy for storage of surface water started in the 1960ies to irrigate 1 million ha, followed by the improvement of access to drinking water and the national plan of liquid sanitation. In the energy issue, Morocco has to face the regular increase of demand that has exceeded 105 tons of oil equivalents (toe). The adoption in 2009 of an ambitious strategy enables Morocco to shift towards other sources of energy, mainly solar and wind sources. The construction of four solar energy plants (Noor 1–4) tends to achieve a ratio of self-generation of energy of 42% by 2020 and of 52% by 2030.

4 Conclusion and Recommendation

Agricultural water footprint in the MENA countries will be increasingly problematic due to relatively high population growth rates. The seriousness of the problem increases as the countries mainly suffer from physical water scarcity, which makes the solution even harder. In conclusion, in almost all MENA countries, sustainable management of water resources is required. Based on these findings, following issues require attention [2].

Demand–Supply Policy

Countries in these regions are experiencing localized water stress, coupled with an increasing inability of traditional water sharing systems to cope with growing demand, with the new users, and with competition between sectors and individual users. The general solution is the introduction of government-managed water sharing systems, where water rights are formalized and better specified. This will raise a significant challenge for comprehensive regulations and good field management for water allocation operation, which will be able to provide equity, sustainability and economic efficiency. Demand management is a key instrument and should be preferred. Water footprint methodology can be used as a Decision Support Tool for managing valuable water resources.

Water Quality

Most MENA countries are facing increasing evidence of surface as well as groundwater pollution. There is a strong challenge for strengthening policy and institutional framework, improving compliance and enforcement capabilities, and developing public participation. Several countries have developed water quality monitoring and classification systems. However, it has been observed that statistically sound monitoring designs, sound long term data storage facilities, and standardized reporting systems are lacking or do not exist. One main reason identified is the fragmentation of the water quality related activities within institutional framework in most countries. This poses a significant challenge for cooperation in water quality management especially in transboundary river systems.

Water data issues

Statistics on water data reported by several agencies as FAO and World Bank differ from each other to some extent. The situation is quite common in the world and can usually be explained by different authors reporting different sources and years. But for the MENA Region, there is considerable work required to present a consistent and complete picture of its water resources management. The existence of good data is an absolute requirement for reasonable water resources management. In order to develop a water resources management strategy useful, consistent and credible datasets supporting the regional water policy and water management are required.

Water for Food

In almost all MENA countries, most of available water is used for irrigation. Given the demands on water resources coming from other sectors and the environment, and considering the growing population, it is important to limit withdrawals yet to grow more foods. Thus, there is a general opinion that the target should be for a zero percentage increase in irrigation water demand for an increase in water productivity. This is posing a clear challenge for the region.

Drinking Water and Sanitation

Millennium Development Goals Progress Report of UN for the year 2014 indicates differences across MENA countries in respect to both access to safe drinking water and improved sanitation.

Water Related Disasters and Climate Change

Many countries in the MENA Region are vulnerable to floods, causing damages to agricultural production, urban infrastructure and touristic industry. Coupled with the growing number of population and the economic development, the potential damage values are increasing. It has been indicated as well that climate change has made the situation worse in many areas. Concerned governments are applying structural and non-structural measures to cope with their problems. It is now also considered important to develop capacity building in integrated management and climate change adaptation.

Water Governance

Currently, MENA countries are facing not only with irrigation matters, but with a multitude of water related problems, such as urban flash floods, landslides and erosion, water scarcity, polluted water ways, and contaminated ground water. Implementing strategies for integrated water resources management (IWRM) will play a crucial role in managing all these water related problems. Institutional reforms, policy changes, good water governance, and trans-boundary cooperation are the essential parts of IWRM that will ensure management of water resources in a sustainable way.

Acknowledgements This literature review was prepared mainly referring to papers published from scientists being members of the Exceed Swindon network or supported through this project. The authors would like to acknowledge DAAD financial support for Exceed Swindon project which made an invaluable contribution to water related issues in the MENA countries as well as in other regions in the world through providing the opportunity to cooperate within such broad science community.

References

1. http://siteresources.worldbank.org/INTMNAREG TOPWATRES/Overview/20536158/WATER-ENG.pdf
2. Şahin H, Karadirek EI, Topkaya B. Maintaining agricultural water security in MENA region. In: Mahjoub B, Aydin ME, editors. Water perspectives in emerging countries, water scarcity: risks and solutions. Cuvillier Verl., Göttingen; 2018. pp. 12–23. ISBN 978-3-7369-9841-4. eISBN 978-3-7369-9841-5.
3. FAO. Crop statistics; 2009. Retrieved from the web site of FAO. http://faostat3.fao.org/faostat-gateway/go/to/download/Q/QC/E.
4. FAO. AQUASTAT; 2013. Retrieved from the web site of FAO. http://www.fao.org/nr/water/aquastat/data/query/index.html.
5. Al-Omari A. Challenges to the water sector in the MENA region. In: Mahjoub B, Aydin ME, editors. Water perspectives in emerging countries, water scarcity: risks and solutions. Cuvillier Verl., Göttingen; 2018. pp. 2–11. ISBN 978-3-7369-9841-4. eISBN 978-3-7369-9841-5.
6. AFED. Arab environment climate change: impact of climate change on Arab Countries. In: Tolba M, Saab N, editors. 2009 report of the Arab forum for environment and development. Technical Publications and Environment & Development Magazine; 2009. pp. 47–62. ISBN: 9953-437-28-9.
7. Arab Center for the Studies of Arid Zones and Dry Lands (ACSAD). Water resources and their uses in the Arab world. In: Proceedings of the first Arab symposium on water resources and their uses in the Arab world, organized by ACSAD, Arab Fund for Economic and Social Development (AFESD) and Kuwait Fund for Arab Economic Development (KFAED), Kuwait; 1997. pp. 25–121.
8. Intergovernmental Panel on Climate Change (IPCC). Impacts, adaptation and vulnerability. In: Parry ML, Canziani OF, Palutikof JP, van der Linden PJ, Hanson CE, editors. Contribution of working group II to the fourth assessment report of the intergovernmental panel on climate change. Cambridge, United Kingdom and New York, NY, USA: Cambridge University Press; 2007. ISBN: 978 0521 88009-1 Hardback; 978 0521 70596-7 Paperback, 2007.
9. Chenoweth J, Hadjinicolaou P, Bruggeman A, Lelieveld J, Levin Z, Lange M, Xoplaki E, Hadjikakou M. Impact of climate change on the water resources of the eastern mediterranean and middle east region: modeled 21st century changes and implications. Water Resour Res. 2011;47:W06506. https://doi.org/10.1029/2010WR010269.
10. Zyadin A. Water shortage in MENA region: an interdisciplinary overview and a suite of practical solutions. J Water Resourc Protect. 2013;5:49–58.

11. Mirkin B. Population levels, trends and policies in the Arab region: challenges and opportunities. Arab human development report, report. Research Paper Series. New York: UNDP Regional Bureau for Arab States; 2010. pp. 6–39. https://www.undp.org/content/dam/rbas/report/Population%20Levels, Trends.pdf.
12. Roudi-Fahimi F, Kent M, Challenges and opportunities—the population of the Middle East and North Africa. Popul Bull. 2007;62(2):24.
13. World Bank. A water sector assessment report on the countries of the cooperation council of the Arab States of the Gulf. Report No. 32539-MNA, vol. 1, 2005. http://documents.worldbank.org/curated/en/415761468052156362/A-water-sector-assessment-report-on-the-countries-of-the-Cooperation-Council-of-the-Arab-States-of-the-Gulf.
14. United Nations Development Program (UNDP). Arab development challenges report 2011: towards the developmental state in the Arab Region; 2011. https://www.undp.org/content/undp/en/home/librarypage/hdr/arab-development-challenges-report-2011.html.
15. Wade A, Black E, Brayshaw D, El-Bastawesy M, Holmes P, Butterfield D, Nuimat S, Jamjoum K. A model-based assessment of the effects of projected climate change on the water resources of Jordan. Philos Trans Ser A, Math Phys Eng Sci. 2010;368(1931):5151–5172.
16. Kannan A. Reallocation of water resources in the Arab region: an emerging challenge in water governance. Eur J Sustain Develop. 2014;3(3):283–98.
17. Beaumont P. Water policies for the Middle East in the 21st century: the new economic realities. Int J Water Resour Dev. 2002;18(2):315–34.
18. Zetlanda D, Gasson C. A global survey of urban water tariffs: Are they sustainable, efficient and fair? Int J Water Resour Dev. 2013;29(3):327–42.
19. AFED. Arab environment: water-sustainable management of a scarce resource. In: El-Ashry M, Saab N, Zeitoon B, editors. Annual Report of the Arab Forum for Environment and Development (AFED). Beirut, Lebanon: Technical Publications; 2010. ISBN: 978-9953-437-55-2. https://www.sekem.com/wp-content/uploads/2016/12/ENglish2015L.pdf.
20. Abu-Zeid M, Hamdy A. Water crisis and food security in the Arab world: where we are and where do we go. Arab Water Council Meeting, 14 April, Cairo, 76 p., 2004. Nasr, M.: Assessing Desertification and Water Harvesting in the Middle East and North Africa: Policy Implications, Policy Paper no. 10, Center for Development Research, Bonn University, Germany; 1999.
21. League of Arab States (LAS). United Nations environment program (UNEP), Center for environment and development for the Arab region and Europe (CEDARE): environmental outlook for the Arab region; 2010.
22. FAO. Global food price monitor: with focus on the Horn of Africa. Rome; 2011. ISBN 978-92-5-106927-1. http://www.fao.org/3/a-i2330e.pdf
23. Mehmet O, Bicak H. Modern and traditional irrigation technologies in the eastern Mediterranean. Ottawa: International Development Research Centre; 2002.
24. Rached E, Brooks D. Water governance in the Middle East and North Africa: an unfinished agenda. Int J Water Resour Dev. 2010;26(2):141–55.
25. UNESCO. World water assessment program. World water development report 4: managing water under uncertainty and risk, vol. 1; 2012. p. 210.
26. Kavvas M, Chen Z, Anderson M, Ohara N, Yoon J, Xiang F. A study of water balances over the Tigris-Euphrates watershed. Phys Chem Earth. 2011;36(5–6):197–203.
27. Abu-Hashim M, Lilienthal H. Management water resources—can deficit irrigation adapt water scarcity in MENA? In: El Hajjaji S, Dahchour A, Dichtl N, editors. Water perspectives in emerging countries, water use in MENA countries. Cuvillier Verl., Göttingen; 2017. p. 276–87. ISBN 978-3-7369-9749-3. eISBN 978-3-7369-8749-4.
28. Hassan A, Ismail S, Elmoustafa A, Khalaf S. Estimation of evaporation losses in high Aswan dam (HAD) reservoir using remote sensing techniques. In: Mahjoub B, Aydin ME, editors. Water perspectives in emerging countries, water scarcity: risks and solutions. Cuvillier Verl., Göttingen; 2018. p. 209–20; ISBN 978-3-7369-9841-4. eISBN 978-3-7369-9841-5.
29. https://fr.slideshare.net/KhalilAmiri/tunisie-recherche-scientifique-20172022-priorits-orientations-futures-et-initiatives-cls.
30. http://www.watchwater.tn/fr/node/11.
31. Elaboration de la Vision et de la Strategie Eau 2050 de la Tunisie, Banque Africaine de Developpement, FAD, Novembre 2010. http://www.africanwaterfacility.org/fileadmin/uploads/awf/Projects/AWF-Project-appraisal-report-TUNISIA-VISION2050.pdf.
32. Ulusal Havza Yönetim Stratejisi, Türkiye Cumhuriyeti, Tarım ve Orman Bakanlığı; 2014. http://suyonetimi.ormansu.gov.tr/Libraries/su/uhys_belgesi_3.sflb.ashx, https://www.tarimorman.gov.tr/SYGM/Belgeler/havza%20koruma%20eylem%20planlar%C4%B1/Konya_Kapali_Havzasi.pdf.
33. Havza Koruma Eylem Planlarının Hazırlanması Projesi - Konya Kapalı Havzası, Türkiye Bilimsel ve Teknolojik Araştırma Kurumu, Marmara Araştırma Merkezi, Cevre ve Temiz Üretim Enstitüsü, Tarım ve Orman Bakanlığı, Kasım; 2013.
34. Ulusal kuraklık yönetimi strateji belgesi ve eylem planı, Tarım ve Orman Bakanlığı; 2017. http://suyonetimi.ormansu.gov.tr/Libraries/su/Ulusal_Kurakl%C4%B1k_Y%C3%B6netimi_Strateji_Belgesi_ve_Eylem_Plan%C4%B1.sflb.ashx.
35. http://www.dsi.gov.tr/docs/hizmet-alanlari/tarim-sulama.pdf?sfvrsn=2.

36. Pimentel D, Houser J, Preiss E, White O, Fang H, Mesnick L, Barsky T, Tariche S, Schreck J, Alpert S. Water resources: agriculture, the environment, and society. Bioscience. 1997;47(2):97–106.
37. http://www.ormansu.gov.tr/osb/haberduyuru/guncelhaber/14-09-06.
38. http://www.fao.org/docrep/S2022E/s2022e02.htm.
39. http://www.fao.org/faostat/en/#data/PP.
40. Güneş K. Agricultural water resources management, environmental and financial evaluation—a case study. In: Abu-Hashim M, Aydin ME, editors. Water perspectives in emerging countries, water–energy–food NEXUS in MENA Region. Cuvillier Verl., Göttingen; 2018. p. 199–205. ISBN 978-3-7369-9943-5. eISBN 978-3-7369-8943-6.
41. Özdemir A, Aydın ME. Problems caused by improper water and land management and global climate change effects in Karapinar Basin (Konya, Turkey). In: El Hajjaji S, Dahchour A, Dichtl N, editors. Water perspectives in emerging countries, water use in MENA countries. Cuvillier Verl., Göttingen; 2017. p. 100–10. ISBN 978-3-7369-9749-3. eISBN 978-3-7369-8749-4.
42. Aydin ME, Kumcu SY. Assessment of water management in irrigation applications in the Karapinar Basin, Konya, Turkey. In: Abu-Hashim M, Aydin ME, editors. Water perspectives in emerging countries, water–energy–food NEXUS in MENA region. Cuvillier Verl., Göttingen; 2018. p. 141–48. ISBN 978-3-7369-9943-5. eISBN 978-3-7369-8943-6.
43. Dahchour A, El Hajjaji S. The challenge of management of water in Moroccan cities. In: El Hajjaji S, Dahchour A, Dichtl N, editors. Water perspectives in emerging countries, water use in MENA Countries. Cuvillier Verl., Göttingen; 2017. p. 2–8. ISBN 978-3-7369-9749-3. eISBN 978-3-7369-8749-4.
44. Rochdane S, Reichert B, Messouli M, Babqiqi B, Jacobi Khebiza M. Climate change impacts on water supply and demand in Rheraya watershed (Morocco), with potential adaptation strategies. Water. 2012;4:28–44.
45. Choukr-Allah R, Hamdy A. Wastewater treatment and reuse as potential water resource for irrigation. In: Hamdy A, editors. The use of non-conventional water resources. Proceedings of the international workshop Alger, Algeria, 12–14 June 2005. Options Méditerranéennes, Séries A n. 66; 2005. p. 101–24.
46. Makhokh M, Bourziza M. Country report for the expert. Consultation on wastewater management-Morocco-SEEE. 22–24 May 2011, Dubai, UAE; 2011.
47. UN: Millennium Development Goals Report; 2014. Retrieved from the website of UN. http://www.un.org/millenniumgoals/reports.

Reservoir Sedimentation Under Changing Climate

S. Yurdagul Kumcu and M. Ali Kokpinar

Abstract

During the past decades, Earth's river systems have shown significant changes, particularly concerning their flow characteristics with time and space, and fluvial sediment loads which are carried by them. These changes are most likely due to climate change. As climate change is uncertain and its effects are not uniform in both temporally and spatially, such variations lead to increasing concerns about the hydrological regimes of river basins. The rise in Earth's near-surface air temperature and changes in precipitation patterns are prominent features of climate change. Although climate change is primarily considered as an increase in temperatures and global warming, its most important effects are those that will occur due to the change in precipitation regime. The hydrological system is directly and/or indirectly affected by the climatic conditions in the world. Changes in temperatures affect evapotranspiration rate, cloud characteristics, soil moisture, storm intensity, and snowfall and melting regimes. As a result, changes in precipitation cause changes in temporal, spatial and severity of flood and drought events. In addition, it causes changes in the surface flow regime, the amount of water seeping underground, plant pattern and growth rates. Reservoirs are essential in managing floods and droughts, but their effectiveness can be influenced by climate change. Sediment yield, which may increase due to climate change, can cause a rapid decrease in the reservoir's storage capacity and shorten its design life. The accumulation of sediment at the reservoirs is a significant problem for many aspects of water management, like flood control, energy production, irrigation and drainage systems, domestic water supplies, river transportation systems, and recreational activities. Sediment accumulation can lead to shortening economic lifespan of the dam by resulting in severe consequences. In this study, the effect of climate change on reservoir sedimentation is investigated and it is depicted that the amount of sedimentation is considerably increasing by the effect of climate change.

Keywords

Climate change · Loss of reservoir storage · Reservoir sedimentation · Sediment yield

S. Y. Kumcu (✉)
Department of Civil Engineering, Necmettin Erbakan University, Konya, Turkey
e-mail: yurdagulkumcu@gmail.com

M. A. Kokpinar
Department of Civil Engineering, TED University, Ankara, Turkey

A. M. Bahadir et al. (eds.), *Hydrology and Urban Water Supply*, Water and Wastewater Management,
https://doi.org/10.1007/978-3-031-72589-0_6

1 Introduction

Scientists have identified close links between water issues and climate change and identified climate change as the main cause of disruption to the water cycle and changes in water quantity and quality. Despite the constancy of the water supply on Earth due to the water cycle, managing water resources in terms of quantity and quality is becoming increasingly challenging due to the variability of water resources at different locations and over time, leading to potential difficulties in the near future.

Climate change refers to a broad concept causing alterations in various climate factors, mostly in temperature and precipitation. Global warming is considered the main reason behind these changes. Over the past few decades, numerous researchers have focused on working on global warming due to its far-future impacts on various aspects of our lives. The primary factor responsible for global warming is the rise in the planet's temperature, which, according to NASA, increased by approximately 1 °C during the twentieth century. Even though rise in may be considered minor, its consequences are significant and can result in lasting damage to the planet [1]. The accelerated melting of glaciers in Antarctica and Greenland is having a significant impact on ocean currents, contributing to rising sea levels and more frequent occurrences of destructive storms and hurricanes globally [2]. The climate change increases average temperatures of both on land and sea. It increases extreme weather events. Additionally, changes in precipitation patterns have been observed, with previously rainy areas experiencing more rainfall, particularly in the winter time, while arid regions have become even drier, especially during the summer months. The rise in sea levels, the loss of ice mass in arctic and mountainous regions, and the occurrence of extreme weather events, such as waves, floods, and droughts, have resulted in an increase in their frequency and intensity. Heavy rainfall and frequent thunderstorms resulting from climate change are the contributing factors to floods. Flood is one of the most important results of changes and is also expected to have noticeable effects on soil since rainfall and runoff are the factors controlling soil erosion and sediment transport within landscapes. It is worth noting that heavy rainfall and floods play a vital role in erosion and sediment transport towards the dam reservoirs, which are crucial structures that provide reliable water supply, hydropower, and flood control. Reliable predictions of the quantity and rate of runoff, and sediment transport are needed to help decision makers in developing watershed management plans for better soil and water conservation measures.

Numerous research studies have predicted a diverse range of impacts on streamflow, soil erosion, and sediment load resulting from climate change. The aim of this study is to offer a concise overview of how climate change affects sediment accumulation in reservoirs.

2 Climate Change

Commonly, climate change means global warming which is continually increasing in average global temperature. The current rise in global average temperature is more rapid than previous changes, and is primarily caused by humans burning fossil fuels [3, 4]. Using fossil fuels, deforestation, increasing industrial development, and agricultural works increase the amount of greenhouse gasses especially carbon dioxide (CO_2) and methane gasses at the atmosphere [5]. If the earth surface warms up, some part of the back-emitted heat as infrared radiation is absorbed by these gases, which causes global warming by trapping more heat at Earth's lower atmosphere. Because of climate change, deserts are enlarging, while heat waves and wildfires are encountered more commonly [6]. Increased warming in the Arctic has given rise to melting permafrost, glacial melting and sea ice loss [7]. Higher temperatures are also causing more severe storms, droughts, and other weather extremes [8]. Rapid environmental change in mountains, coral reefs, and the Arctic is forcing many

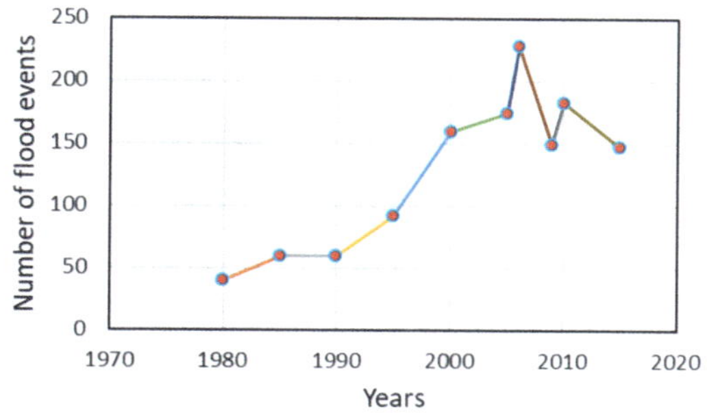

Fig. 1 General increase in the change of flood events in the world [10]

species to relocate or to become extinct. Some studies may help with reducing warming up of the Earth, preventing oceans from ongoing heating, acidification and rising of water level [9].

Climate change has wide effects on water resources depending on the region. Water related extremes such as runoff, flooding, or sea level rise are more critical in some regions than water shortages. While increased evaporation rate might dry out some regions, excessive precipitation can put some other areas in danger. Furthermore, warmer air can hold more water vapor, which can lead to more intense rainstorms, causing major problems like extreme flooding. With climate change, the flood disasters are increasing rapidly in the world, as seen in Fig. 1 (Emergency Events Database (EMDAT), 2019) [10]. As in the world, flood events are seen in Turkey due to climate change. Table 1 summarizes the number of flood events, loss of life and economic losses in the last 45 years. When Table 1 is inspected, it is seen that the total number of 15-year flood events is 1528 between 2006 and 2020. This value corresponds to 73.3% of total number of 2084 observed during the total number of 45-year flood events. This situation, clearly shows the effect of the factors of global warming, industrialization, etc. on the floods in recent years [11].

When Table 1 is inspected, it is seen that the total number of 15-year flood events is 1528 between 2006 and 2020. This value corresponds to 73.3% of total number of 2084 observed during the total number of 45-year flood events. This situation, clearly shows the effect of the factors of global warming, industrialization, etc. on the floods in recent years.

Climate change threatens people with extreme heat and more frequent and severe rain events and so floods, can increase erosion and result in greater amounts of sediment washing into rivers, streams, lakes and reservoirs.

3 Soil Erosion

Stronger storms, higher river levels, and faster stream velocity can increase the rate of erosion and result in higher suspended sediment load in water bodies. These parameters affect normal distribution of bed load along river, lake and stream beds as well. The sediment load carried by streams comes from various sources. The sediment load transported by streams varies from very coarse material to very fine material, and the major portion of the sediment load carried by streams comes from the erosion of material in the drainage basin as seen in Fig. 2 [13].

Table 1 Flood events and their effects in Turkey during 1975–2020 [12]

Time interval	Number of flood	Loss of life	Effected area (ha)
1975–1985	263	126	293.167
1986–1995	143	271	241.403
1996–2005	150	120	175.068
2006–2015	683	212	184.295
2015–2020	845	57	6.064
Total	2.084	786	959.997

Fig. 2 **a** Variation of size of sediment particles along the streams. **b** Sediment erosion

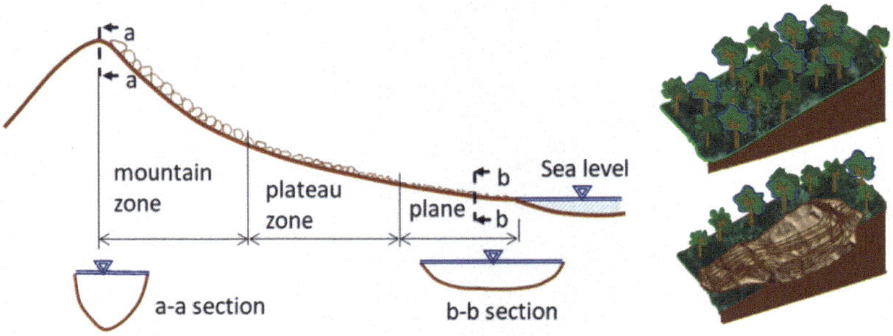

4 Floods

Among the most destructive natural hazards are the floods caused by river overflows. A warmer climate with its increased climate variability will increase the risk of floods [14, 15]. Flooding is an overflowing of water onto land that is normally dry. Flooding may occur as an overflow of water from water bodies, such as a river, lake, or ocean, in which the water overtops or breaks levees, resulting in some of that water escaping its usual boundaries [16]. Floods include river floods, flash floods, urban floods and sewer floods, and can be caused by intense and/or long-lasting precipitation, snowmelt, dam break, or reduced conveyance due to ice jams or landslides. Floods depend on precipitation intensity, volume, timing, antecedent conditions of rivers and their drainage basins (e.g., presence of snow and ice, soil character, wetness, urbanization, and existence of dikes, dams, or reservoirs). Floods carry sediment particles and these particles are deposited at the reservoir as the flow velocity slows down. Finer particles get deposited in front of/upstream of the dam structures. Since sediment accumulation decreases the reservoir capacity, the operational life of reservoirs gets negatively affected.

5 Sediment Transport and Sediment Yield

Factors such as precipitation, water flow in the form of layers on the ground surface, winds and glaciers cause erosion by removing the grains on the surface of the ground. These removed grains are dragged into the river by the water flow [17]. The amount of this material depends on the size of the river basin, soil structure, topography, vegetation, wind as well as the intensity and duration of precipitation. This load is called suspended load. Frequent and intense rain events can increase suspended sediment loading coming from storm water runoff. In a stream, in addition to the sediment that comes from the river basin by soil erosion, there may also be grains removed by the stream from its own bottom by channel erosion. These sediment particles are the source of bed load. While suspended load depends on hydrological characteristics of river basin, bed load depends on hydraulic characteristics of river flow. Although suspended load does not directly depend on river flow characteristics, it generally increases with discharge capacity of the flow as a result of increase in discharge capacity of river and erosion. It is clear that, if the velocity of the flow increases, the sediment carrying capacity of the river increases, too. In many rivers, a large amount of the sediment load is suspended load with a rate of 80–95% (Fig. 3).

Sediment is the final product that emerges from the erosion of the soil surface getting affected by the factors such as water, wind and glacier. Water resource development projects are mostly affected by solid particles carried by water. However, not all eroded sediment reaches the stream completely. Some of these materials are deposited in natural or man-made barriers or in channels and floodplains. The part of the whole amount of eroded material reaches a downstream stream measurement location is called Sediment Yield (SY). The value of SY in

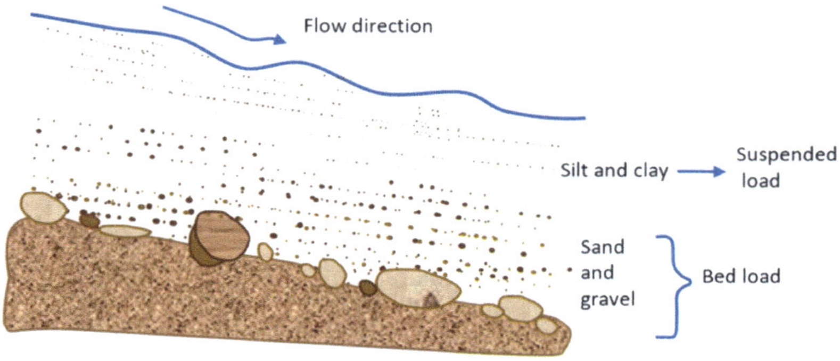

Fig. 3 Type of sediment load transport along the flow

unit watershed is called Sediment Yield Ratio (SYR), and it is expressed in terms of "sediment weight/area/year" (ton/km^2/year). It is estimated that 14 billion tons of the sediment particles that goes to the oceans in the world in a year is suspended sediment and 1 billion tons is the bed material. The transportation of this amount of sediment particles to the oceans means that the entire land surface is eroded by 57.5 mm in 1000 years. This amount of erosion does not have a uniform distribution over the world and shows regional differences [18].

Almost half of the suspended solids which is poured into the oceans originate from the Asian continent. On the other hand, just 8% comes from African and European continents. The highest sediment transport rates are in the South Pacific islands of Japan, Taiwan, the Philippines, Indonesia, New Guinea and New Zealand.

Obtaining reliable sediment data along the rivers requires long-term continuous monitoring. The methods used to determine SY depend either directly on measurement results or indirectly on calculation methods. Sediment quantity obtained by direct measurements is the most reliable SY determination method. To achieve this, it is necessary to measure solids transport either by mapping in reservoirs or by sampling solids from rivers.

6 Reservoir Sedimentation

Most of the sediment deposited in the reservoirs of dams is composed of suspended sediment load caused by the basin erosion. Suspended sediment rate is of great importance and to be taken into account in the dead volume calculation of dam reservoirs. In many basins, most of the erosion occurs in the upper parts of the basin where the slope is steep. By the way, the most important magnitude affecting erosion is precipitation, especially precipitation that causes floods.

Reservoir siltation due to soil erosion is among the critical environmental, scientific, and engineering challenges currently. Reservoirs help to slows water runoff, alter the microclimate of the landscape, mitigate flash flooding, aid in erosion control, accumulate water in the landscape, and filter out waste materials. Dams can serve as a significant local source of drinking and industrial water supply as well as for energy provision and fish breeding. They also provide a habitat for water plants and animals, promoting biodiversity and serving as essential elements of landscape's beauty. Dams are also frequently used for recreational activities and sports facilities. However, the efficiency of reservoirs for water management and other purposes gradually

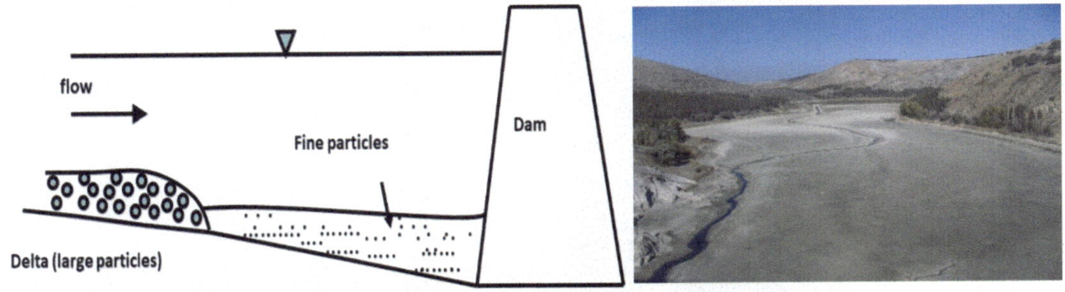

Fig. 4 **a** Sediment load accumulation along the reservoir base, **b** siltation in Cubuk 1 Dam (Turkey)

decreases with time due to the continuous emergence of deposits in watersheds, leading to the gradual siltation of water reservoirs (Fig. 4).

Catchment management strategies such as erosion control plans, drainage management, and runoff control must be put in place to minimize sediment transport via rivers and drainage networks. This will help with reducing sediment yield and to prevent high sediment loads from accumulating.

7 Effects of Climate Change on Reservoir Sedimentation

Scientists have established a close relationship between climate change and water, as the former is considered the leading cause of disruptions in the water cycle and alterations in both the quantity and quality of water. Despite the constancy of the water supply on Earth due to the water cycle, managing water resources in terms of quantity and quality is becoming increasingly challenging due to the variability of water resources at different regions and with time, leading to potential difficulties in the near future.

The rapid melting of ice and glaciers, caused by the rising temperatures, is responsible for the increase in sea levels, which, in turn, can lead to floods. Based on the latest report published by World Meteorological Organization (WMO) in 2022, the average sea level across the world, also known as the global mean sea level (GMSL), has shown ongoing increase. Another type of flood occurs, when water evaporated from one region falls in another region as unexpected amounts and at unexpected times and places [19].

The siltation of reservoirs is a natural phenomenon that has been observed to have significantly changed in recent years. Numerous authors from different parts of the world have reported an unprecedented increase in the siltation of water reservoirs [20–26]. Siltation has become a significant issue, leading to a considerable build-up of sediments in almost all reservoirs across the World. According to some authors, for example, the storage capacity of European reservoirs has been reduced by up to one-third over the last 50 years due to sediment accumulation [27]. The process of siltation is closely associated with soil erosion, which primarily depends on factors such as rain fall, soil erodibility, vegetation cover, topographic features of the land, and measures taken to protect against soil erosion. Changes in climate, land use, and agricultural practice can have a significant impact on the siltation of water reservoirs. As a result, these factors can contribute to the rapid accumulation of sediments in reservoirs and alter their capacity for water storage and other essential functions. Some authors estimate that sediment accumulation in the reservoirs will be faster in the future by the effect of climate change. The influence of climate change on suspended sediment transport in Danish rivers was studied by Thodsen et al. (2008) using the HIRHAM regional climate model. Results incorporating projected changes in land use/land cover for the period 2071–2100 showed an increase in suspended sediment transport in the winter months as a result of the increase in river discharge, and decreases during summer and early autumn months [28, 29] used the Soil and Water Assessment Tool SWAT model to simulate the impacts of climate change on stream discharge and sediment yield from the Song Cau watershed in Vietnam. The results showed that the highest changes in stream discharge (up to 11.4%) and sediment load (up to 15.3%) can be expected in wet seasons in the 2050s according to the high-emission scenario (A2), while for the low-emission scenario the corresponding changes are about 8.8% and 12.6%. A similar study of a quantitative estimate of changes in sediment loads (1991–2007) in response to climate change in eight large Chinese rivers was also conducted by Lu et al. 2013 [30]. Precipitation change coupled with rising temperatures has played a significant role in influencing the sediment delivery dynamics. They found that, lower precipitation along with rising temperatures has significantly reduced sediment loads delivered into the sea in semi-arid climates (4–61%) and increasingly warmer and wetter climates in subtropical zones has yielded more sediment (0.4–11%), although the increase was offset by human impact. It is clear that sediment yield is significant parameter in the reduction of reservoir capacity. Azari et al. (2016) tried to assess the impacts of climate change on streamflow and sediment yield from the Gorganroud River basin in the North of Iran. In order to study the effects of climatic variations, they implemented the SWAT model for simulating the hydrological regime and the SUFI-2 algorithm for parameter optimization. The climate change scenarios were constructed using the outcomes of three general circulation models for three emission scenarios. The study results for 2040–2069 showed an increase in annual streamflow of 5.8%, 2.8% and 9.5%, and an increase in sediment yield of 47.7%, 44.5% and 35.9% for the A1F1, A2 and B1 emission scenarios, respectively. This implies that the impact of climate change on sediment yield is greater than on streamflow. Monthly variations show that the increase in discharge and sediment yield is more pronounced in wet seasons, and the decrease is more pronounced in the summer period (July to September). The results highlighted the strong impact of climate change and reflected the importance of incorporating such analysis into adaptive management [31]. It is obvious that more intense rainfalls may lead to more soil erosion, which can be further fostered by changes in land use. Increasing in the sediment load, carried by soil erosion to the water body, accelerates sediment accumulation in the reservoir. Thus, the reservoir operation becomes worsening and dam infrastructure is rapidly damaged. In 2019, Zhang et al. conducted a case study on the Zhenjiangguan watershed in order to estimate the relation between climate change and sediment yield. They used 1980–1999 as base period for comparison the estimated data. They used three downscaled general circulation models with two emission scenarios to generate possible climatic conditions for three future periods. These periods are P1 (2020–2039), P2 (2050–2069) and P3 (2080–2099). The temperature and precipitation projections showed a significant increasing trend compared with the baseline condition. Climate change was expected to increase the average annual runoff by 15–38% compared with the base period, and the average annual sediment yield will increase by 4%–32% [3]. Chen et al. (2020) also analysed the influence of climate change on sediment yield variation, sediment transport and erosion deposition distribution at the watershed scale of Gaoping River basin, which is among the largest basins in southern Taiwan. It is implemented that there has been a gradual increase of precipitation and typhoon intensity in the last 55 years. Both the frequencies of occurrence and the severity of flooding were showing an increasing trend. They utilized the Physiographic Soil Erosion Deposition (PSED) model and found a general increase in soil erosion and deposition volume under the A1B-S climate change scenario. The situation is even worsened with increasing return periods. Total erosion volume and sediment yield in the watershed were increased by 4–25% and 8–65%, respectively, and deposition volumes increased by 2–23%. The study showed how climate change variability would influence the watershed through increased sediment yields, which might even worsen the impacts of natural disasters [5]. Sirisena et al. (2021) conducted a study on Irrawaddy River to figure out the effects of climate change and reservoir impact on streamflow and sediment load. They used SWAT program as a tool with two end-member Representative Concentration Pathways; RCP 2.6 and RCP 8.5. They used the streamflow and sediment load data belonging to baseline period 1991–2005 as reference to estimate the future data. They made a projection for the mid-century of 2046–2065 and end-century of 2081–2100. Their results showed that streamflow and sediment amounts are increased as 8–17% and 13–26% under RCP 2.6, and 9–45% and 18–75% under RCP 8.5, respectively [20]. The relationship between reservoir sediment aggradation and rainfall, temperature and inflow is also studied by Mazhar et al. (2021) for the period of 1980–2017. They found that the inflow in the Tarbela reservoir is the strongest variable in sediment generation. The increase of sedimentation in the reservoir is due to the fact that the sediment load gets transported downstream and most of it ends up settling in Tarbela reservoir due to its high trap efficiency [32]. Ahmad et al. (2021) examined the impact of reservoir sedimentation and climate change by 2050 on water supply of Indus Basin Irrigation Systems (IBIS). As climate change will likely result in increasing temperatures across the basin, particularly at higher altitudes, the results show that this will result in more glaciers melting in upcoming decades. Although the overall impacts of climate change on flows are uncertain, the general expectation is for greater flows in the next few decades. They indicated that the volume and timing of the water supply of IBIS rely on reservoirs which are accumulating sediment continuously resulting in reduced water storage volume. So, they conducted some model studies using the scenarios of reservoir sedimentation and their interaction with and without climate change. Their scenarios were current condition and nine potential future scenarios that represent reservoir sedimentation, climate change, and combined reservoir and climate change impacts on water balances of the IBIS. They described the reservoir sedimentation scenarios assuming that current sedimentation rates keep continuing to 2050. Reservoir sedimentation will diminish the water supply in the Rabi season by 2050 to Sindh-Balochistan and sedimentation combined with climate change in Sindh-Balochistan exacerbates the decrease in water supply in Rabi. They concluded that no single strategy will be sufficient. Water supply, demand policies and management should be taken into consideration all together. Water supply management options include reservoir sedimentation management [33]. Abdul Razad et al. (2021) studied the impacts of climate change on sediment influx rate in Ringlet Reservoir. As climate change causes more frequent and intense rainfall events, leading to severe erosion in the catchment and sediment transferred into rivers and reservoirs, long-term term sediment load accumulation resulted in by rainfall. The studied catchment receives high amount of sediment throughout the year. Although the rainfall projections in the future showed a decreasing trend, the comparison of annual sediment loads for the historical period of 2000 to 2016 and the future projection period of 2017–2030 showed an increase of 12%. This is because of extreme high rainfalls [34].

Do et al. (2022) predicted the long-term period of Pleikrong Reservoir sediment deposition by SWAT model with 14 bias-corrected GCM/RCM models under two emissions scenarios, They showed that the reduction of reservoir storage capacity because of sedimentation varied from 25 to 62% by 2050 depending on the different climate change models. The model shows that the reservoir capacity losing in considering the impact of climate change is much faster than the one with compared to when there is no climate change considered. They also noted that sedimentation rate of reservoirs is much higher than that in the designed process. The used assumptions for dead volume of reservoirs are general but incorrect assumptions. Therefore, there is a need to better understand of the spatial distribution of deposition of sediments in reservoirs [32]. Li et al. (2022) studied the impacts of changes in climate and land use/land cover on hydrological processes and sediment yields in the Xin'anjiang Reservoir basin. They used SWAT model and performed GCM projections for the future period of 2068–2100. The results showed that all scenarios of climate changes, land use/land cover alterations, and combined changes projected an increase in sediment yields in the basin. Under three representative concentration pathways (RCP), climate change significantly increased the annual sediment yields (by 41.03–54.88%). Deforestation may also increase the annual sediment yield (by 1.1–1.2%) in the future. The comprehensive influence of changes in climate and land use/land cover on sediment yields was 97.33–98.05% (attributed to climate change) and 1.95–2.67% (attributed to land use/land cover change) at the annual scale, respectively. This means that during the period 2068–2100, climate change will exert a much larger influence on the sediment yields than land use/land cover alteration in Xin'anjiang Reservoir Basin if the future land use/land cover remains unchanged after 2015 [35].

As a result of these studies it is to be emphasized that climate change has a strong and negative effect on the reservoir sedimentation due to precipitation variation, land erosion and temperature increase.

8 Conclusion

The process of siltation of dams and natural reservoirs is closely associated with soil erosion, which primarily depends on factors such as rain erosivity, soil erodibility, vegetation cover, and topographic features of the land. Changes in climate can have a significant impact on rain erosivity and soil erosion, which can affect the siltation of water reservoirs. The siltation of reservoirs is a natural phenomenon that has been observed to have significantly changed in recent years. The rapid accumulation of sediments in reservoirs, resulting in decrease in water storage capacity, is important for reservoir operation systems and the aim of the reservoir use. In future scenarios, this requires practicable and result-oriented management strategies, and their applications are to be applied in action plans accordingly.

References

1. NASA. Global climate change, vital signs of the planet, climate spiral (1880–2022). 21 Feb 2023.
2. Pattyn F, Ritz C, Hanna E, Assay-Davis X, DeConto R, Durand G, Favier L, Fettweis X, Goelzer H, Golledge NR, Munneke PK, Lenaerts JTM, Nowicki S, Payne AJ, Robinson A, Seroussi H, Trusel LD, van den Broeke M. The Greenland and Antarctic ice sheets under 1.5 °C global warming. Nat Clim Change. 2018;8:1053–61.
3. Zhang S, Li Z, Lin X, Zhang C. Assessment of climate change and associated vegetation cover change on watershed-scale runoff and sediment yield. Water. 2019;11(7):1373.
4. Fluixá-Sanmartín J, Altarejos-García L, Torres AM, Ignacio Escuder-Bueno I. Review article: climate change impacts on dam safety. Nat Haz Earth Syst Sci. 2018;18(9):2471–88.
5. Chen CN, Samkele ST, Tsai CH. Climate change impacts on soil erosion and sediment yield in a watershed. Water. 2020;12(8):2247.
6. IPCC SRCCL. Summary for policymakers; 2019.
7. Pörtner H-O, Roberts DC, Masson-Delmotte V, Zhai P, Tgnor M, Poloczanska E, Mintenbeck K, Alegria A, Nicolai M, Okem A, Petzold J, Rama B, Weyer NM. The ocean and cryosphere in a changing climate, a special report of the intergovernmental panel on climate change. IPCC; 2019.
8. Seneviratne SI, Zhang X, Adnan M, Badi W et al. Chapter 11: weather and climate extreme events in a changing climate. IPCC AR6 WG1; 2021.
9. Rompaey AV, Govers G, Puttemans C. Modelling land use changes and their impact on soil erosion and sediment supply to rivers. Earth Surf Process Landf. 2002;27:481–92.
10. EMDAT. 2019. www.emdat.be. Accessed 20 March 2023.
11. Li H, Yu C, Qin B, Li Y, Jin J, Luo L, Wu Z, Shi K, Zhu G. Modeling the effects of climate change and land use/land cover change on sediment yield in a large reservoir basin in the east Asian Monsoonal region. Water. 2022;14:2346.
12. Kutsal C. Problems encountered on flood control facilities. MS Thesis (in Turkish), Konya, Turkey; 2020.
13. Yanmaz M, Kumcu SY. Sediment problems and properties lecture notes of post-graduate course in sediment transport technology; 2009. p. 27–39.
14. Wetherald RJ, Manabe S. Simulation of hydrologic changes associated with global warming. J Geophys Res. 2003;107:1–15.
15. Bernstein L, Bosch P, Canziani O, Chen Z, Christ R, Riahi K. IPCC, 2007: climate change 2007: synthesis report; 2007.
16. Kumcu SY. Flood management under changing climate, water and wastewater, global problems and measures. Switzerland: Springer; 2022. p. 35–40.

17. Technical Report, TUBITAK. The effect of sedimantation problem of HES in energy planning (in Turkish), 105G098, Ankara, Turkey; 2011.
18. Kumcu SY, Kökpınar MA. Measurement of sediment transport, lecture notes of post-graduate course in sediment transport technology; 2009. p. 65–89.
19. Hamlet AF, Lettenmaier DP. Effects of 20th century warming and climate variability on flood risk in the western U.S. Water Resourc Res. 2007;43(6).
20. Sirisena TAJG, Maskey S, Bamuawala J, Ranasinhe R. Climate change and reservoir impacts on 21th century sreamflow and fluvial sediment loads in the Irrawaddy River, Myanmar. Front Earth Sci. 2021;9: 644527.
21. FD2628 impact of climate change on dams & reservoirs, final guidance report, ATKINS. https://assets.publishing.service.gov.uk/media/6033dfa3e90e076605eab4de/Review_of_indirect_Impacts_of_climate_change_on_dams_and_reservoirs_final_report.pdf. Accessed 28 March 2023
22. Choukri F, Raclot D, Naimi M, Chikhaoui M, Nunes P, Huard F, Herivaux C, Sabir M. Distinct and combined impacts of climate and land use scenerios on water availability and sediment loads for a water supply reservoirs in Northern Morocco. Int Soil Water Conserv Res. 2020;8:141–53.
23. Shrestha B, Maskey S, Babel GA, Uhlenbrook S. Sediment related impacts of climate change and reservoir development in the lower Mekong River Basin: a case study of the Nam Ou Basin, Lao PDR. Clim Change. 2018;149:13–27.
24. Jusco V, Sedbak R, Kudela P. Siltation of small water reservoir under climate change: a case study from forested mountain landscape of western Carpathians, Slovakia. Water. 2022;14(17):2606.
25. Altinbilek D. The role of dams in development, international energy symposium. Ossiach; 2001. p. 59–63.
26. Nerantzaki SD, Giannakis GV, Nikolaidis NP, Zacharias I, Karatzas GP, Sibetheros IA. Assessing the impact of climate change on sediment loads in a large Mediterranean watershed. Soil Sci. 2016;181 (7):306–14.
27. Novak P, Moffat AIB, Nalluri C, Narayanan R. Hydraulic structures. 4th ed. Oxon: Taylor and Francis; 2006.
28. Thodsen H. Climate change induced changes on suspended sediment transport in Danish Rivers. EGU General Assembly Conference Abstracts, 10624; 2010.
29. Phan DB, Wu CC, Hsieh SC. Impact of climate change on stream discharge and sediment yield in Northern Viet Nam. Water Resour. 2011;38:827–36.
30. Lu C, Dong X, Tang J, Lju G. Spatio-temporal trends and causes of variations in runoff and sediment load of the Jinsha River in China. J Meteorol Sci. 2019;16:2361–2378.
31. Azari M, Moradi HR, Saghafian B, Faramarzi M. Climate change impacts on streamflow and sediment yield in the North of Iran. Hydrol Sci J. 2016;61 (1):123–33.
32. Mazhar N, Mirza AI, Abbas S, Akram MAN, Ali M, Javid K. Efects of climatic factors on the sedimentation trends of Tarbela Reservoir, Pakistan; 2021. p. 3122.
33. Ahmad MA, Pena-Arancibia JL, Yu Y, Stewart JP, Podger GM, Kirby JM. Cimate change and reservoir sedimentation implications for irrigated agriculture in the Indus basin Irrigation System in Pakistan. J Hydrol. 2021;603(126967):1–18.
34. Abdul Razad AZ, Shamsuddin SH, Setu A, Sidek LM. Future impacts of limate change on sediment influx rate in hydropower reservoir using SWAT, STACLIM 2021. In: IOP conference on series, earth and environmental science, vol. 880 (012024); 2021. p. 1–11.
35. Do XK, Nguyen TH, Ngo LA, Felix ML, Jung K. Prediction of reservoir sedimentation in the long term period due to the impact of climate change: a case study of Pleikrong reservoir. J Disaster Res. 2022;17 (4):552–60.

Role of Dams on Climate Change Adaptation, Mitigation and Resilience

Esra Siltu, A. Murat Hatipoglu, S. Yurdagul Kumcu, and Bulent Selek

Abstract

Water resources in Turkey are not sufficiently well distributed and need to be better managed to meet and ensure current and future needs. Turkey is becoming more urbanized while population increases, industry grows, and the demand for agricultural production increases. Due to global warming and climate change, freshwater is increasingly becoming a scarce resource. Water resources should be managed efficiently to optimise the benefits of water, e.g. for energy supply, irrigation and drinking water supply, and to minimise negative environmental impacts such as erosion and flooding. In recent decades, Turkey has made great strides in developing and managing water resources for irrigation, flood control, power generation and other purposes through the construction of dams. Through the construction of dams, Turkey can store water during the rainy season for use throughout the year for irrigation, power generation, drinking water supply and sanitation. Especially in times of floods and droughts, the reservoirs play an important role. As a result of climate change dry seasons become drier and drier and heavy rainfall to cause sudden and increasingly frequent flooding, the importance of dams cannot be ignored. Currently, 992 dams are in operation in Turkey. This study examines the role of dams in climate change adaptation using Turkey as an example.

Keywords

Climate change · Climate change adaptation · Dam reservoir · Mitigation · Resilience

1 Introduction

Despite the fact that there are adequate water resources globally, distribution of available freshwater resources varies both spatially and temporally [1]. Moreover, climate change has incontrovertible impacts on water resources and it is essentially a water crisis. While the world gets warmer, the hydrological cycle intensifies and pressure on freshwater ecosystems increases as a result of changes in precipitation regimes [2]. Due to climate change, the amount of saline water is increasing compared to freshwater reserves, land uses are changing, glaciers and snow masses are melting, groundwater use is excessively increasing, continents are becoming drier and sea levels are rising [3]. Some parts of

E. Siltu · A. M. Hatipoglu · B. Selek
General Directorate of State Hydraulic Works, Ankara, Turkey

S. Y. Kumcu (✉)
Department of Civil Engineering, Necmettin Erbakan University, Konya, Turkey
e-mail: yurdagulkumcu@gmail.com

A. M. Bahadir et al. (eds.), *Hydrology and Urban Water Supply*, Water and Wastewater Management,
https://doi.org/10.1007/978-3-031-72589-0_7

the world receive less rainfall than in the past, while others receive more, and rainfall and temperature regimes are changing unpredictably. As a result of these, significant increases in the frequency and severity of water-related disasters are observed worldwide. According to United Nations, 74% of the disasters occurred worldwide between 2001 and 2018 were water-related, and in the last 20 years, more than 166,000 people have lost their lives due to the floods and droughts and the resulting economic damage approximates to 700 billion dollars [3]. In addition, 2.3 billion people, including 733 million at high and critical levels, live in countries experiencing water stress [4]. On the other hand, the quality of water resources is deteriorating, with around 1.4 million people losing their lives each year due to the use of drinking water contaminated with pathogens and lack of proper treatment of wastewater [5]. All these facts reveal that, development and management of water resources needs to be given more importance than ever before since development and management of water resources may provide effective means of combating climate change. Sustainable management of water resources is the key to improve the resilience of communities to the impacts of climate change as well as reducing carbon emissions. Although the sustainable water resources management comprehends various measures and approaches, construction and operation of dams have a central role in climate change adaptation and mitigation as they improve the storage capacity and ensure water supply during drought periods, regulating flows in streams and preventing floods as well as generating clean energy by hydroelectric power plants and reducing carbon emissions. Through these features, dams make a major contribution to food, water and energy security, disaster risk reduction, protection of human health and welfare, continuity of industrial development and economic growth and achieving 2030 sustainable development goals.

In this study, the role of dams on climate change adaptation, mitigation and resilience will be assessed through the case of Turkey. The status of water resources in Turkey is presented firstly and the role of Turkey's dams in climate change adaptation, mitigation and resilience efforts is assessed.

2 Status of Water Resources and Dams of Turkey

Turkey's total annual precipitation is 450 billion m^3 and the average annual areal precipitation is 574 mm. The average annual surface runoff (natural flow) is 185 billion m^3. However, the annual amount of usable surface water is 94 billion m^3, secure exploitable groundwater reserve is 18 billion m^3, and the total amount of annual usable water is 112 billion m^3. Accordingly, Turkey's per capita water availability in 2022 is 1,322 m^3. According to the Falkenmark index (Falkenmark, 1989 as cited in Brown and Matlock, 2011), shows that Turkey is among the countries experiencing water stress [6]. Falkenmark indicator limits are given in Table 1.

The change of Turkey's annual available freshwater per capita from 1960 to 2022 is shown in Fig. 1.

As can be interpreted from the graph in Fig. 1, Turkey's amount of annual available freshwater per capita has decreased significantly over the years as a result of population growth, industrialization, increase in agricultural production,

Table 1 Falkenmark indicator limits [5]

Annual per capita available freshwater (m^3/year/capita)	Status
>1700	No stress
1000–1700	Stress
500–1000	Scarcity
<500	Absolute scarcity

Fig. 1 The change of Turkey's annual available freshwater per capita [6]

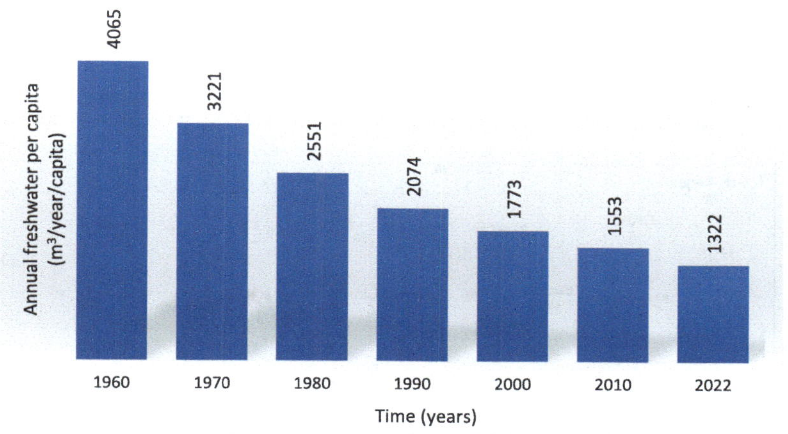

changes in water use habits and decrease in precipitation due to climate change.

In addition to that, precipitation and water potential across the country are not distributed evenly. Although the precipitation and water potential are high in the Eastern Black Sea Region where population, agricultural and industrial activities are not intense, water potential is low in the Aegean and Marmara Regions where production activities and population are very intense. The distribution of the amount of water per capita in Turkey's 25 water basins is shown in Fig. 2 [7].

Figure 2 demonstrates that Marmara and Kücük Menderes Basins are in the category of absolute water scarcity and the amount of freshwater per capita is at very low levels. In Burdur, Gediz, Susurluk, Akarçay, Büyük Menderes, Asi and Sakarya Basins, which are important regions in terms of industrial activities and agricultural production, the amount of freshwater per capita is at the water scarcity level. This situation explicitly reveals the need to develop water resources to improve the storage capacity and improve the climate change adaptation and resilience. According to General Directorate of State Hydraulic Works (DSI) Annual Reports of 2022 [8], there are 992 dams under operation in Turkey. Most of the largest dams of Turkey; Keban, Karakaya and Atatürk are constructed on Euphrates River. Optimum operations of these dams depend on forecasting incoming water in early summer season. Snowmelts in spring and early summer season are the main source of water in Euphrates River and snowmelt constitutes the 65–70% of the annual flow in Euphrates. Moreover, snowmelts recharge the groundwater resources in the area.

Besides these, the total amount of groundwater allocated across Turkey is approximately 17.80 billion m^3 in 2022. 64% of allocated groundwater is used for agricultural irrigation, 24% for domestic purposes and 9% for industrial purposes, and almost all of the secure exploitable reserve has already been allocated. On the other hand, 43% of the available surface water potential has been allocated. In 2022, the amount of surface water used is approximately 41 billion m^3. Of the total allocated surface and groundwater, 77% is used for agricultural irrigation, while 23% is used for domestic purposes and industrial sectors.

3 Impacts of Climate Change on Turkey's Water Resources and Turkey's Climate Change Adaptation Policy

Turkey is located in Mediterranean climate zone which is expected to be impacted by climate change intensively. The annual average

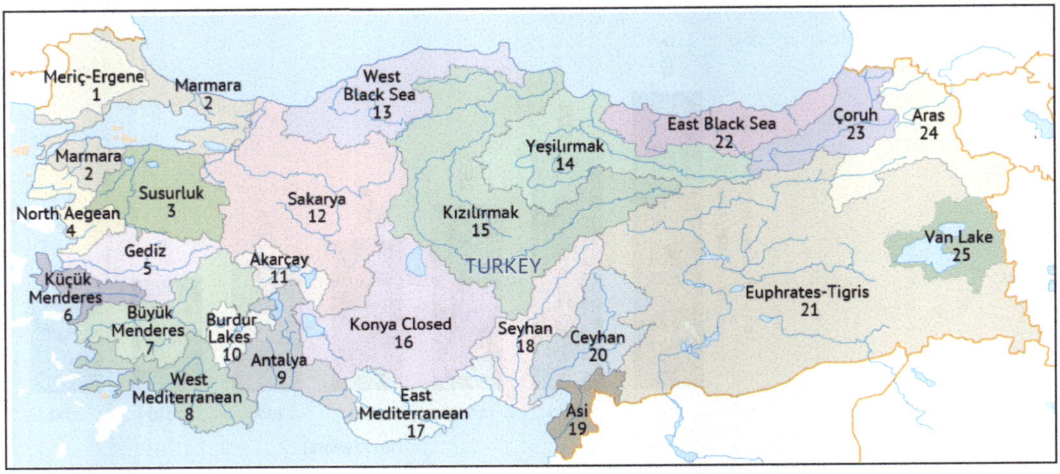

(a)

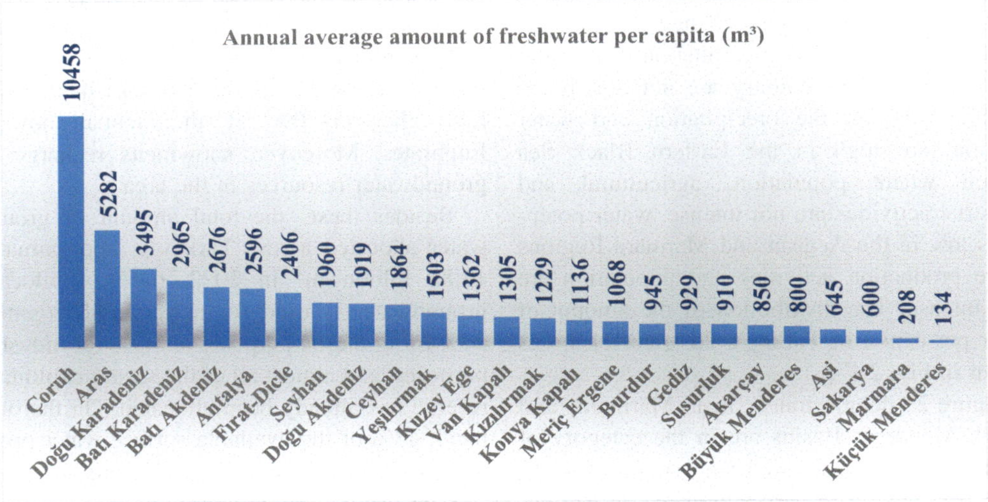

(b)

Fig. 2 **a** Turkey's hydrological basins (*source* Fanack Water), **b** distribution of annual average available freshwater per capita according to basins [7]

temperatures are forecasted to increase by 2.5–4 °C in Turkey while up to 4 °C increase in Aegean and East Anatolian regions and up to 5 °C increase in inner regions are forecasted to occur [9]. "Baseline water stress" is defined as the ratio of total annual freshwater withdrawals to the annual amount of available freshwater [10]. The Middle East and North Africa, including Turkey, are estimated to be the regions of the world where baseline water stress will be the most severe [11]. Figure 3 shows the level of baseline water stress in the Middle East and North Africa.

Water stress in Turkey is predicted to worsen further with the impact of climate change. River basins are also estimated to face many risks due to changing climate conditions. One of these basins is the Euphrates-Tigris Basin. The amount of precipitation is decreasing in the interior and southern parts of the country, especially in the

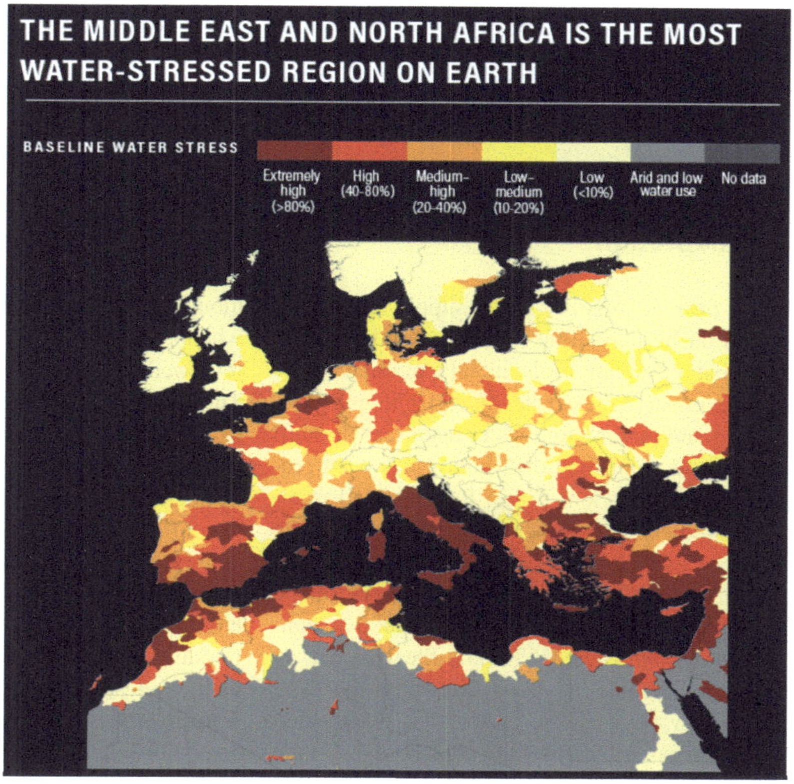

Fig. 3 Baseline water stress in Middle East and Africa [11]

Euphrates-Tigris Basin. Another projected risk is the increase in average temperatures, especially in winter. As a result of the increase in average temperatures in winter, precipitation during the winter occurs in the form of rain rather than snow. This leads to a decrease in snow cover which is very critical for groundwater recharge. The rise in temperatures also causes the snow to melt earlier in the spring months and spring floods occur earlier. Another impact of climate change is the increase in precipitation in the summer period in Northern and Western Anatolia. Due to increased precipitation, the possibility of floods in these regions increases [12]. On the other hand, the increase in the duration and severity of droughts due to climate change increases the importance of improving the water storage capacity and the need for storage facilities. In a study investigating the effects of climate change on the inflow flows and stored water volume of Demirköprü Dam in the Gediz Basin, it is estimated that although the inflow is expected to increase compared to previous years, they will decrease due to climate change throughout the twenty-first century [13]. In addition to these, in a study evaluating the probability of drought occurrence across Turkey based on SPI (Standard Precipitation Index) values, it was found that the probability of SPI values of the Mediterranean coastal area and the Syrian border line being less than zero, in other words, the probability of these areas being arid was higher than 50% [14]. All of these facts reveals that climate change adaptation is mainly an issue of water resources management.

In this context, water resources management is one of the main pillars of Turkey's National Climate Change Adaptation Strategy and development of water resources to increase water storage capacity and to reduce the risk of floods are defined as major adaptation measures. Besides contributing to climate change adaptation, dams in Turkey are contributing to climate change mitigation efforts substantially through hydroelectric energy generation. Share of hydroelectric energy in total annual energy generation is 31% in 2022 with a total installed capacity of 31.712 MW.

4 Role of Dams on Climate Change Adaptation

Dams provide regular water supply and flood protection by regulating flow in streams and are the most powerful tools in adaptation climate change impacts. The role of dams in reducing drought and maintaining water resources through storage and controlled distribution has been recognized since ancient times. Recent flood modelling studies have shown that dams reduce the exposure of the world's population to floods by up to 20.6%. According to the outcomes of the projections made under the representative concentration pathways RCP 6.0, the number of people vulnerable to flooding would be 15.3 million per year by the end of the twenty-first century in case no dams were constructed. On the other hand, 12.9% fewer individuals are subjected to flooding by the protection of existing dams [15].

As previously mentioned, river basins in Turkey are vulnerable to the devastating impacts of climate change. These impacts are far more observable in South and Southeast Anatolia, especially in Euphrates-Tigris Basin. The hydrological regime of the Euphrates-Tigris river system which is mainly fed from snow precipitation is characterized by variable precipitation and sharp flow fluctuations within years, with significant floods resulting from spring snowmelt. Intense spring floods that occur in April and May also cause the loss of crucial water required for irrigation and electricity production, and therefore drought in summer and autumn months. The variation can be clearly observed when the historical discharge values are reviewed. Annual discharge of Euphrates at Turkish-Syrian border range from minimum flow of 14 km^3/a (1961) to a maximum of 57 km^3/a (1969). The discharge values for the Tigris at the Turkish border dropped to 7 km^3/a in 1961 and rose to 34 km^3/a in 1969 [16]. On the other hand, historic flood events have recorded in Euphrates in 612 B.C.E., and in Tigris River in 1896. The entire area of Nineveh city was destroyed by the flooding of Euphrates in 612 B.C.E. and Baghdad was destroyed by the flood in Tigris resulting from rise of water level up to 3 m in 1896. The river flow was regulated, these variations were compensated and impacts of floods and droughts were reduced by the large dams constructed on Euphrates and Tigris rivers starting from the 1970s and extending into the twenty-first century [17].

Figure 4 shows that the impacts of projected increase in frequency and intensity of extreme wet seasons and drought events in the twenty-

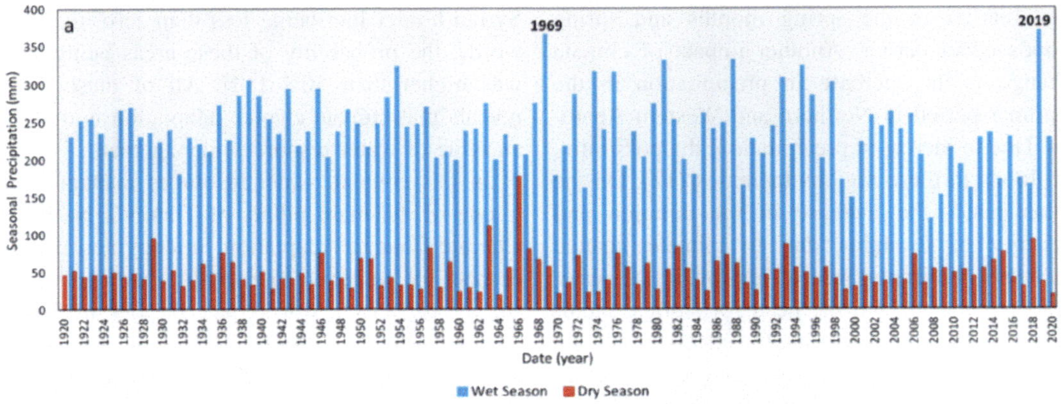

Fig. 4 Precipitation time series of Tigris Euphrates Basin derived from GPCC data (1920–2020) [16]

first century due to global warming over the Tigris Euphrates Basin. Extreme drought and wet periods were identified for the period of 1920–2020. It is depicted from this figure that, in case of no dams are constructed in river basins, most of the flow is carried out of the watershed by river system and water cannot be utilized effectively. Starting from the 1970s, major dams were constructed on Euphrates-Tigris river system and significant amount of water was impounded in reservoirs of dams. Thanks to the storage capacity created by these dams, flooding and the resulting destructive impacts of a 1 in 100 extreme rainfall event in 2019 could be prevented. Prior to this extreme precipitation event, a prolonged (2007–2018) and intense drought was experienced in Euphrates-Tigris Basin. These extreme conditions are concrete implications of climate change. Although extreme weather events occurring as a result of climate change will cause disruptive social and economic impacts, many of the well-engineered designed watersheds can defense it selves to these anticipated impacts by taken precautions. Similarly, in Euphrates-Tigris Basin, the river flow was regulated by dams in Turkey, Syria, Iraq and Iran and flood risks were minimized [17].

Mohamed et al. (2021) [18] had an assessment on the impacts of climate change on Tigris-Euphrates Basin and they found out that this watershed is well prepared for modulating the projected climatic variabilities of the twenty-first century and reducing hazards associated with flooding events.

5 Role of Dams on Climate Change Mitigation

Hydroelectric power plants generate clean and renewable energy. By the use of hydropower, 200 million metric tons of carbon emission was avoided in the U.S. each year which equals to the emission of more than 38 million passenger cars [19]. As compared to conventional coal power plants, hydropower provides 3 GT per year decrease in CO_2 emissions annually and that amount corresponds to 9% of global annual CO_2 emissions.

Share of electricity production from hydropower was 16.6% in 2016 throughout the world [20]. In 2021, approximately 55% of renewable energy generated in world was hydropower [21]. Hydropower can be tool to mitigate climate change by contributing to reducing of greenhouse gas emissions. However, it can also be impacted by climate change due to the variations in flow and amount of stored water. Nevertheless, it is expected that the impacts of climate change on hydropower generation would be limited even though there may be variations between different regions of the world. Hydropower is still recognized as a safe, cheap and clean energy source and has a crucial role in global climate change mitigation efforts [22]. It is now generally accepted that hydropower generation is an important alternative to generate electricity. Its greenhouse gas emissions are significantly less compared with fossil fuels and is able to stabilize renewable energy systems based on wind and solar power [23]. With about half of the world's electricity currently coming from hydropower, it is the foundation of low-carbon electricity production (Fig. 5). The contribution of hydropower is more than that of nuclear power by 55% and greater than that of all other renewable energy sources combined, including wind, solar PV, biofuels, and geothermal. According to data of 2020, hydropower constitutes the 17% of the world's electricity production, making it the third-largest source of energy after coal and natural gas. Global hydropower capacity increased by 70% over the past 20 years, but its proportion to overall output remained same as a result of the expansion of wind, solar PV, coal, and natural gas [24].

In 2022, 34.6% of our electricity is produced from coal, 22.2% from natural gas, 20.6% from hydraulic energy, 10.8% from wind, 4.7% from sun, 3.3% from geothermal energy and 3.7% from other sources. Distribution of Turkey's installed capacity by resources as of the end of February 2023; 30.3% from hydraulic energy, 24.4% from natural gasses, 20.9% from coal,

Fig. 5 Low-carbon electricity generation by technology, 2020 [25]

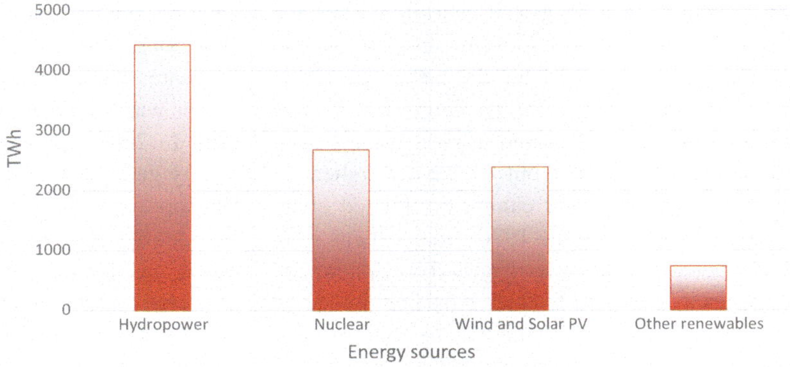

11% from wind, 9.3% from solar energy, 1.6% geothermal energy and 2% 5% of it is in the form of other resources. Turkey's hydroelectric energy potential corresponds to 5% of the world and 16% of Europe [25]. The dams, having the highest installed capacities under operation are constructed on the Euphrates and Tigris. Among them, Atatürk Dam and HEPP, Karakaya Dam and HEPP and Keban Dam and HEPP, have constructed on the Euphrates by having installed capacities of 2405 MW, 1800 MW and 1330 MW, respectively. Ilısu Dam and HEPP is located on the Tigris with installed capacity of 1209 MW.

6 Conclusion

Today, water reservoirs such as dams are an element that can help reduce greenhouse gas emissions in the context of renewable energy. Dams are also water reservoirs that help bridge periods of drought. In this context, cost–benefit analyses must be carefully analysed at the planning stage, taking into account possible environmental impacts and the costs involved.

References

1. Yildiz V, Hatipoglu MA, Kumcu SY. Climate change impacts on water resources, water and wastewater management. Springer; 2022. p. 17–25.
2. Oki T, Kanae S. Global hydrological cycles and world water resources. Science. 2006;13(5790).
3. Bates BC, Kundzewicz ZW, Wu S, Palutikof JP. Climate change and water. Technical paper of the intergovernmental panel on climate change. IPCC Secretariat, Geneva; 2008.
4. UNESCO, UN-Water. United nations world water development report 2020: water and climate change. Paris: UNESCO; 2020.
5. UNEP. UN environment programme. Annual Report; 2019.
6. Brown A, Matlock MD. A review of water scarcity indices and methodologies. White Paper, Sustain Consortium. 2011;106(1).
7. Fanack Water; 2023. https://water.fanack.com/summer-of-drought-lessons-from-iran-and-turkey.
8. DSI. Annual Report, Ankara, Turkey; 2020.
9. Ministry of environment, urbanization and climate change, climate change national action plan 2011–2023. Ankara; 2012.
10. Hofste RW, Reig P, Schleifer L. Aqueduct 3.0: updated decision relevant global water risk indicators. Technical Note. Washington, DC: World Resources Institute; 2019.
11. Hofste RW, Reig P, Schleifer L. 17 countries, home to one-quarter of the world's population, face extremely high water stress. World Resources Institute; 2019.
12. Demircan M, Gürkan H, Eskioğlu O, Arabacı H, Coskun M. Climate change projections for Turkey: three models and two scenarios. Turk J Water Sci Manage. 2017;1(1):22–43.
13. Gorguner M, Kavvas ML. Modeling impacts of future climate change on reservoir storages and irrigation water demands in a Mediterranean Basin. Sci Total Environ. 2020;748: 141246.
14. Turkes M. Climate and drought in Turkey, water resources of Turkey. Cham: Springer; 2020. p. 85–125.
15. Boulange J, Hanasaki N, Yamazaki D, Pokhrel Y. Role of dams in reducing global flood exposure under climate change. Nat Commun. 2021;12:417.
16. Altınbilek D. Development and management of the euphrates-tigris basin. Water Resourc Develop. 2004;20(1):15–33.

17. Abdelmohsen K, Sultan M, Save H, Abotalib A, Yan E, Zahran K. Buffering the impacts of extreme climate variability in the highly engineered Tigris Euphrates river system. Sci Rep. 2022;12:4178.
18. Mohamed S, Karem A, Himanshu S, Abotalib F, Eugene Y, Zahran KH. Assessment of the impacts of climate variability on highly engineered watersheds: a case study from the Tigris Euphrates watershed. AGU fall meeting 2021, New Orleans, LA, H21A-02; 2021. p. 13–7.
19. National Hydropower Association, Clean & Sustainable; 2023.
20. Our world in data, share of electricity production from hydropower. Retrieved on 11 March 2023 from: https://ourworldindata.org/grapher/share-electricity-hydro?tab=chart. 2023.
21. Ritchie H, Roser M, Rosado P. Energy. Published online at OurWorldInData.org; 2022. Retrieved from: https://ourworldindata.org/energy.
22. Berga L. The role of hydropower in climate change mitigation and adaptation: a review. Engineering. 2016;2:313–8.
23. Biswas A, Tortajada C. Roles of hydropower and dams in society, including mitigation and adaptation to climate change. World Atlas: Hydropower and Dams; 2021. p. 8–9.
24. International Energy Agency. Hydropower special market report analysis and forecast to 2030, report; 2021.
25. Republic of Turkey Ministry of Energy and Natural Resources. Executive summary—hydropower special market report—analysis. IEA, Hydropower Special Market Report; 2023.

Drought Management and Planning for Water Resources

Veysel Yildiz, Mustafa Onur Onen, Cemil Arslan, and Murat Ali Hatipoglu

Abstract

Droughts have serious consequences for ecosystems, agriculture, water supply and energy, requiring proactive measures to mitigate long-term and potentially irreversible damages. Droughts are divided into four types: meteorological, agricultural, hydrological and socio-economic. Each of these categories is found in arid, semi-arid and tropical regions of the world. Drought monitoring and forecasting studies help decision-makers and planners to adapt to future water management challenges. The objective of this chapter is to provide a comprehensive analysis of the impacts of droughts on water resources. It highlights how changes in the climate influence drought frequency and severity, emphasizing the growing uncertainties surrounding water availability. It also discusses adaptation strategies, contrasting crisis management with risk management for sustainable water use. Special attention is given to Turkiye's drought management, including basin-level plans, infrastructure improvements, and agricultural water conservation initiatives. These insights aim to enhance planning and policy for mitigating drought risks and securing water resources.

Keywords

Climate change · Global water cycle · Uncertainty · Water resources

V. Yildiz (✉)
Department of Civil and Structural Engineering, The University of Sheffield, Sheffield, UK
e-mail: vyildiz1@sheffield.ac.uk

M. O. Onen · C. Arslan · M. A. Hatipoglu
General Directorate of State Hydraulic Works, Ankara, Turkey
e-mail: moonen1@sheffield.ac.uk

C. Arslan
e-mail: cemil.arslan@metu.edu.tr

M. A. Hatipoglu
e-mail: mhatipoglu@dsi.gov.tr

C. Arslan
Department of Civil Engineering, Middle East Technical University, Ankara, Türkiye

1 Introduction

Droughts lead to substantial damage to ecosystems and various sectors that guaranty civilisation (agriculture, water supply, energy, etc.). Classifying, quantifying and estimating the onset and end of droughts have long been a challenge to monitor and understand the impacts of droughts, plan and manage water resources, take early action so that severe damage can be prevented and sustainable development can be promoted. [23, 27, 46]. Starting from this point, droughts are first classified into 4 main

A. M. Bahadir et al. (eds.), *Hydrology and Urban Water Supply*, Water and Wastewater Management,
https://doi.org/10.1007/978-3-031-72589-0_8

categories. These are meteorological, agricultural, hydrological and socio-economic [27, 40, 78].

Meteorological droughts occur when there is unusually little precipitation over a long period of time or other weather-related factors such as high temperatures or strong winds. Agricultural droughts occur when a lack of water affects crops and livestock, while hydrological droughts impact water resources such as rivers, lakes, and groundwater. Socioeconomic droughts occur when a lack of water affects the economy and livelihoods of people in a given region. As drought is a condition that is measured relative to the long term average of a hydro meteorological variable for a particular region, each of these drought categories can be observed both in arid, semi-arid and tropical regions of the world. To be fair in space, time and magnitude in the determination of droughts, several indices have been invented. Comprehensive reviews document the calculation, strengths, and weaknesses of drought indices that help characterize the severity of droughts in terms of their magnitude, duration, and intensity [17, 27, 46, 82]. There is no single drought index that meets the requirements of all applications. In addition, all drought indices use precipitation either alone or coupled with other hydro meteorological variables.

The objective of this paper is to provide a thorough analysis of the impacts of drought on water resources, including an examination of observed trends and future projections, as well as approaches for modelling and managing water resources during drought conditions, with a specific focus on Turkey's adaptation strategies.

2 Impacts of Drought on Water Resources

Droughts can cause a significant reduction in the amount of water available in rivers, lakes, and groundwater sources. In fact, studies have pointed out that decreases in stream flows are generally observed several times more than decreases in precipitation due to non-linear hydrological factors [60, 69]. This gap in reductions between precipitation and surface water may reach further levels with the effect of decreases in groundwater levels. On the other hand, it was determined that a decreasing trend in precipitation was not observed in some basins where there was a significant decrease in flows [53]. Catchment runoff is largely influenced by the balance of precipitation and evapotranspiration. Changes in vegetation or soil water and groundwater storage may affect runoff response. Climate-soil-vegetation interactions within a particular location determine how rainfall becomes runoff. Due to the cumulative influence of droughts on streamflow generation, dry conditions may persist until a sufficiently long wet time that will restore the catchment function.

Infrastructural developments in water supply such as reservoirs and conveyance systems allow storing water to mitigate hydrological fluctuations in time and space. Yet, due to management decisions and infrastructure limits, catastrophic disruptions in water supply may not occur during the most severe droughts. A dry winter, for example, can impede water transfers from a reservoir to another place the following summer. According to a study conducted in England and Wales that combined climatic, hydrological, and water resource models on a national scale, if a reservoir's storage volume is extremely low compared to long term average, nearby catchments have a 40% chance of having storage volumes at extremely dry conditions as well. In the study, they find that since low reservoir storage correlates with the storage of neighbouring reservoirs and stabilizes after 100–150 km, inferring that interbasin transfers of at least this length are required to improve drought resistance [14].

Droughts can have severe impacts on aquatic ecosystems, including fish and other aquatic life. Reduced summer flows and temperature rise can lead to changes in the composition and distribution of species, and in extreme cases, may lead to the collapse of entire ecosystems [9, 77]. In fact, water temperature, along with flow, is one of the most critical factors on ecosystem status in surface water resources. Water temperature has a direct influence on the metabolisms and lifecycles of aquatic organisms, as well as on water quality, with most chemical and bacterial processes running quicker at higher temperatures, leading to an increase in the prevalence of algal blooms, for example [8, 34, 76, 77]. Rising water temperatures alter lake hydrodynamics, lengthen the thermal stratification period, and deepen the thermocline. These changes then boost nutrient release from sediments and modify nutrient circulation [5, 25, 34] which will increase the available nutrient necessary for the algal proliferation. Phosphorus release from the sediments on the lake bottom can be further enhanced by repeated drying and wetting due to changes in water storage and wetted perimeter. It is also found this effect can be stronger with higher degree of sediment drying [61]. According to long-term predictions, increased water temperature in lakes due to global warming will generate more trophic lake conditions, boosting algae growth and modifying aquatic ecosystems [34].

Dilution of point source effluents, particularly dissolved phosphorus, nitrate, biochemical oxygen demand, and micro-organic pollutants, are reduced as a result of lower river flows during prolonged dry periods. In these periods, lower river flow velocities and increased water residence periods can also increase the risk of algal blooms [31, 33, 43, 51].

Droughts can have significant economic impacts on communities that rely on water resources for their livelihoods. For example, reduced water availability can lead to crop failures, reduced agricultural production, and higher food prices. In addition, droughts can increase the cost of water treatment and distribution, and may require additional investments in water infrastructure. The Millennium Drought that was experienced in Australia between 1996 and 2010 is known as the most severe drought in the last and the current century. Despite the improved water use efficiency, fall in agricultural water supply by 27% in 2002/2003 and 64% in 2008/2009 compared to the annual average of 1993–2002, resulted in a 99% and 84% decrease in cotton and rice production in Australia, respectively. On the other hand, the volume of wheat product taken from the unit area decreased by 12% compared to the normal period, and this decrease was compensated by increasing the cultivated areas [69].

Droughts can also cause problems in maintaining power supply/demand balance. In drought periods while hydropower is primarily affected, water scarcity can also cause shortages and temperature rise in cooling water supplied for the demands of thermoelectric power plants [70]. W. Wan et al. [74] assessed the risk of reduction in global hydroelectricity generation due to droughts at both the 0.5° spatial resolution and country levels. They also demonstrated that 18 nations experience a decline of more than 40% and half of the 134 countries with hydropower experience a reduction of more than 20% of mean annual hydropower per 10 years (Fig. 1).

Eutrophic lake conditions due to harmful algal bloom and their toxins make drinking water treatment difficult causing disturbances in clean water supply [9, 54]. In addition, harmful algal blooms create chemicals that give drinking water a disagreeable taste and odor, resulting in customer complaints. They pose a costly problem for both drinking water utilities and recreational concessionaires since water utilities must utilize expensive treatment technologies to eliminate these compounds [13, 15, 16, 24, 65, 79]. Environmental degradation caused by droughts can also have unfavorable economic outcomes. These financial losses become more noticeable in aquatic ecosystems. [54] classified various costs associated with freshwater eutrophication under 2 major groups named as damage and policy response costs; then estimated the total damage costs and policy response costs of freshwater eutrophication per year in the UK as up to 160 million $ and 77 million $, respectively.

Droughts can also have social impacts, particularly on vulnerable populations such as poor and marginalized communities. Water scarcity can lead to social unrest and conflict, and may

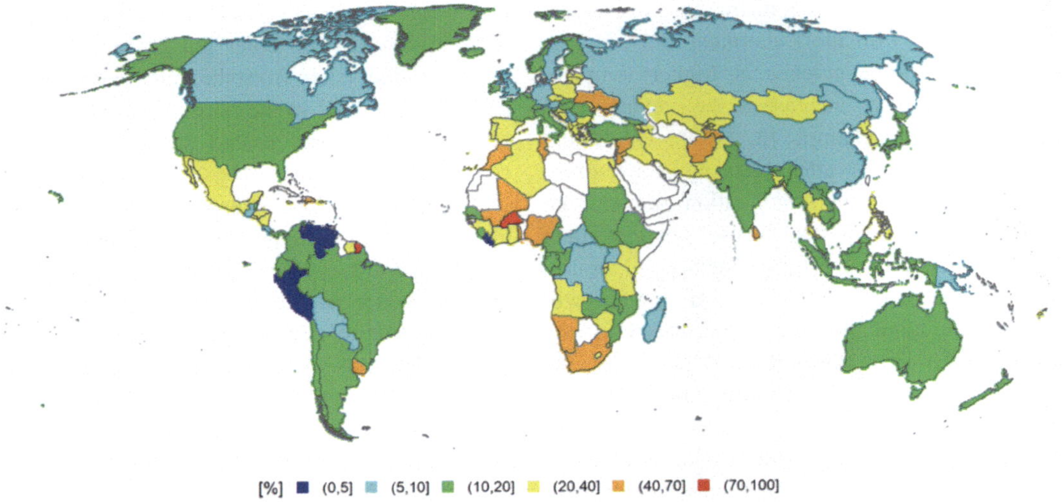

Fig. 1 Country level estimated hydropower reduction in percent of mean annual generation that occurs in 1 out of every 10 years based on simulation for the period 1975–2016 [74]

exacerbate existing social inequalities. D. A. Musolino et al. [47] represents the opinion that drought only causes economic loss or damage. They demonstrate that, while the average impact of drought is negative, economic effects vary in sign and extent depending on social and regional factors. Contrary to popular belief, some farmers can profit significantly from droughts. According to the findings, the number and variety of key study locations and drought episodes, crop type, and geographical area should be examined further. The policy response should focus on drought preparedness and resilience, which brings significant economic benefits. They suggest taking a proactive approach to drought risk reduction by investing in water supply management techniques that increase water availability for irrigation farming, as well as demand-side management strategies that minimise susceptibility by internalising the cost of risk. Discriminating based on agricultural households' income levels, farm turnover, and profit margins can help tailor support policies. Policy interventions should also be aimed at strengthening resilience and reducing vulnerability to droughts be considered collective subsidies, because final consumers are likely to suffer the lion's share of the entire cost of the drought [47].

3 Drought Projection and Modelling Approaches

Changes in climate significantly influence the hydrological cycle, evaporation, evapotranspiration, precipitation, surface runoff, infiltration, etc. This induces the disruption of the hydrological cycle by altering when, where, and how much precipitation falls. Hence it affects the quantity and availability of water resources that causes change in the frequency and magnitude of droughts. Hydro meteorological observations from the Global Precipitation Climatology Center (GPCC) have disclosed positive trends in the frequency, length, and intensity of meteorological droughts in many regions including Mediterranean, Central Africa and Amazonia between 1951 and 2010 [66]. Furthermore, Climatic Research Unit (CRU) measurements have revealed a substantial positive tendency in land regions affected by meteorological drought [49]. Projections of General Circulation Models (GCMs) are routinely used to estimate potential drought impact in the future. Although increasing trends in temperature are predicted to continue into the foreseeable future [42], there is no clear signal on precipitation which varies significantly with different climate scenarios [3, 48, 71]. Increasing trend in temperature might change the precipitation pattern in winter (from snow to rain) and cause early snow melting in spring and more evaporation. These possible changes will likely result in an increasing trend of extreme hydrological events [80]. There is a general consensus that high emission scenarios of greenhouse gases will rise the degree of meteorological, agricultural and ecological and hydrological droughts more than low emission scenarios [2, 11, 37, 73, 75, 84]. Indeed, projections of climate change and its impact on water resources are inherently uncertain [72], and this is likely to increase as a result of climatic, technological, economic and sociopolitical changes [1, 38, 41]. Therefore, droughts are expected to be more frequent, severe, and long-lasting in the future than in prior decades globally [4, 18, 67, 68]. Recent studies by [11, 83] concludes that streamflow declines in many places of the world including Mediterranean and other regions with Mediterranean-climates in the future, posing a major danger to existing fresh-water supplies. For example, according to the study by (SYGM, 2016) which perform hydrological impacts study in Turkey by using the projections of three global climate models to assess potential future changes in streamflow, a decrease around 20% in precipitation also, decrease around 60% in gross water potential is projected for the period of 2041- 2070. Increase intensity of drought conditions are expected with large uncertainty in the region for the near future over Turkey. Long-term forecasts are not very heartwarming in regions that are more fortunate in terms of precipitation compared to the Mediterranean climate either. B. Dobson et al. [14] calculated that the probability of a year with water use restrictions in England and Wales doubles by 2050 and is four times worse by 2100.

Planners and decision-makers in the water sector are concerned about the potential impact of droughts, balancing the cost of infrastructure investment against the high potential social and environmental costs of inaction [56, 64]. The water resources sector therefore faces the challenge of planning and operating systems under uncertainty. Historically, conventional engineering approaches to water systems planning have best been summarized as a predict-then-act framework [36], where optimization of a design objective under the assumption of a best-estimate (i.e., most likely) predictions of the future, suggests the"best" course of action. In this framework, rainfall-runoff models are routinely used to simulate future runoff by using climate model projections of dynamically downscaled global climate models (GCMs) which are driven by greenhouse gas emission scenarios [10, 52] or modified historical hydro-climatic conditions. Despite improvements in climate science, the GCMs developed to project climate futures generate a wide range of projections that often disagree on both the direction and magnitude of precipitation changes [45] which results in various hydrological projections that used to define droughts. What is more, rainfall-runoff models have generally been calibrated and validated under historical conditions, with no assurance that calibrated parameters would still be valid in different hydro-climatic conditions [52]. Instead, a number of studies have found that rainfall-runoff models' predictive skill decreases with changed climatic conditions especially when predicting drier climates [19, 59, 62]. Although recent methodological developments aim to increase performance of hydrological models to climate change [e.g. 20, 50, 57], a recent study by [58] demonstrated that optimal values of model parameters considerably changed through time mainly because of climate change as well as changes in land use and river structure. Besides, the significant resources and experience needed for model calibration mean that water resource assessments routinely fail to explore the uncertainty space associated with climate change, with far-reaching consequences for planning. These are the main challenges to their suitability for drought impact assessments.

Predict-then-act is appropriate when uncertainties can be well characterised [63]. If it turns out that the future predictions differ from the assumed future(s), the optimal solution is unlikely to succeed [21, 22]. Accordingly, decision makers aim to find design and operation alternatives which are robust to uncertain and changing conditions [39]. A robust alternative is one in which the expected solution is insensitive to variations in the parameters, i.e. the parameter influences are few sensitive [28]. Some bottom-up analytical methods have been developed to provide plausible projections. This requires that uncertainties are described using scenarios that consider coherent projected pathways based on different assumptions [38]. Examples include Robust Decision Making [7], Decision Scaling [6], Info-Gap [30], Robust Decision Making [35], Dynamic Adaptive Policy Pathways [21] and Many-Objective Robust Decision Making [32]. A methodological flowchart of bottom-up approaches can be found in Fig. 2 [28]. Bottom-up approaches have so far tended to rely on multipliers, also known as the "delta change" approach, where they are applied uniformly along a time series [e.g. 28, 29] or to climate variables such as temperature and precipitation [e.g. 56] to investigate the impacts of climate change. An upgraded form is introduced by [55] in which a range of multipliers of flow characteristics is applied to the stationary synthetic time series to capture a diverse suite of nonstationary monsoonal dynamics. In this approach, the hydrograph is modified instead of shifting the flows uniformly by focusing specifically on the river of interest. However, modifying flows by multipliers fails to describe catchment response under changing climate in a coherent way across the full range of hydrological conditions. Therefore, there is a need for approaches that consider coherent shifts across the full range of flow conditions. Recently, [81] proposed a novel approach based on a near-universal parameterization of flow duration curves (FDCs), relating these parameters with streamflow statistics that are of interest to water resource management and on the perturbation of them to simulate a broad range of hypothetical future stream flows. This method can be used in bottom-up approaches to generate future flows to represent hydrological variability and extreme dry conditions.

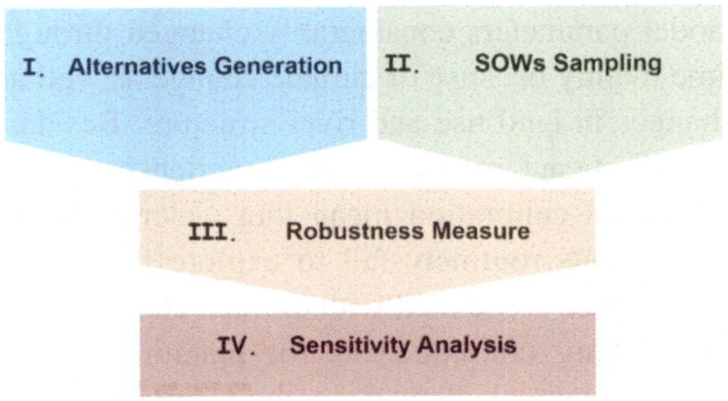

Fig. 2 Methodological flowchart of the bottom-up approaches, based on: (1) the generation or selection of decision alternatives; (2) the sampling of the uncertain factors which make up plausible future states of the world; (3) quantification of robustness measure to rank alternatives; (4) the identification the most critical uncertain factors controlling robustness using sensitivity analysis

4 Adaptation Strategies; How to Reduce the Risk?

Water resources are particularly susceptible to the impacts of drought, so regular monitoring and examination are required to define their current state, evaluate the effects of drought conditions, and ensure their long-term utilization. Based on the findings of this process, action plans must be developed and implemented to mitigate the effects of drought, lowering the risks of depletion and degradation of these vital resources. Drought management can be divided into two approaches: crisis management, which involves studies conducted after the onset of the drought, and risk management, which is a systematic process that evolves gradually and includes proactive measures implemented before the onset of the drought. In a crisis management, attempts are made to reduce harm by acting as soon as a crisis develops. The following measures are implemented to protect water resources in crisis management:

- Increasing the amount of water stored by constructing reservoirs on underground and surface water resources,
- Desalination and purification to use resources like seawater or brackish water to use in daily life,

- Application of cloud seeding in regions where the atmosphere and topography are suitable,
- Providing reliable water supply in regions where water resources are insufficient, maximum benefit should be obtained from precipitation waters by applying the water harvesting method.

Risk management includes the preparation of long-term drought action policies and plans, the establishment of drought monitoring centers, and methods to ensure the effective and rational use of water resources. Risk management plan is an applied administrative plan created to keep the water shortage at a minimum level against possible negative effects when faced with drought risks and to solve the drought problem as soon as possible [44]. The following measures are implemented to protect water resources in risk management:

- Integrated water resources management,
- Ensuring efficient use of water and implementation of sectoral water allocation plans,
- Ensuring the protection of groundwater in terms of quantity and quality and regulating against unsustainable groundwater exploitation,
- Improving the efficiency of water systems through engineering improvements,
- Investing in active maintenance, data analytics, quick repair and the use of robust materials to reduce wasted water,
- Effective and efficient use of water in agricultural production,
- Improving the efficiency of water systems through engineering improvements,
- Dissemination of the basin-based production model to expand the cultivation of plants with low water needs,
- Reusing or recycling water from the wastewater treatment process for other uses, such as energy generation or agriculture,
- Measuring rainwater and storm water harvesting, which can be used locally as well as connected with centralized systems,
- Managing land use and pollution sources to protect watersheds and improve water quality, making use of nature-based solutions,
- Desalination and purification to use resources like seawater or brackish water.

4.1 Adaptation Strategies Followed in Turkey

The precipitation regime of Turkey, which is located in a semi-arid region, varies greatly according to the seasons and regions, and it is seen that the water needs of some river basins exceed the potential of the resources. In addition, the effects of climate change in Turkey such as the increase in temperatures, decrease in precipitation, loss of surface waters, and increase in droughts, soil degradation, erosion, and flooding on the coasts directly threaten the existence of water resources. In Fig. 3, the annual precipitation of Turkey is presented. It is aimed to reduce the uncertainties to be encountered in the amount

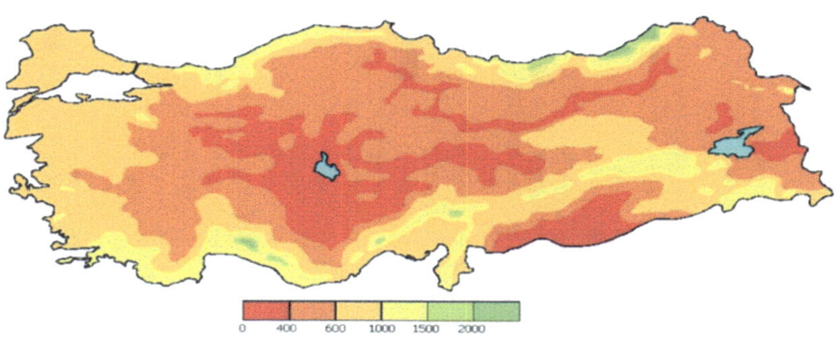

Fig. 3 The annual precipitation map of Turkey

of water resources in a more balanced way under climate change and drought conditions by increasing the capacities of storage facilities. Within the scope of combating drought in water resources, Drought Management Plans are being prepared at the basin scale in Turkey.

Drought management plans ensure that studies such as drought analysis, water balance and sectoral vulnerability are planned and managed to reduce the negative impacts of potential drought risks and to improve and solve problems before, during and after drought. Drought management plans identify drought-prone areas in the basin so that institutions/organisations in the regions know what is sensible to do in terms of drinking water supply, agriculture, industry and water supply for the ecosystem. It gives information on measures that will enable producers to continue production, such as turning to agricultural products that require less water, reducing irrigated areas, or changing irrigation techniques when faced with decreasing underground and surface water resources. Drought Management Plans guide the industrial establishments in the basin to direct their investments according to the drought situation and current water situation of the basin and to implement the measures that will ensure that they are least affected by drought. Moreover; Drought Management Plans ensure that the public has information about drought and creates social awareness.

Figure 4 shows the drought assessment of the country based on the Standard Precipitation Index (SPI) for the almost forty-year period. SPI which only needs precipitation values is used to analyze the monthly rainfall data [26]. SPI is spatially consistent because it enables comparison between several locations in various climates [12]. In order to use water efficiently in the fight against drought, it is necessary to use water more effectively and efficiently in irrigation areas, to bring renewal projects to the forefront to eliminate drainage problems and to expand the use of modern closed irrigation systems. Studies on the effective use of water in Turkey are evaluated under two main headings as structural and non-structural studies. These are:

- Completion of all water storage facilities to develop the economically usable water potential in the country,
- Construction of irrigation facilities with pressure pipes,

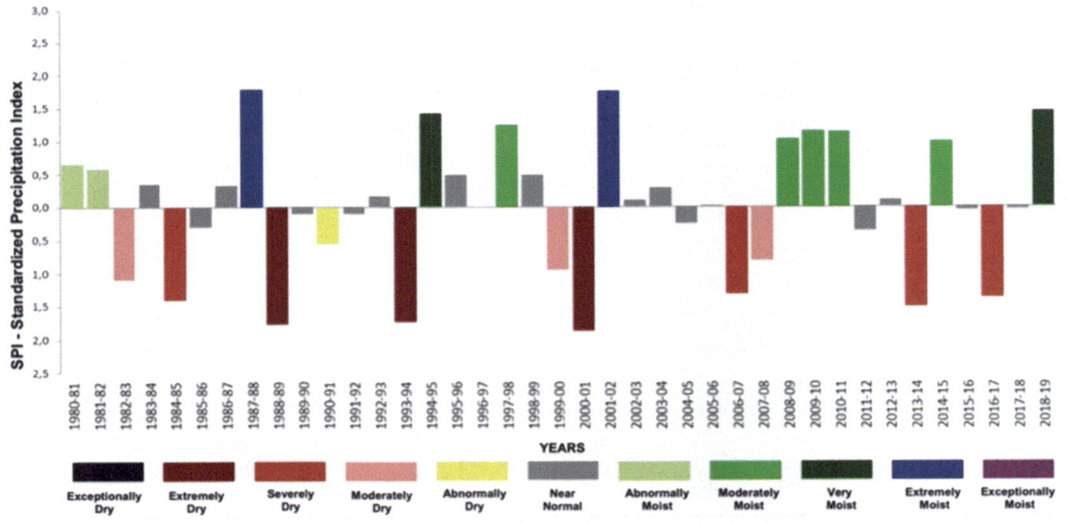

Fig. 4 The drought assessment of Turkey based on the Standard Precipitation Index for the period between 1980 and 2019

- Conversion and modernization of the irrigation facilities in operation to the pressure pipe system within the scope of the" Renewal Project",
- Meeting the maintenance and repair needs of the irrigation facilities in operation, together with the water user organizations,
- Establishing measurement facilities in the irrigation facilities in the operation to ensure the fair and reliable transmission and distribution of water and to detect losses and leaks,
- Construction of necessary wastewater facilities within the scope of studies for the reuse of used water in irrigation.

Classical irrigation systems have been abandoned by DSI in recent years. In new irrigation and renovation projects, the high pressure network systems are preferred in terms of minimum loss of land, high transmission efficiency, ease of placing water metering facilities, the possibility of applying irrigation water fees based on volume, ease of operation, length of facility life, low maintenance and repair costs, farmer's inability to intervene in the system, and water saving.

Non-Structural Studies covers the studies:

- Water management studies, which include hydroelectric energy production based on the needs of the irrigation season and the distribution of water by taking into account the needs of all sectors,
- Irrigation management, which includes gradual pricing studies, in which those who use a lot of water pay more and those who use less water pay less in water usage service fees.

The basis of these studies is the operation hydrology study in all the storage facilities in operation, the comparison of the existing water and the water needed, and the planning of the sectoral sharing of the water according to the result. The use of the Flow Estimation and Basin Optimization Model (ATHOM) for the dissemination of closed systems, the realisation of renewal projects, and the measurement and good management of water provides significant water savings in irrigation.

Before the irrigation season, "general irrigation planning studies" are carried out for each irrigation facility, according to the area to be irrigated, the type of crop to be planted, and the plant water needs of this type of product. Within the scope of planning studies, weekly and daily water distribution programs are made throughout the irrigation season.

Considering the results of long-term monitoring and evaluation, it is seen that there is an increase in economical irrigation methods due to factors such as the increase in closed irrigation systems, the increase in the importance of effective use of water due to droughts, and the awareness of water users.

5 Conclusion

Drought observation and prediction studies are critical for water resource planning and management. Decision-makers and planners of water resources can thus make informed decisions about water allocation, storage, conservation and management strategies. This chapter underscores the need of proactive drought management strategies to ensure sustainable water resource management while mitigating the adverse effects of drought. To effectively navigate the growing uncertainties related to water availability and demand, it is crucial to employ robust methodologies that quantify and address these uncertainties. Techniques such as scenario analysis, robust decision-making frameworks, and adaptive management approaches can provide valuable insights into potential outcomes under uncertain conditions. By incorporating these methodologies into drought management practices, decision-makers can create more resilient systems that respond effectively to evolving challenges. Furthermore, fostering collaboration among stakeholders and investing in innovative technologies will be vital for adapting to future challenges. Türkiye's Drought Management Plans shows a successful approach that integrates crisis and risk management strategies, focusing on infrastructure improvements and water conservation techniques. Ultimately, a coordinated approach to drought management, combined with comprehensive monitoring systems and uncertainty mitigation strategies, will not only protect water resources but also enhance ecological balance and strengthen community resilience against prolonged droughts.

References

1. AghaKouchak A, Chiang F, Huning LS, Love CA, Mallakpour I, Mazdiyasni O, Moftakhari H, Papalexiou SM, Ragno E, Sadegh M. Climate extremes and compound hazards in a warming world. Annu Rev Earth Planet Sci. 2020;48:519–48.
2. Almazroui M, Saeed F, Saeed S, Ismail M, Ehsan MA, Islam MN, Abid MA, O'Brien E, Kamil S, Rashid IU, et al. Projected changes in climate extremes using cmip6 simulations over srex regions. Earth Syst Environ. 2021;5(3):481–97.
3. Araya-Osses D, Casanueva A, Roman-Figueroa C, Uribe JM, Paneque M. Climate change projections of temperature and precipitation in chile based on statistical downscaling. Clim Dyn. 2020;54(9):4309–30.
4. Ault TR. On the essentials of drought in a changing climate. Science. 2020;368(6488):256–60.
5. Blenckner T, Omstedt A, Rummukainen M. A Swedish case study of contemporary and possible future consequences of climate change on lake function. Aquat Sci. 2002;64:171–84.
6. Brown C. Decision-scaling for robust planning and policy under climate uncertainty. World Resources Report Uncertainty Series, vol. 14; 2011.
7. Bryant BP, Lempert RJ. Thinking inside the box: A participatory, computer-assisted approach to scenario discovery. Technol Forecast Soc Chang. 2010;77(1):34–49.
8. Caissie D. The thermal regime of rivers: a review. Freshw Biol. 2006;51(8):1389–406.
9. Charlton MB, Bowes MJ, Hutchins MG, Orr HG, Soley R, Davison P. Mapping eutrophication risk from climate change: future phosphorus concentrations in English rivers. Sci Total Environ. 2018;613:1510–26.
10. Chen C, Kalra A, Ahmad S. Hydrologic responses to climate change using downscaled gcm data on a watershed scale. J Water Clim Change. 2019;10(1):63–77.
11. Cook BI, Mankin JS, Marvel K, Williams AP, Smerdon JE, Anchukaitis KJ. Twenty-first century drought projections in the cmip6 forcing scenarios. Earth's Future. 2020;8(6):e2019EF001461.
12. Dhawale R. Moving from crisis management to risk assessment for drought planning using standardized precipitation index (spi) and standardized groundwater level index (swi): case study of Marathwada, India.
13. Dietrich AM. Aesthetic issues for drinking water. J Water Health. 2006;4(S1):11–6.
14. Dobson B, Coxon G, Freer J, Gavin H, Mortazavi-Naeini M, Hall JW. The spatial dynamics of droughts and water scarcity in England and wales. Water Resourc Res. 2020;56(9):e2020WR027187.
15. Dodds WK, Bouska WW, Eitzmann JL, Pilger TJ, Pitts KL, Riley AJ, Schloesser JT, Thornbrugh DJ. Eutrophication of us freshwaters: analysis of potential economic damages; 2009.
16. Dunlap CR, Sklenar KS, Blake LJ. A costly endeavor: addressing algae problems in a water supply. J Am Water Works Ass. 2015;107(5):E255–62.
17. Eslamian S, Ostad-Ali-Askari K, Singh VP, Dalezios NR, Ghane M, Yihdego Y, Matouq M. A review of drought indices. Int J Constr Res Civ Eng. 2017;3:48–66.
18. Fang Z, Zhang W, Brandt M, Abdi AM, Fensholt R. Globally increasing atmospheric aridity over the 21st century. Earth's Future. 2022;10(10):e2022EF003019.
19. Fowler R, Knoben W, Peel M, Peterson T, Ryu D, Saft M, Seo K-W, Western A. Many commonly used rainfall-runoff models lack long, slow dynamics: implications for runoff projections. Water Resourc Res. 2020;56(5):e2019WR025286.
20. Guilpart E, Espanmanesh V, Tilmant A, Anctil F. Combining split-sample testing and hidden markov modelling to assess the robustness of hydrological models. Hydrol Earth Syst Sci. 2021;25(8):4611–29.
21. Haasnoot M, Kwakkel JH, Walker WE, ter Maat J. Dynamic adaptive policy pathways: a method for crafting robust decisions for a deeply uncertain world. Glob Environ Chang. 2013;23(2):485–98.
22. Hamarat C, Kwakkel JH, Pruyt E. Adaptive robust design under deep uncertainty. Technol Forecast Soc Chang. 2013;80(3):408–18.
23. Hao Z, AghaKouchak A. Multivariate standardized drought index: a parametric multi-index model. Adv Water Resour. 2013;57:12–8.
24. Harris TD, Graham JL. Predicting cyanobacterial abundance, microcystin, and geosmin in a eutrophic drinking-water reservoir using a 14-year dataset. Lake Reservoir Manage. 2017;33(1):32–48.
25. Hassan H, Hanaki K, Matsuo T. A modeling approach to simulate impact of climate change in lake water quality: phytoplankton growth rate assessment. Water Sci Technol. 1998;37(2):177–85.
26. Hayes M. Comparison of major drought indices: introduction. Available at drought. unl. edu/Planning/Monitoring/ComparisonofIndicesIntro. Aspx (verified 24 Dec. 2011). Natl. Drought Mitigation Ctr., Univ. of Nebraska, Lincoln, 2011.

27. Heim RR Jr. A review of twentieth-century drought indices used in the United States. Bull Am Meteor Soc. 2002;83(8):1149–66.
28. Herman JD, Reed PM, Zeff HB, Characklis GW. How should robustness be defined for water systems planning under change? J Water Resour Plan Manag. 2015;141(10):04015012.
29. Herman JD, Zeff HB, Reed PM, Characklis GW. Beyond optimality: Multistakeholder robustness tradeoffs for regional water portfolio planning under deep uncertainty. Water Resour Res. 2014;50 (10):7692–713.
30. Hipel KW, Ben-Haim Y. Decision making in an uncertain world: Information-gap modeling in water resources management. IEEE Trans Syst Man Cybern Part C (Appl Rev). 1999;29(4):506–17.
31. Hutchins M, Williams R, Prudhomme C, Bowes M, Brown H, Waylett A, Loewenthal M. Projections of future deterioration in UK river quality are hampered by climatic uncertainty under extreme conditions. Hydrol Sci J. 2016;61(16):2818–33.
32. Kasprzyk JR, Nataraj S, Reed PM, Lempert RJ. Many objective robust decision making for complex environmental systems undergoing change. Environ Model Softw. 2013;42:55–71.
33. Knapp AS, Milewski AM. Spatiotemporal relationships of phytoplankton blooms, drought, and rainstorms in freshwater reservoirs. Water. 2020;12 (2):404.
34. Komatsu E, Fukushima T, Harasawa H. A modeling approach to forecast the effect of long-term climate change on lake water quality. Ecol Model. 2007;209 (2–4):351–66.
35. Lempert RJ. A new decision sciences for complex systems. Proc Natl Acad Sci. 2002;99(suppl 3):7309–13.
36. Lempert RJ, Popper SW, Groves DG, Kalra N, Fischbach JR, Bankes SC, Bryant BP, Collins MT, Keller K, Hackbarth A et al. Making good decisions without predictions: robust decision making for planning under deep uncertainty. RAND Corporation, Santa Monica, California. Available: rand. org/pubs/research briefs/RB9701; 2013.
37. Li H, Li Z, Chen Y, Xiang Y, Liu Y, Kayumba PM, Li X. Drylands face potential threat of robust drought in the cmip6 ssps scenarios. Environ Res Lett. 2021;16(11): 114004.
38. Maier HR, Guillaume JH, van Delden H, Riddell GA, Haasnoot M, Kwakkel JH. An uncertain future, deep uncertainty, scenarios, robustness and adaptation: How do they fit together? Environ Model Softw. 2016;81:154–64.
39. Marchau VA, Walker WE, Bloemen PJ, Popper SW. Decision making under deep uncertainty: from theory to practice. Springer Nature; 2019.
40. McKee TB, Doesken NJ, Kleist J et al. The relationship of drought frequency and duration to time scales. In: Proceedings of the 8th conference on applied climatology, vol. 17. Boston; 1993. p. 179–83.
41. McPhail C, Maier H, Kwakkel J, Giuliani M, Castelletti A, Westra S. Robustness metrics: how are they calculated, when should they be used and why do they give different results? Earth's Future. 2018;6(2):169–91.
42. Meehl GA, Stocker TF, Collins WD, Friedlingstein P, Gaye AT, Gregory JM, Kitoh A, Knutti R, Murphy JM, Noda A et al. Global climate projections. Chapter 10; 2007.
43. Meis S, Thackeray S, Jones I. Effects of recent climate change on phytoplankton phenology in a temperate lake. Freshw Biol. 2009;54(9):1888–98.
44. Mengü GP, Serü A, Özcçakal E. Kuraklık yönetim stratejileri. Ege Üniversitesi Ziraat Fakültesi Dergisi. 2011;48(2):175–81.
45. Miralles-Wilhelm F, Clarke L, Hejazi M, Kim S, Gustafson K, Munoz-Castillo R, Graham N. Physical impacts of climate change on water resources. World Bank; 2017.
46. Mishra AK, Singh VP. A review of drought concepts. J Hydrol. 2010;391(1–2):202–16.
47. Musolino DA, Massarutto A, de Carli A. Does drought always cause economic losses in agriculture? An empirical investigation on the distributive effects of drought events in some areas of southern Europe. Sci Total Environ. 2018;633:1560–70.
48. Naschen K, Diekkrüger B, Leemhuis C, Seregina LS, van der Linden R. Impact of climate change on water resources in the kilombero catchment in Tanzania. Water. 2019;11(4):859.
49. Nasrollahi N, AghaKouchak A, Cheng L, Damberg L, Phillips TJ, Miao C, Hsu K, Sorooshian S. How well do cmip5 climate simulations replicate historical trends and patterns of meteorological droughts? Water Resour Res. 2015;51(4):2847–64.
50. Nicolle P, Andréassian P, Royer-Gaspard P, Perrin C, Thirel G, Coron L, Santos L. Technical note–rat: a robustness assessment test for calibrated and uncalibrated hydrological models 2. Hydrol Earth SystSci Discuss. 2021;2021:1–22.
51. Paerl HW, Gardner WS, Havens KE, Joyner AR, McCarthy MJ, Newell SE, Qin B, Scott JT. Mitigating cyanobacterial harmful algal blooms in aquatic ecosystems impacted by climate change and anthropogenic nutrients. Harmful Algae. 2016;54:213–22.
52. Peel MC, Blöschl G. Hydrological modelling in a changing world. Prog Phys Geogr. 2011;35(2):249–61.
53. Petrone KC, Hughes JD, Van Niel TG, Silberstein RP. Streamflow decline in southwestern Australia, 1950–2008. Geophys Res Lett. 2010;37(11).
54. Pretty JN, Mason CF, Nedwell DB, Hine RE, Leaf S, Dils R. Environmental costs of freshwater eutrophication in England and wales; 2003.
55. Quinn JD, Reed PM, Giuliani M, Castelletti A, Oyler JW, Nicholas RE. Exploring how changing monsoonal dynamics and human pressures challenge multireservoir management for flood protection, hydropower production, and agricultural water supply. Water Resour Res. 2018;54(7):4638–62.

56. Ray PA, Bonzanigo L, Wi S, Yang Y-CE, Karki P, Garcia LE, Rodriguez DJ, Brown CM. Multidimensional stress test for hydropower investments facing climate, geophysical and financial uncertainty. Glob Environ Chang. 2018;48:168–81.
57. Royer-Gaspard P, Andréassian V, Thirel G. Pmr–a proxy metric to assess hydrological model robustness in a changing climate. Hydrol Earth Syst Sci. 2021;25(11):5703–16.
58. Ruijsch J, Verstegen JA, Sutanudjaja EH, Karssenberg D. Systemic change in the rhine-meuse basin: quantifying and explaining parameters trends in the pcr-globwb global hydrological model. Adv Water Resour. 2021;155: 104013.
59. Saft M, Peel MC, Western AW, Perraud J-M, Zhang L. Bias in streamflow projections due to climate-induced shifts in catchment response. Geophys Res Lett. 2016;43(4):1574–81.
60. Scanlon BR, Duncan I, Reedy RC. Drought and the water–energy nexus in texas. Environ Res Lett. 2013;8(4): 045033.
61. Schönbrunner IM, Preiner S, Hein T. Impact of drying and reflooding of sediment on phosphorus dynamics of river-floodplain systems. Sci Total Environ. 2012;432:329–37.
62. Seibert J, van Meerveld H, et al. Hydrological change modeling: challenges and opportunities. Hydrol Process. 2016;30(26):4966–71.
63. Singh R, Reed PM, Keller K. Many-objective robust decision making for managing an ecosystem with a deeply uncertain threshold response. Ecol Soc. 2015;20(3).
64. Singh VP. Hydrologic modeling: progress and future directions. Geosci Lett. 2018;5(1):1–18.
65. Smith VH, Sieber-Denlinger J, deNoyelles Jr F, Campbell S, Pan S, Randtke SJ, Blain GT, Strasser VA. Managing taste and odor problems in a eutrophic drinking water reservoir. Lake Reservoir Manage. 2002;18(4):319–23.
66. Spinoni J, Naumann G, Carrao H, Barbosa P, Vogt J. World drought frequency, duration, and severity for 1951–2010. Int J Climatol. 2014;34(8):2792–804.
67. Spinoni J, Vogt JV, Naumann G, Barbosa P, Dosio A. Will drought events become more frequent and severe in Europe? Int J Climatol. 2018;38 (4):1718–36.
68. Sreeparvathy V, Srinivas V. Meteorological flash droughts risk projections based on cmip6 climate change scenarios. NPJ Clim Atmosp Sci. 2022;5 (1):77.
69. Van Dijk AI, Beck HE, Crosbie RS, De Jeu RA, Liu YY, Podger GM, Timbal B, Viney NR. The millennium drought in southeast Australia (2001–2009): Natural and human causes and implications for water resources, ecosystems, economy, and society. Water Resour Res. 2013;49(2):1040–57.
70. Van Vliet MT, Sheffield J, Wiberg D, Wood EF. Impacts of recent drought and warm years on water resources and electricity supply worldwide. Environ Res Lett. 2016;11(12): 124021.
71. Varotsos KV, Karali A, Lemesios G, Kitsara G, Moriondo M, Dibari C, Leolini L, Giannakopoulos C. Near future climate change projections with implications for the agricultural sector of three major mediterranean islands. Reg Environ Change. 2021;21 (1):1–15.
72. Vicente-Serrano SM, Domínguez-Castro F, McVicar TR, Tomas-Burguera M, Peña-Gallardo M, Noguera I, Lopez-Moreno JI, Pena D, El Kenawy A. Global characterization of hydrological and meteorological droughts under future climate change: the importance of timescales, vegetation-CO_2 feedbacks and changes to distribution functions. Int. J. Climatol. 2020;40(5):2557–67.
73. Vicente-Serrano SM, Peñna-Angulo D, Beguería S, Domínguez-Castro F, Tomas-Burguera M, Noguera I, Gimeno-Sotelo L, El Kenawy A. Global drought trends and future projections. Philos Trans Roy Soc A. 2022;380(2238):20210285.
74. Wan W, Zhao J, Popat E, Herbert C, Döll P. Analyzing the impact of streamflow drought on hydroelectricity production: a global-scale study. Water Resourc Res. 2021;57(4):e2020WR028087.
75. Wang T, Tu X, Singh VP, Chen X, Lin K. Global data assessment and analysis of drought characteristics based on cmip6. J Hydrol. 2021;596: 126091.
76. Webb BW, Hannah DM, Moore RD, Brown LE, Nobilis F. Recent advances in stream and river temperature research. Hydrol Processes: Int J. 2008;22(7):902–18.
77. Whitehead PG, Wilby RL, Battarbee RW, Kernan M, Wade AJ. A review of the potential impacts of climate change on surface water quality. Hydrol Sci J. 2009;54(1):101–23.
78. Wilhite DA, Glantz MH. Understanding: the drought phenomenon: the role of definitions. Water Int. 1985;10(3):111–20.
79. Wurtsbaugh WA, Paerl HW, Dodds WK. Nutrients, eutrophication and harmful algal blooms along the freshwater to marine continuum. Wiley Interdiscip Rev Water. 2019;6(5): e1373.
80. Yildiz V, Hatipoglu MA, Kumcu SY. Climate change impacts on water resources. Water and wastewater management: global problems and measures; 2022. p. 17.
81. Yildiz V, Milton R, Brown S, Rougé C. Statistical generation of climate-perturbed flow duration curves. EGUsphere; 2023. p. 1–14.
82. Zargar A, Sadiq R, Naser B, Khan FI. A review of drought indices. Environ Rev. 2011;19(NA):333–49.
83. Zhang Y, Zheng H, Zhang X, Leung LR, Liu C, Zheng C, Guo Y, Chiew FH, Post D, Kong D et al. Future global streamflow declines are probably more severe than previously estimated. Nat Water. 2023;1–11.
84. Zhao T, Dai A. Cmip6 model-projected hydroclimatic and drought changes and their causes in the twenty-first century. J Clim 35(3):8.

Groundwater

Groundwater Quality for Drinking Purposes and Potential Health Risk Assessment

Sumon Ganguli, Md. Akter Hosen Rifat, Md. Sohel Mahmud, Sabbir Howlader, Dipta Das, Md. Nazrul Islam, Sujan Kanti Das, Md. Nur E. Alam, and Ashok Kumar Chakraborty

Abstract

In Feni, a south-eastern district of Bangladesh, the rapid growth of industry, agricultural activities, gas deposits and lack of waste disposal could pose a threat to groundwater. We investigate the comprehensive pollution and health risk assessment due to trace metals found in groundwater for the first time. The major parameters including physico-chemical to trace metals were studied using data from 39 locations. It was found, the mean value of total dissolved solids (TDS), electrical conductivity (EC), turbidity, manganese (Mn), and iron (Fe) exceeded the prescribed limits stipulated by WHO. WQI (BSTI) standards found 20.51% and 56.41% (total = 76.92%) of water samples to be excellent and good, whereas WQI (WHO) standards found only 10.25% and 17.94% (total = 28.19%) in the same category. About 72% were graded as 'unsuitable for drinking' based on heavy metal pollution index (HPI). Contamination degree (Cd) and Nemerrow pollution index (NPI) values were largely coincided with HPI. Principal component analysis (PCA) indicated that geological and anthropogenic sources were responsible for pollution which subsequently largely consistent with cluster analysis (CA) and correlation matrix (CM). Ca-Mg-HCO_3 type water was found from Chadha diagram and mostly supported by Scholler diagram. Gibbs plot indicated that groundwater samples fell mostly in the rock-dominant zone, while the Ficklin-Caboi diagram demonstrated groundwater is of near neutral high metal class. Analysis showed that there was a significant non-carcinogenic risk due to arsenic (As), Mn and cobalt (Co) exposure through oral ingestion. Additionally, cadmium (Cd), As and chromium (Cr) were associated with high lifetime total cancer risks

S. Ganguli (✉) · Md. A. H. Rifat · Md. S. Mahmud · S. Howlader · D. Das · Md. N. Islam
Department of Applied Chemistry and Chemical Engineering, University of Chittagong, Chittagong, Bangladesh
e-mail: sumonganguli@cu.ac.bd

S. Ganguli · Md. A. H. Rifat · Md. S. Mahmud · S. Howlader · D. Das
Biomaterials Research Laboratory (BRL), Department of Applied Chemistry and Chemical Engineering, University of Chittagong, Chittagong, Bangladesh

Md. N. Islam
School of Pharmacy, University of Queensland, Brisbane, Australia

S. K. Das
BCSIR Laboratories, Chattogram, Bangladesh

Md. N. E. Alam
Atomic Energy Centre, Bangladesh Atomic Energy Commission, Dhaka, Bangladesh

A. K. Chakraborty
Department of Applied Chemistry and Chemical Engineering, Islamic University, Kushtia, Bangladesh

A. M. Bahadir et al. (eds.), *Hydrology and Urban Water Supply*, Water and Wastewater Management,
https://doi.org/10.1007/978-3-031-72589-0_9

(TCR) for adults and children. The results strongly suggest that groundwater consumption may pose non-carcinogenic and carcinogenic health risks to children and adults. It is prudent that more effective administrative, social, and control measures should be taken to minimize the contaminant exposures to groundwater.

Keywords

Carcinogenic risk · Groundwater · Hazard index · Non-carcinogenic risk · Trace metals

1 Introduction

Groundwater is a natural resource that is essential for life [1]. Today, however, groundwater is the most abused valuable resource [2, 3]. Globally, groundwater provides 43% of irrigation water and 40% of drinking water [4]. The demand for water has continuously increased with the development of socio-economic status [5]. In Bangladesh, there are between 6 and 11 million wells and 98% of the population uses groundwater as their main source of drinking water [6]. Therefore, sustainable groundwater management is the biggest challenge for Bangladesh [7]. There are several factors that influence the quality and quantity of water in Bangladesh either directly or indirectly, including geological and anthropogenic sources [6].

Feni, a southeastern district of Bangladesh under Chattogram division, with an area of 928.34 km^2 of flat areas, has 1,437,371 populations making it ninth-most populous district in Chattogram (population density 1500 per km^2). There are 2 industrial areas in this district including 4 heavy industries, 17 medium industries, and 826 small industries all over the areas. In addition, there is a gas field in the Feni sadar area. Total area of agriculture land is 75,922 ha and arable land is 74,720 ha [8]. Therefore, groundwater pollution and subsequent human health effects are a serious concern in Feni because of the presence of industries and improper waste disposal practices. Furthermore, ample pesticide/fertilizer uses, atmospheric deposition, seawater intrusion, and pollution of river/seawater estuaries from the nearest chemical industrial activities could be harmful to the groundwater (Bhatiari/Sitakund/Mirsharai of Chattogram; about 86, 67 and 45 km from Feni).

Health effects range from cancer to non-cancer when significant amounts of metals in water are ingested [9, 10]. It is worthwhile to mention that the Bangladesh national drinking water survey (BNDWQS) found that 97.8% of Bangladeshi residents drink safe water. In spite of this, numerous studies in recent decades have shown that drinking groundwater is a health hazard in the long run, including cancer-causing as well as non-cancer-causing effects. In Bangladesh, Pabna, Kushtia, Chuadanga, Meherpur, and Jessore districts were surveyed by Chakraborti et al. (2015) and reported that As pollution in those areas [11]. Cancer and non-cancer risks were found for As in the groundwater of Satkhira district [12]. On the other hand, trace metals induced carcinogenic and non-carcinogenic effects for As and Mn, As, Fe, barium (Ba), zinc (Zn), and aluminium (Al) in Rangpur district respectively whether in Gopalganj, As induced carcinogenic effect and As, Fe, Mn, and B induced non-carcinogenic effect were reported recently [13]. In another study from Haripur gas blowout area of Sylhet district showed cancer risk due to lead (Pb), Cd, and nickel (Ni) [14]. Cd and Mn in the groundwater did not induced cancer risk in Chuadanga district, whereas carcinogenic effect was originated due to Pb [15]. Furthermore, in the capital city of Bangladesh, Dhaka, non-carcinogenic health effects were found for Fe, Mn and Zn [16]. Fe and Mn played a key role in non-carcinogenic health risks in Khulna and Jessore district [17, 18].

In case of carcinogenic effect, Cr in the groundwater of coal mine area, Dinajpur [19–21], Cd in Hazaribagh tannery, Dhaka [22], Ni, Cr in Manikgonj [23], Cd in Rajshahi city [24], Ni, Cd in Chuadanga [15], As in southwest coastal region [25], As in central region [26], As and Pb in southwestern region [27] are found to show carcinogenic effect. However, all the studied metals did not show any carcinogenic effect in the southeast coastal zone [28]. Hence, great concern to develop cancer and non-cancer in the people's lifetime exists by drinking the groundwater of those areas.

To our knowledge, out of 64 districts in Bangladesh, groundwater from few areas was investigated in terms of health risk assessment due to the presence of trace metals. Of note, no groundwater quality monitoring studies with health risk assessment were yet to be done. Taken together, we aimed a comprehesive pollution charaterization and health risk assessment of groundwater from Feni district for the first time.

2 Materials and Methods

2.1 Study Area and Sample Collection

The study area extends from 91°15′18.4″ to 91° 32′02.8″ E (longitude) and 22°02′22.1″ to 23°49′ 56.7″ N (latitude). About 49% of incomes coming from agriculture, industry and commerce. Groundwater is used for drinking purpose at the Feni district where generally deep tube wells are used. The pumped groundwater is directly used by the populations. The depth of the wells is varied from 10 to 290 m with a mean value of 71 m. Sterilized plastic and glass bottles are used to collect water according to APHA protocol. The map of the study area and the sampling locations (n = 39) were presented in Fig. 1.

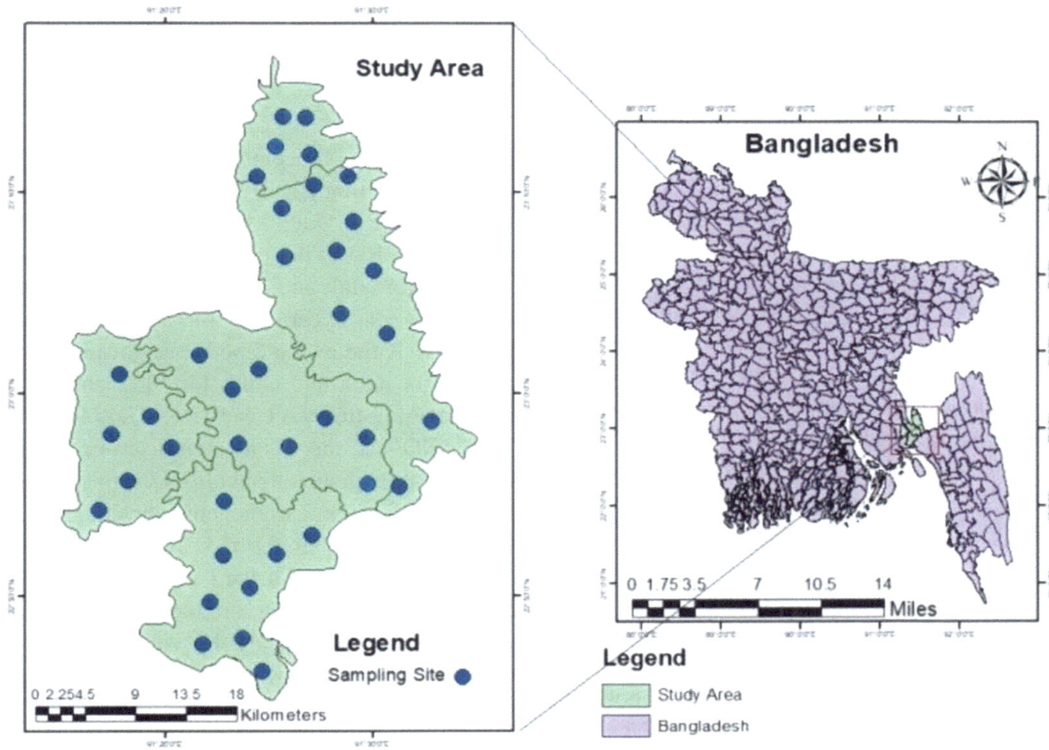

Fig. 1 Map of the study area (Feni District)

2.2 Physico-Chemical Parameter Measurements

Physico-chemical parameters such as temperature, total dissolved solids (TDS), electrical conductivity (EC), pH, turbidity, salinity, total hardness, chloride, bicarbonate and carbonate were measured according to the established protocol reported elsewhere [29, 30].

2.3 Ion and Trace Metals Determination

Na, K, Mg, and Ca were measured by Inductively Coupled Plasma Optical Emission Spectrometry (ICP-OES: Shimadzu 9820) following a method explained elsewhere [31]. On the other hand, NO_3 and SO_4 were determined by UV-spectrophotometer following APHA (2017) protocol [32]. Trace metals such as Cr, Mn, Fe, Zn, As, Cd, Pb, Ni and Co were monitored by Inductively Coupled Plasma Mass Spectroscopy (ICP-MS: 7900 Agilent ICP-MS, Agilent Technologies Inc.) following the procedure published by our group [29, 30].

2.4 Quality Indices

The WQI was calculated by applying the weighted arithmetic method reported earlier [33]. In addition, weighted arithmetic mean method was followed to calculate the heavy metal pollution index (HPI) where two basic steps were developed [34]. Maximum admissible concentration (MAC) values were collected from Siegel (2002) [35] and standard values were taken from DoE (1997) [36]. Moreover, contamination degree (Cd) was calculated by following the equations and standard values reported elsewhere [36–38]. Furthermore, the Nemerow's pollution index (NPI) refers to the pollution calculation which was developed by Nemerow and Sumitomo (1970) [39] and the index was calculated by following the equations developed by them [39].

2.5 Health Risk Assessment

Health risk assessment was done following USEPA (1989) [40] guidelines. Hazard Quotient (HQ) was estimated for all trace metals determined from groundwater. Trace metals contaminated groundwater by ingestion (oral) for adult and child via drinking deep wells water was considered. HQ is calculated by the following equation:

$$HQ = \frac{CDI}{RfD}$$

RfD is the reference dose (mg/kg/day) of trace metals as suggested by US EPA [41] [Fe = 0.700, As = 0.0003, Cr = 0.003, Mn = 0.024, Zn = 0.300, Pb = 0.0015, Cd = 0.0005, Ni = 0.02 and Co = 0.0003]. CDI stands for chronic daily intake (mg/kg/day) which is derived by dividing total dose by body weight of a person by using the formula:

$$CDI = \frac{CxIRxEFxED}{BWxAT}$$

where, C is the concentration of trace metal in groundwater (mg/L); IR is the human water ingestion rate (L/day; 2.0 and 1.0 for adult and child) [42]; ED is the exposure duration (years; 70 and 6 for adult and child); EF is the exposure frequency (days/year; 365 for adult and child) [40]; BW is the average body weight in (kg; 70 and 15 for adult and child) [40, 43], and AT is the averaging time (AT = 365 × ED) [40, 43].

For HQ < 1, there is no significant health risk of non-carcinogenic effects. If HQ value exceeds 1 (HQ > 1), non-carcinogenic health risk is anticipated [18]. For the risk assessment of the drinking water, a hazard index (HI) was employed by summing up the calculated HQ values of all trace metals. HI value less than 1 (HI < 1) indicates a lower possibility for adverse non-cancer health effects over a lifetime of exposure. However, for HI > 1, an adverse non-cancer health effects are evident [44].

Cancer risk (CR) assessment was done with the following equation [45]:

$$CR = CDI \times CSF$$

CSF (Cancer slope factor) value for As, Cr, Pb and Cd are 1.5 [46], 0.5 [46], 0.0085 [47] and 15.0 [47] $(mg/kg/day)^{-1}$. CR value between 10^{-4} and 10^{-6} is considered as the acceptable carcinogenic risk [45].

2.6 Statistical Analysis

Descriptive statistical analysis, Pearson's correlation matrix, PCA, and CA were calculated by SPSS-26.0. Gibbs, Chadha and Schoeller diagrams were plotted using Grapher 14. Fracklin-Cobai diagram was plotted by MS Excel-16.

3 Results and Discussion

3.1 Physico-Chemical Parameters

A comparison of physico-chemical parameters in groundwater from Feni district with WHO, BSTI (Bangladesh) and USEPA standards is shown in Table 1.

During the period of measurements, the mean temperature was 24.26 ± 0.25 °C, varying from 29.10 to 30.06 °C, indicating that all samples remained within the range of WHO, BSTI and USEPA limits (20–30 °C). Water pH was measured as 6.92 ± 0.47 which is very close to neutral water as well as within the prescribed value by the regulatory authorities (Table 1). Our findings are in line with previous reports of nearly neutral water in Chittagong city's supply water [48]. A water's acidity or weak acidity indicates that it contains very little inorganic sediments or domestic and agricultural wastes [49]. Acidic pH is possibly derived mainly from free CO_2 and HCO_3 concentrations [50], and in the present study the HCO_3 concentration is found as 198.46 ± 89.95 mg/L (mean content) which are within the WHO standards (200 mg/L), indicating low acidity of the water. Inorganic and organic matters are characterized by TDS [51]. The range of TDS lay within 59.0 to 3700.00 mg/L with the mean value of 516.79 ± 606.58 mg/L. Mean value of TDS was within the acceptable limits of BSTI (1000 mg/L), however, crossed the WHO and USEPA limits (500 mg/L) (Table 1). Water samples in 46.15%of sampling stations were found to be a freshwater type with a mean concentration of TDS less than 1000 mg/L [52]. The conductivity of pure water is very low, making it a good insulator. The mean EC was 753.82 ± 797.07 μS/cm, ranging from 70.0 to 4560.0μS/cm. EC values were largely varied in the present study area and the mean values were out of the WHO and BSTI limit (500 μS/cm). About 51.28% of samples exceeded the stipulated value. Noteworthy that EC values less than 1000 μS/cm are classified as very weakly mineralized water defined by Detay and Carpentor (1997) [53], and EC values of 0–250 μS/cm indicate low salinity according to WHO [52]. Turbidity varied from 0.44 to 60.00 NTU with the means of 12.60 ± 14.27NTU. There were deviations from the acceptable mean values (5.0, 10.0 and 5.0 NTU for WHO, BSTI and USEPA respectively), possibly due to the aggregation of dissolved constituents. Health risks may arise from consuming highly turbid water [54], and in the present study 53.83% of water samples exceeded the WHO standard. Drinking water with high salinity may cause hypertension [55]. The mean salinity of 0.44 ± 0.42 ppt was found in all of the sampling stations. The low salinity content indicates that saline surface water intrusion has not occurred. The mean values of total hardness (TH) were found to be 142.62 ± 111.31 mg/L and the ranges were 32.0 to 460.0 mg/L which are within the acceptable limits prescribed by WHO (300 mg/L), BSTI and USEPA (500 mg/L). According to the results, water in the Feni district is moderately hard (75–150 mg/L) in nature as per the classification of TH reported elsewhere [56].

Table 1 Descriptive analysis for drinking water quality of Feni district

Parameters	Unit	Range	Mean ± SD	WHO Standards	BSTI Standards	USEPA Standards
Temp	(°C)	29.1–30.6	29.81 ± 0.40	20–30	20–30	20–30
pH	–	5.7–7.6	6.92 ± 0.47	6.5–8.5	6.5–8.5	6.5–8.5
TDS	ppm	59.0–3700.0	516.79 ± 606.58	500	1000	500
EC	μS/cm	70.0–4560.0	753.82 ± 797.07	500	500	–
Salinity	ppt	0.1–2.0	0.44 ± 0.42	–	–	–
Turbidity	NTU	0.44–60	12.60 ± 14.27	5	10	5
TH	ppm	32.0–460.0	142.62 ± 111.31	300	500	500
Na	ppm	16.04–22.96	20.65 ± 1.71	200	200	200
K	ppm	1.11–17.31	4.69 ± 3.35	12	12	–
Ca	ppm	14.38–41.93	26.46 ± 7.00	200	75	–
Mg	ppm	1.83–7.38	3.77 ± 0.96	150	30–35	–
Chloride	ppm	67.88–153.52	111.36 ± 19.91	250	150–600	250
Nitrate	ppm	0.38–3.65	1.70 ± 0.87	50.0	10.0	10.0
Sulphate	ppm	1.30–7.37	3.63 ± 1.58	250.0	400.0	250.0
Carbonate	ppm	0	0	–	–	–
Bicarbonate	ppm	83.0–427.0	198.46 ± 89.95	200	–	–
Cr	ppb	5.4–27.3	10.49 ± 5.16	50	50	100
Mn	ppb	41.4–3951	390.67 ± 672.71	100.0	100.0	50.0
Fe	ppb	79.56–20,284.5	3006.60 ± 3509.15	300	1000	300
Zn	ppb	3.8–63.7	24.28 ± 14.70	3000	5000	5000
As	ppb	BDL-12.40	5.58 ± 5.26	10	50	10
Cd	ppb	BDL-3.0	1.35 ± 1.17	3	5	5
Pb	ppb	BDL-51.7	4.71 ± 8.86	10	50	15
Ni	ppb	BDL-54.10	27.54 ± 12.15	70	100	–
Co	ppb	BDL-61.18	5.20 ± 10.25	50	50	–

3.2 Ion Chemistry

Table 1 shows the content of several cations and anions. A major abundance of Na^+, K^+, Ca^{2+}, and Mg^{2+} was found in the study area. Noteworthy that the mean content of all the cations were within the stipulated value given by the regulatory bodies. Among them, Ca^{2+} contributed the highest concentrations (26.46 ± 7.00) followed by Na^+, K^+ and Mg^{2+}. The content of Ca^{2+} was ranged from14.38 to 41.93 mg/L. Ca^{2+} and Mg^{2+} control the hardness of water [57]. The mean content of Mg^{2+} was 3.77 ± 0.96 mg/L and varied from 1.83 to 7.38 mg/L. The low amounts of Ca^{2+} and Mg^{2+} made the water soft or moderately hard. Na^+ is essential for nerve and muscle function in the human body [58]. The mean content of Na^+ was 20.65 ± 1.71 mg/L and the content was varied from 16.04 to 22.96 mg/L. The values of K^+ were varied from 1.11 to 17.31 mg/L with the mean concentration of 4.69 ± 3.35 mg/L. Anions were also found to be within acceptable limits (Table 1). Here, CO_3^{2-}, NO_3^-, SO_4^{2-}, Cl^-, and HCO_3^- were found as the major anions. Among them, the concentration of HCO_3^- ion was higher. The mean content of HCO_3^- was 198.46 ± 89.95 mg/L which was within the prescribed limit fixed by WHO (200 mg/L). The most likely cause of this could be the weathering of carbonates and the dissolution of carbonic acid [59]. When Cl^- concentration in water exceeds 100 mg/L, the water recognized saline water or brackish [60]. In the study area, the mean Cl^- content was 111.36 ± 19.91 mg/L which has been exceeded 100 mg/L, however it was within the prescribed value given by WHO and USEPA (250 mg/L). NO_3^- is considered a disease causing anion like hypertension in adults and methemoglobinemia in infants [53], however the concentration of nitrate was found within the permissible range. Very insignificant mean concentrations of SO_4^{2-} and no CO_3^{2-} were detected in our study area. The abundance of anions in the drinking water could be sequenced as: HCO_3^-, Cl^-, SO_4^{2-}, NO_3^-, and CO_3^{2-}. It is hypothesized that the comparatively low contents of Na^+, Ca^{2+}, Mg^{2+}, HCO_3^-, and Cl^- in the drinking water are explained by lower extraction, moderate to heavy rainfall due to tropical monsoon climate in the locality, or due to low rock weathering [61].

3.3 Trace Metals Chemistry

Fe, Mn, As, Cr, Pb, Cd, Co, Ni, and Zn contents were measured in this study (Table 1). Among them, Fe (3006.60 ± 3509.15 μg/L) contributed the highest concentrations followed by Mn (390.67 ± 672.71 μg/L), Ni (27.54 ± 12.15 μg/L), Zn (24.28 ± 14.70 μg/L), Cr (10.49 ± 5.16 μg/L), As (5.58 ± 5.26 μg/L), Co (5.20 ± 10.25 μg/L) and Pb (4.71 ± 8.86 μg/L). The mean concentrations of Fe were very high, while the ranges were largely varied (79.56–20,284.5 μg/L). Fe values exceeded the acceptable limits set by WHO, USEPA (300 μg/L) and BSTI (1000 μg/L). Here 89.74% of samples in the study area exceeded the WHO standard. Fe in groundwater is unfavorable due to metallic taste, staining clothes, discoloring plumbing fixtures [62]. Free radicals produced by excessive Fe can damage cell structure and cause cancer [63]. Furthermore, high concentrations of Fe can lead to potential health hazards such as nausea, diarrhea, and cardiovascular diseases; even liver and kidney might be affected [64]. Recently, Fe content out of the acceptable limit had also reported in Chittagong [64]. In Bangladesh, tube well water has variable and elevated Fe levels [65]. Of note, according to the National Drinking Water Quality Survey (NDWQS) in 2009, approximately 70% of drinking water sources in Bangladesh (n = 2896), mainly groundwater, have been found to contain elevated level of Fe (>300 μg/L). Dissolution of rocks and ferruginous minerals, as well as wastewater from domestic and industrial sources, may result in high Fe concentrations [66]. According to the Bangladesh National Drinking Water Quality Survey (NDWQS) in 2009, approximately 61% of drinking water samples (n = 2896), mainly groundwater, contained elevated level of Mn (>100 μg/L) [65]. Groundwater of Bangladesh is contaminated with Mn, which poses a threat to health has been reported recently in different areas of Bangladesh [18]. The British Geological Survey (BGS) of Groundwater Studies— a systematic survey of 61 of the 64 districts of Bangladesh, identified elevated and variable level of Fe and Mn in tubewells water [67] In this study, mean Mn concentration ranges were from 41.4 to 3951 μg/L and the mean content is found out of the allowable limit fixed by WHO, BSTI (100 μg/L) and USEPA (50 μg/L). About 70% if water samples exceeded the WHO standard was found. The present results might indicate possible health hazards due to Mn contamination. Numerous study reported possible health hazards including Parkinson disease, Huntington disease, cardiovascular disease, hyperkeratosis, diabetes mellitus, pigmentation changes, and Alzheimer disease, and kidney, liver, respiratory and neurological disorders not only due to Fe but also Mn [68–70]. In a recent report by Liang et al. (2020), studied the source and mobilization mechanism of Mn and Fe in groundwater, revealed few major reasons including Fe and Mn nodules in clay layer, higher TDS values, reductive condition of soil which results progressive dissolution of geologic materials and residence time of groundwater [71].

Bangladesh's groundwater is contaminated with As, posing a health risk [72]. However, mean As concentrations in this study area were within the recommended value by WHO, USEPA (10 µg/L) and BSTI (50 µg/L). On the other hand, Cr, Pb, Cd, Co, Ni, and Zn were found in small quantities and most importantly within the permissible limits according to WHO, USEPA and BSTI standards (Table 1), which indicates the water might be safe for drinking in terms of those parameters. Taken together, the water is somewhat safe for drinking purposes in terms of ion chemistry, however, giving serious concerns for excess TDS, EC, Turbidty, Fe and Mn concentrations. The standard deviation shows very large fluctuation indicating large variations in the parameters such as TDS (±606.58), EC (±797.07), TH (±111.31), Mn (±672.71), Fe (±3509.15), Cl^- (±19.91), Turbidity (±14.27), HCO_3^- (±89.95), Zn (±14.70), Ni (±12.15), Co (±10.25) among the samples, while the rest samples were less than ± 10.00. To make very clear, we aimed to assess the groundwater quality further using different indices.

3.4 Groundwater Quality Assessment in Terms of HPI, Cd, NPI and WQI

Groundwater quality is evaluated by the HPI based on its heavy metal content [73]. As per the guidelines value, the calculated HPI values and their water grade, description for each sampling location (samples percentage and sampling station) are presented in Table 2. The values of HPI ranged from 7.65 to 256.59 with the mean value of 82.49 for groundwater. In accordance with Edet and Offiong's classification, (2002) [74], among the 39 groundwater samples collected in the study area, 71.79% were graded as 'Unsuitable for drinking', while the rest were suitable [74].

According to Cd, 17.94 and 15.38% of groundwater samples showed medium and high grade of pollution based on the drinking water classified by Backman et al. (1998) [75], while the 66.69% of samples depicted low grade of pollution. Basically, Cd, and HPI reflect the accumulated effects of each harmful metal. In order to obtain a more precise result, it is necessary to quantify all heavy metal concentrations using those indices. However, in this study, only eight (8) heavy metals have been taken into consideration in computing Cd. Therefore, further clarification is warranted. To keep in mind that we aimed again to analyze the groundwater samples using other indices including NPI and WQI. The limitations of HPI, and Cd (not considering all the heavy metals) could be overcome by NPI and WQI. As per the stipulated guidelines for NPI, in the present study, the values 3.0 were obtained for 46.15% groundwater samples indicating severe pollution. Moreover, 10.25% and 25.64% of samples showed moderate and slight pollution because the NPI values were remained in the range of 2.0 to 3.0 and 1.0 to 2.0. Therefore, water quality based on NPI was largely coincided with HPI but not with Cd. In case of water quality assessment based on the extensively studied WQI [76], interesting results were obtained and represented in Table 2. We have analyzed the parameters to determine the WQI based on BSTI and WHO standards. Interestingly, an opposite result was obtained. In one hand, based on the WQI (BSTI standards), 20.51% and 56.41% (total = 76.92%) of water samples were found as excellent and good category, while the only 10.25% and 17.94% (total = 28.19%) were found under the same category based on WQI (WHO standards). On the other hand, in total (poor + very poor + unsuitable category) 23.06% and 71.78% were found based on WQI (BSTI standards) and WQI (WHO standards) respectively. This phenomenon can be explained by that fact that the permissible values were relaxed in case of BSTI standards than WHO standards, therefore, the unit weight were found large for BSTI standards than WHO. Furthermore, unit weight also varied between different parameter.

Table 2 Pollution indices of groundwater of Feni district

Index title	Classification	Grade of pollution	Sample (%)	Sample station
HPI [74]	<100	Suitable for drinking	28.20	S-1, 10–13, 21, 22, 24, 27, 32, 34
	>100	Unsuitable for drinking	71.79	S-2, 3–9, 14–20, 23, 25, 26, 28–31, 33, 35–39
C_d [38]	<1	Low	66.69	S-1–8, 15, 17–21, 23, 25, 26, 28, 29, 31–33, 36–39
	1–3	Medium	17.94	S-9, 11, 14, 16, 22, 30, 35
	>3	High	15.38	S-10, 12, 13, 24, 27, 34
NPI [15]	≤0.5	No pollution	0	–
	0.5–0.7	Safe	10.25	S-8, 18, 25, 29
	0.7–1.0	Precaution	7.69	S-2, 7, 39
	1.0–2.0	Slight pollution	25.64	S-1, 3, 15, 19, 20, 26, 28, 33, 37, 38
	2.0–3.0	Moderate pollution	10.25	S-5, 6, 31, 36
	>3.0	Severe pollution	46.15	S-4, 9–14, 16, 17, 21–24, 27, 30, 32, 34, 35
WQI (based on BSTI standard) [76]	0–25	Excellent	20.51	S-1, 2, 20, 22, 25, 26, 29, 39
	26–50	Good	56.41	S-3, 5–8, 11, 13–15, 17–19, 23, 28, 31–38
	51–75	Poor	12.82	S-4, 9, 10, 16, 21
	76–100	Very poor	5.12	S-24, 30
	>100	Unsuitable water	5.12	S-27, 12
WQI (based on WHO standard) [76]	0–25	Excellent	10.25	S-20, 22, 25, 39
	26–50	Good	17.94	S-1–3, 19, 26, 29, 36
	51–75	Poor	30.76	S-5, 7, 13, 15, 18, 23, 28, 32, 33, 35, 37, 38
	76–100	Very poor	20.51	S-4, 6, 8, 11, 14, 17, 31, 34
	>100	Unsuitable water	20.51	S-9, 10, 12, 16, 21, 24, 27, 30

3.5 Pollutant Sources

3.5.1 Principal Component Analysis (PCA)

PCA was analyzed for the reduction of the data set presented by scree plot approach (Fig. 2) [77]. Here, we report only the R-mode PCA analysis in the present study due to small area and population. Kaiser–Meyer–Olkin (KMO) testing was performed on components to determine their suitability for PCA. In KMO testing, a value of >0.5 indicates an observed correlation magnitude and partial coefficients. KMO was 0.533 in this study, which indicates that heavy metal results are suitable for component analysis.

From PCA, three components with eigenvalues >1 (Fig. 2) were taken to explain 60.711% of total variance. The use of eigenvalue >1 is recommended for verifying three loading factors. From the data set, 23.837%, 21.429%, and 15.445% of variances were explained by PC1, PC2, and PC3 respectively (Table 3). Three

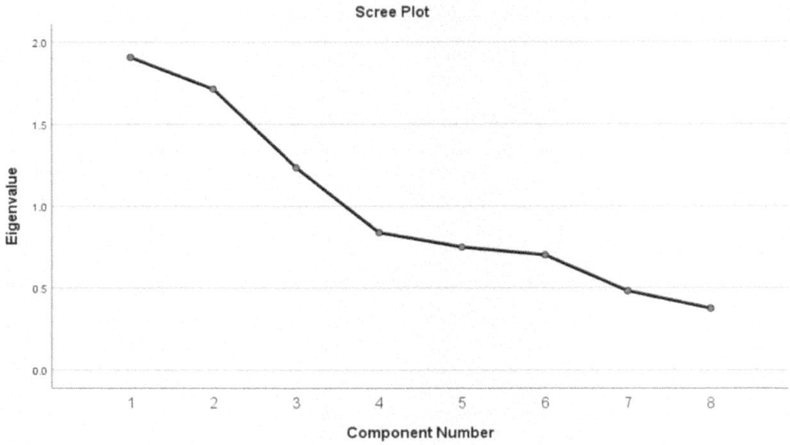

Fig. 2 Scree plot

Table 3 Principle component analysis of drinking water quality parameters of Feni district

Parameters	PC1	PC2	PC3
Pb	0.741	0.357	0.041
Cr	0.713	0.283	−0.027
Cd	0.661	−0.007	0.320
Ni	−0.012	0.851	0.005
Fe	−0.216	0.563	−0.443
Co	−0.391	0.484	−0.253
As	−0.399	0.078	0.664
Mn	−0.232	0.475	0.655
Eigen value	1.907	1.714	1.236
% Total variance	23.837	21.429	15.445
Cumulative % variance	23.837	45.266	60.711

classes of factor loadings are specified here, which are strong, moderate, and weak with the values in that order of >0.75, 0.75–0.50, and 0.50–0.30 [78].

PC1 (eigenvalues = 1.907, most significant variable) was positively dominant with moderate ratio which contains Pb, Cr and Cd. Natural causes like geogenic process along with anthropogenic activities (like industrialization, municipal wastage, automobile discharge) control these enrichments [79–81]. PC2 (eigenvalues = 1.714) was the second significant variable showing strong to moderate loadings dominated by Fe and Ni respectively. In the study area, Ni may be assayed into the groundwater through leaching of metals from industrial activities and burning. Where Fe might be generated in groundwater naturally as a result of the weathering of iron-rich rocks or rock water interaction [82, 83]. In PC3 (eigenvalues = 1.236), moderate positive loadings of As and Mn are related to geogenic and anthropogenic factors. As accumulated in sediments by geological process. Bangladesh comprises of four major geomorphological regions based on As contamination where Feni district fall in Flood Plain region [11, 84]. Besides, As also might be generated from untreated industrial wastage, pesticides and chemical fertilizers [85]. Mn occurs in nature as a mineral from sedimentary rocks or from industrial waste products [16, 21].

3.6 Cluster Analysis

Cluster analysis (CA) is an empirical method for identifying similarities and dissimilarities among different classes by forming clusters. R-mode CA was carried out in this study area (Fig. 3). As an output of cluster analysis, two clusters were found in a dendrogram where cluster 1 has two sub clusters. One sub cluster consisted by As, Cd, Cr, Pd, Co, Ni where other sub cluster contained only Mn. These results revealed the both geogenic and anthropogenic (industrialization, municipal wastage, automobile discharge, agro chemicals) sources which are similar with PCA [86–88]. Cluster 2 included Fe which might be attributed to geogenic sources from the iron containing minerals and rocks [89]. Overall, a good agreement between PCA and CA was observed.

3.7 Correlation Matrix (CM)

CM identified the relation and variation among the water samples through physico-chemical parameters and groundwater pollution sources (Table 4) [90]. Significant and non-significant trends were observed in the correlation matrix. Significant and positive trends indicate the relation between two variables in such a way that they move in the same direction, and when two variables move in inverse directions it indicates a non-significant trend. pH exhibited a significant positive correlation with Salinity ($r = 0.549$, $p < 0.01$) and weak positive correlation with EC ($r = 0.418$, $p < 0.01$), TH ($r = 0.434$, $p < 0.01$), K ($r = 0.453$, $p < 0.01$) and HCO3 ($r = 0.481$, $p < 0.01$) which indicated that pH may be increase with the increase of minerals. Correlation between TDS and EC ($r = 0.978$, $p < 0.01$); TDS and Salinity ($r = 0.854$, $p < 0.01$); TDS and TH ($r = 0.712$, $p < 0.01$); TDS and Pb ($r = 0.833$, $p < 0.01$); EC and Salinity ($r = 0.818$, $p < 0.01$); EC and TH ($r = 0.733$, $p < 0.01$); EC and Pb ($r = 0.762$, $p < 0.01$); Salinity and TH ($r = 0.676$, $p < 0.01$), Salinity and HCO_3^- ($r = 0.563$, $p < 0.01$); Salinity and Pb ($r = 0.704$, $p < 0.01$); TH and HCO3 ($r = 0.603$, $p < 0.01$); K and Mg ($r = 0.624$, $p < 0.01$) explained the similar sources which might be due to sea water intrusion with groundwater, mineral dissolution, carbonate

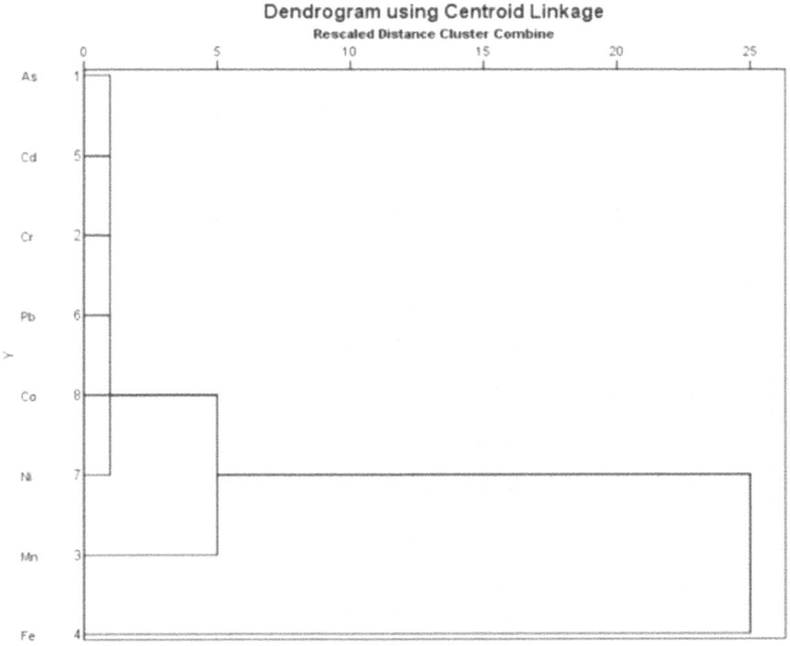

Fig. 3 Cluster analysis

Table 4 Pearson correlation matrix

Parameters	pH	TDS	EC	Salinity	Turbidity	TH	Na	K	Ca	Mg	Bicarbonate	Chloride
pH	1											
TDS	0.361*	1										
EC	0.418**	0.978**	1									
Salinity	0.549**	0.854**	0.818**	1								
Turbidity	−0.047	0.140	0.100	0.121	1							
TH	0.434**	0.712**	0.733**	0.676**	0.406*	1						
Na	0.146	0.004	−0.040	0.199	−0.005	−0.055	1					
K	0.453**	0.346*	0.378*	0.407*	-0.138	0.364*	0.075	1				
Ca	−0.060	0.005	−0.029	0.029	0.258	0.084	−0.141	0.364*	1			
Mg	0.350*	0.280	0.283	0.430**	−0.026	0.384*	0.462**	0.624**	0.174	1		
Bicarbonate	0.481**	0.403*	0.405*	0.563**	0.374*	0.603**	0.227	0.281	0.145	0.433**	1	
Chloride	0.261	0.208	0.164	0.340*	0.085	0.408**	−0.242	0.310	0.221	0.220	0.238	1
Nitrate	−0.110	−0.132	−0.123	0.018	−0.092	−0.081	0.223	−0.169	0.023	0.049	0.042	−0.122
Sulphate	−0.043	0.047	0.060	−0.013	0.429**	0.281	−0.098	−0.059	0.278	0.026	0.354*	−0.016
As	−0.052	0.048	0.105	0.016	0.036	0.130	−0.117	−0.167	−0.199	−0.150	0.112	−0.083
Cr	−0.064	0.402*	0.406*	0.239	0.141	0.250	−0.093	0.213	0.110	0.124	−0.006	0.036
Mn	−0.128	0.042	0.043	0.065	0.418**	0.396*	−0.165	−0.192	−0.060	0.021	0.187	0.264
Fe	−0.430**	−0.152	−0.200	−0.250	0.309	−0.206	−0.243	−0.197	0.154	−0.214	−0.211	−0.036
Zn	0.047	0.233	0.240	0.100	−0.128	0.215	−0.175	0.426**	0.069	0.177	0.223	−0.103
Cd	0.214	0.277	0.220	0.410**	0.002	0.227	0.092	0.410**	0.373*	0.360*	0.261	0.255
Pb	0.203	0.833**	0.762**	0.704**	0.162	0.570**	0.028	0.229	−0.031	0.178	0.233	0.167
Ni	−0.099	0.244	0.259	0.139	0.235	0.128	−0.054	−0.193	−0.089	−0.020	−0.107	−0.153
Co	−0.104	−0.073	−0.105	−0.116	0.043	−0.182	−0.501**	−0.183	0.146	−0.188	−0.137	0.176
Parameters	Nitrate	Sulphate	As	Cr	Mn	Fe	Zn	Cd	Pb	Ni	Co	
pH												
TDS												

(continued)

Table 4 (continued)

Parameters	pH	TDS	EC	Salinity	Turbidity	TH	Na	K	Ca	Mg	Bicarbonate	Chloride
EC												
Salinity												
Turbidity												
TH												
Na												
K												
Ca												
Mg												
Bicarbonate												
Chloride												
Nitrate	1											
Sulphate	−0.032	1										
As	0.388*	−0.077	1									
Cr	−0.086	0.186	−0.111	1								
Mn	−0.087	0.305	0.281	−0.099	1							
Fe	−0.073	0.308	−0.062	0.057	0.082	1						
Zn	−0.273	0.128	−0.053	0.260	−0.147	−0.099	1					
Cd	0.024	−0.002	−0.112	0.299	0.055	−0.172	0.048	1				
Pb	−0.207	0.064	−0.181	0.446**	−0.022	−0.044	0.227	0.361*	1			
Ni	−0.139	−0.004	0.029	0.176	0.323*	0.296	−0.236	−0.092	0.257	1		
Co	−0.140	−0.162	0.058	−0.167	0.067	0.237	−0.059	−0.125	−0.080	0.296	1	

Correlation is significant at the 0.05 level (2-tailed)*
Correlation is significant at the 0.01 level (2-tailed)**

weathering and industrial wastage [13, 91]. Besides, Na and Co made significant negative correlation ($r = -0.501$, $p < 0.01$) which generated from different sources.

3.8 Gibbs Diagram

To gain a deeper understanding of hydrochemical reactions, Gibbs plots were plotted [30]. Plots drawn with Na/(Na + Ca) and Cl/(Cl + HCO_3) versus TDS indicated that the groundwater samples fell mostly in the rock-dominant zone, which might be governed by carbonate dissolution and hydrolysis of silicate within the aquifers (Fig. 4) [92].

3.9 Chadha Diagram

The hydrochemical analysis of natural water was proposed by Chadha in a classification diagram (Fig. 5a) [30]. From the Chadha diagram in Fig. 5a, Ca-Mg-HCO_3 (Field 1) and Ca–Mg–Cl, Ca–Mg–SO_4 (Field 4) type of water was likely to be dominant in our study area [30]. However, none of the sample were belongs to field 2 (Na–HCO_3 and K-HCO_3) and 3 (Na-Cl/Na-SO_4 and K-Cl/K-SO_4).

3.9.1 Schoeller Diagram

Schoeller diagrams are semi-logarithmic graphs representing the concentrations of ions in groundwater samples which represent the visual

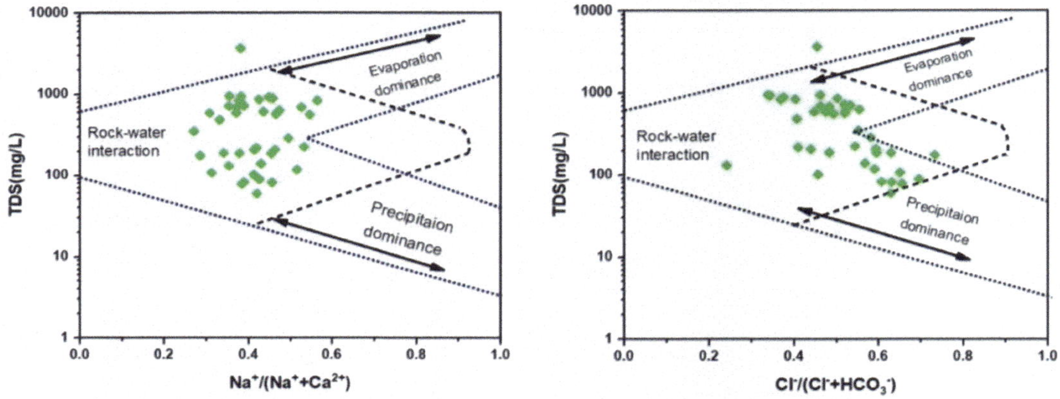

Fig. 4 Gibbs plot

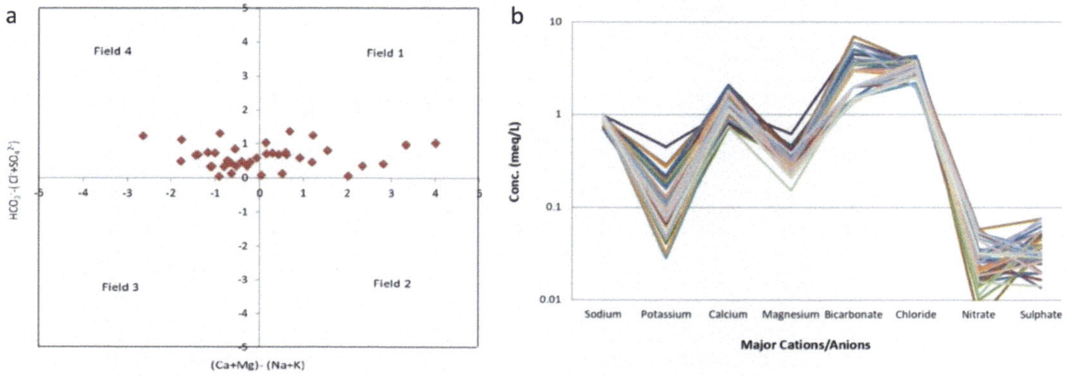

Fig. 5 **a** Chadha plot, **b** scholler diagram

comparisons among the contents of ions in different sample areas [30]. In this study, the Schoeller diagrams revealed the dominance of HCO_3, Cl, Ca and Na (Fig. 5b) in the study area.

3.9.2 Ficklin-Caboi Diagram

The Ficklin-Caboi diagram is used to classify Feni groundwater based on pH and metal load (Fig. 6). The method was introduced by Ficklin et al. (1992) [99] and later modified by Caboi et al. (1999) [93], which allows us to differentiate the various geological aspects of water chemistry [94].

The metal load for each sample was calculated as follows:

$$Metalload(mg/\mathrm{L}) = \sum_{i=1}^{n} Cmetalconc. = Conc.As + Conc.Cr + Conc.Fe + Conc.Pb + Conc.Cd + Conc.Ni + Conc.Co + Conc.Zn$$

Figure 6 presents the plotted data in the diagram. It is observed 15.38% of the groundwater samples were classified as 'near neutral low metal' class, while the 82.05% are classified as 'near neutral high metal' class. Hence, metals present in groundwater of the study area ranged from 'low' to 'high' class and were less mobile in aqueous solutions. Water of this type may be discharged from industrial and urban sources [95].

3.9.3 Health Risk Assessment

Non-Carcinogenic Effect

CDI, HQ, HI, and CR were measured for the assessment of health risks. For child, the average HQ values of As, Mn, and Co were detected higher than unity, which indicated the non-carcinogenic health risk, however, Cr, Fe, Zn, Cd, Pb and Ni were not (Table 5).

For adult, none of the trace metals showed the average HQ value larger than unity, which indicated no non-carcinogenic health risk. Among the maximum values, highest values were obtained for Co (13.59560 and 5.8267) followed by Mn (10.9750 and 4.7036) for child and adult respectively (Table 5). Interestingly, for As, Mn and Co, 41.02, 28.21 and 53.85% of samples showed the HQ values greater than 1.0 for children where the values were 20.51, 10.26 and 5.13% for adult (data not shown). From the trend, it was evident that HQ values were the highest for children denoting children as more susceptible than adults. HI was estimated by summing up HQ values of all trace metals and observed that child (4.4849) and adult (1.9221) were both having higher values than unity

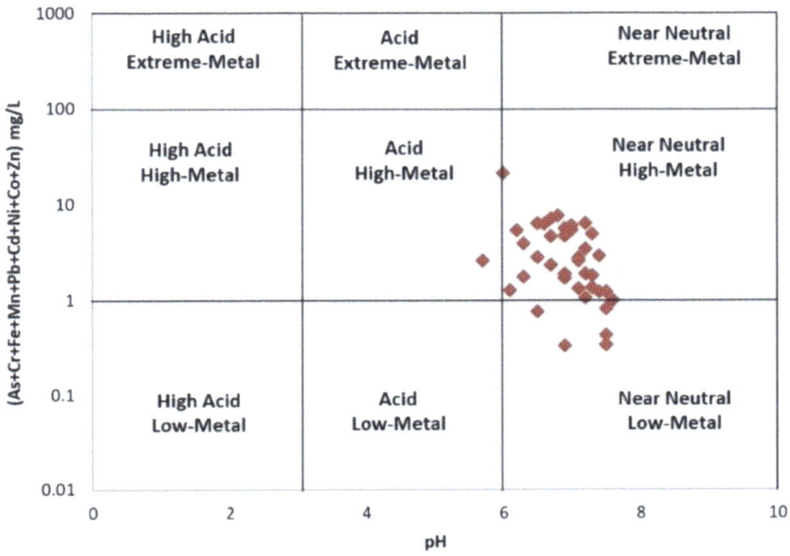

Fig. 6 Ficklin-Caboi diagram

Table 5 Calculated CDI, HQ, HI, CR and TCR values for adult and child through water ingestion

		Child			Adult		
		Average	Min	Max	Average	Min	Max
CDI	As	3.72E−04	0	1.23E−03	1.59E−04	0	5.27E−04
	Cr	7.00E−04	3.60E−04	1.82E−03	3.00E−04	1.54E−04	7.80E−04
	Mn	2.60E−02	2.76E−03	2.63E−01	1.12E−02	1.18E−03	1.13E−01
	Fe	2.00E−01	5.30E−03	1.35E+00	8.59E−02	2.27E−03	5.80E−01
	Zn	1.62E−03	2.53E−04	4.25E−03	6.94E−04	1.09E−04	1.82E−03
	Cd	8.99E−05	0	2.00E−04	3.85E−05	0	8.57E−05
	Pb	3.14E−04	0	3.45E−03	1.35E−04	0	1.48E−03
	Ni	1.84E−03	0	3.61E−03	7.87E−04	0	1.55E−03
	Co	3.46E−04	0	4.08E−03	1.48E−04	0	1.75E−03
HQ	As	1.2389	0.0000	4.1000	0.5310	0.0000	1.7571
	Cr	0.2332	0.1200	0.6067	0.1000	0.0514	0.2600
	Mn	1.0852	0.1150	10.9750	0.4651	0.0493	4.7036
	Fe	0.2863	0.0076	1.9319	0.1227	0.0032	0.8279
	Zn	0.0054	0.0008	0.0142	0.0023	0.0004	0.0061
	Cd	0.1799	0.0000	0.4000	0.0771	0.0000	0.1714
	Pb	0.2094	0.0000	2.2978	0.0897	0.0000	0.9848
	Ni	0.0918	0.0000	0.1803	0.0393	0.0000	0.0773
	Co	1.1547	0.0000	13.5956	0.4949	0.0000	5.8267
HI		4.4849	0.8440	16.6445	1.9221	0.3617	7.1333
CR	As	5.58E−04	0.00E+00	1.85E−03	2.79E−04	0.00E+00	9.23E−04
	Cr	3.50E−04	1.80E−04	9.10E−04	1.75E−04	9.00E−05	4.55E−04
	Pb	2.67E−06	0.00E+00	2.93E−05	1.33E−06	0.00E+00	1.46E−05
	Cd	1.35E−03	0.00E+00	3.00E−03	6.75E−04	0.00E+00	1.50E−03
TCR		2.26E−03	2.19E−04	4.07E−03	3.89E−04	1.10E−04	2.04E−03

(HI > 1 indicates significant health risks) (Table 5). This reveals that there were significant non-carcinogenic risks and importantly children were more prone to CDI, HQ and HI, with higher values than adults.

3.9.4 Carcinogenic Effect

CR for As, Cr, Pb and Cd are also presented in (Table 5). Different types of cancer can be resulted in human body due to the long term consumption of heavy metals [96]. In this study As, Cr, Pb and Cd are considered to calculate the cancer risk. According to USEPA, CR value less than 1×10^{-6} can be considered as insignificant event to cause disease where the values of CR more than 1×10^{-4} can be considered as significant to cause cancer like disease [14]. In this study significant carcinogenic risk was observed for both child and adults presented in Table 5. Mean values of total cancer risk were 2.26E−03 and 3.89E−04 for child and adults respectively where the ranges were 2.19E−04 to 4.07E−03 and 1.10E−04 to 2.04E−03. Besides this, except Pb all other three metals (As, Cr and Cd) have individual significant cancer risk. Cd showed more significant value rather than As and Cr. CR value of Cd varied from 0.00E+00 to 3.00E−03 with mean value 1.35E−03 for child where for adult the values varied from 0.00E+00 to 1.50E−03 with 6.75E−04 mean value. Carcinogenic risk of these elements can be ranked as Cd > As > Cr > Pb for both child and adult. Children are more prone to carcinogenic health risk than adult in this region. The reason is lower body weight of children than the adults. In Bangladesh few other studies also observed cancer risk as like as this study [13, 14, 97].

4 Conclusion

The TDS, EC, turbidity, Mn, and Fe exceeded the prescribed limits stipulated by WHO. Based on the WQI (BSTI standards), 76.92% of water samples were found as excellent as well as good category, while the only 28.19% were found under the same category based on WHO standards. About 72% were graded as 'unsuitable for drinking' based on HPI which is also supported by Cd and NPI values. Geological and anthropogenic sources were responsible for pollution as directed by PCA, CA and CM. Ca-Mg-HCO3 type water was found from Chadha diagram and Scholler diagram. Groundwater samples fell mostly in the rock-dominant zone indicated by Gibbs plot while the Ficklin-Caboi diagram showed the groundwater is of near neutral high metal class. There was a significant non-carcinogenic risk due to As, Mn and Co exposure. Moreover, very high lifetime total cancer risk (TCR) for adults and children were found due to Cd, As and Cr. Children were found more vulnerable to carcinogenic and non-carcinogenic health effect. The water purification process should be addressed with appropriate administrative and social measures.

References

1. Moe CL, Rheingans RD. Global challenges in water, sanitation and health. J Water Health. 2006;4:41–57. https://doi.org/10.2166/wh.2005.039.
2. Arumugam K, Elangovan K. Hydrochemical characteristics and groundwater quality assessment in Tirupur region, Coimbatore District, Tamil Nadu, India. Environ Geol. 2009;58:1509–20. https://doi.org/10.1007/s00254-008-1652-y.
3. Chaudhary V, Satheeshkumar S. Assessment of groundwater quality for drinking and irrigation purposes in arid areas of Rajasthan, India. Appl Water Sci. 2018;8:1–17. https://doi.org/10.1007/s13201-018-0865-9.
4. Salman SA, Shahid S, Mohsenipour M, Asgari H. Impact of landuse on groundwater quality of Bangladesh. Sustain Water Resourc Manage. 2018;4:1031–6. https://doi.org/10.1007/s40899-018-0230-z.
5. Ravikumar P, Somashekar RK. Principal component analysis and hydrochemical facies characterization to evaluate groundwater quality in Varahi river basin, Karnataka state, India. Appl Water Sci. 2017;7:745–55. https://doi.org/10.1007/s13201-015-0287-x.
6. Islam ARMT, Al MA, Rahman MM, Zahid A. Simultaneous comparison of modified-integrated water quality and entropy weighted indices: implication for safe drinking water in the coastal region of Bangladesh. Ecol Ind. 2020;113: 106229. https://doi.org/10.1016/j.ecolind.2020.106229.
7. Saha N, Rahman MS. Groundwater hydrogeochemistry and probabilistic health risk assessment through exposure to arsenic-contaminated groundwater of Meghna floodplain, central-east Bangladesh. Ecotoxicol Environ Saf. 2020;206: 111349. https://doi.org/10.1016/j.ecoenv.2020.111349.
8. Banglapedia. The national encyclopedia of Bangladesh; 2021. https://en.banglapedia.org/index.php/Feni_District
9. Karim Z. Risk assessment of dissolved trace metals in drinking water of Karachi, Pakistan. Bull Environ Contam Toxicol. 2011;86:676–8. https://doi.org/10.1007/s00128-011-0261-8.
10. Kavcar P, Sofuoglu A, Sofuoglu SC. A health risk assessment for exposure to trace metals via drinking water ingestion pathway. Int J Hyg Environ Health. 2009;212:216–27. https://doi.org/10.1016/j.ijheh.2008.05.002.
11. Chakraborti D, Rahman MM, Mukherjee A, Alauddin M, Hassan M, Dutta RN, Pati S, Mukherjee SC, Roy S, Quamruzzman Q, Rahman M, Morshed S, Islam T, Sorif S, Islam R, Hossain M. Groundwater arsenic contamination in Bangladesh—21 years of research. J Trace Elements Med Biol. 2015;1–12. https://doi.org/10.1016/j.jtemb.2015.01.003
12. Rahman MR, Safiqul Islam M, Prodhan MDH, Khabir Uddin M. Organochlorine pesticide residue status in dry fish of Bangladesh and their risk assessment: a review. Int J Sci Res Publ. 2019;9:9480. https://doi.org/10.29322/ijsrp.9.10.2019.p9480
13. Rahman MM, Islam MA, Bodrud-Doza M, Muhib MI, Zahid A, Shammi M, Tareq SM, Kurasaki M. Spatiotemporal assessment of groundwater quality and human health risk: a case study in Gopalganj, Bangladesh. Exposure Health. 2018;10:167–88. https://doi.org/10.1007/s12403-017-0253-y.
14. Howladar MF, Hossain N, Anju KA, Das D. Ecological and health risk assessment of trace metals in

water collected from Haripur gas blowout area of Bangladesh. Sci Rep. 2021;11:1–15. https://doi.org/10.1038/s41598-021-94830-0.
15. Bodrud-Doza M, Islam SMDU, Hasan MT, Alam F, Haque MM, Rakib MA, Asad MA, Rahman MA. Groundwater pollution by trace metals and human health risk assessment in central west part of Bangladesh. Groundw Sustain Dev. 2019;9: 100219. https://doi.org/10.1016/j.gsd.2019.100219.
16. Bodrud-Doza M, Islam SMD, Rume T, Quraishi SB, Rahman MS, Bhuiyan MAH. Groundwater for sustainable development groundwater quality and human health risk assessment for safe and sustainable water supply of Dhaka City dwellers in Bangladesh. Groundw Sustain Dev. 2020;10: 100374. https://doi.org/10.1016/j.gsd.2020.100374.
17. Hossain R, Hassan KM. Geospatial, statistical and human health risk assessment of groundwater contamination around waste disposal site at Khulna in Bangladesh. Int J Innov Sci Res. 2020;49:124–36.
18. Ghosh GC, Khan MJH, Chakraborty TK, Zaman S, Kabir AHME, Tanaka H. Human health risk assessment of elevated and variable iron and manganese intake with arsenic-safe groundwater in Jashore, Bangladesh. Sci Rep. 2020;10:1–9. https://doi.org/10.1038/s41598-020-62187-5.
19. Khan MHR, Seddique AA, Rahman A, Shimizu Y. Heavy metals contamination assessment of water and soils in and around Barapukuria Coal Mine Area, Bangladesh. Am J Environ Protect. 2017;6:80–6. https://doi.org/10.11648/j.ajep.20170604.11
20. Habib A, Reza A, Islam T, Mukta FA, Khan R, Siddique AB, Phoungthong K, Techato K. Simultaneous appraisals of pathway and probable health risk associated with trace metals contamination in groundwater from Barapukuria coal basin, Bangladesh. Chemosphere. 2019;242: 125183. https://doi.org/10.1016/j.chemosphere.2019.125183.
21. Bhuiyan MAH, Islam MA, Dampare SB, Parvez L, Suzuki S. Evaluation of hazardous metal pollution in irrigation and drinking water systems in the vicinity of a coal mine area of northwestern Bangladesh. J Hazard Mater. 2010;179:1065–77. https://doi.org/10.1016/j.jhazmat.2010.03.114.
22. Karim MR, Manshoven S, Islam MR, Gascon JA, Ibarra M, Diels L, Rahman MM. Assessment of an urban contaminated site from tannery industries in Dhaka City, Bangladesh. J Hazard Toxic Radioactive Waste. 2013;17:52–61. https://doi.org/10.1061/(asce)hz.2153-5515.0000139.
23. Halim MA, Majumder RK, Rasul G, Hirosiro Y, Sasaki K, Shimada · J., Jinno K. Geochemical Evaluation of Arsenic and Manganese in Shallow Groundwater and Core Sediment in Singair Upazila, Central Bangladesh. Arab J Sci Eng. 2014;39:5585–5601. https://doi.org/10.1007/s13369-014-1111-x
24. Mostafa MG, Uddin SMH, Haque ABMH. Assessment of hydro-geochemistry and groundwater quality of Rajshahi City in Bangladesh. Appl Water Sci. 2017;7:4663–71. https://doi.org/10.1007/s13201-017-0629-y.
25. Rahman MA, Islam MR, Kumar S, Al-Reza SM. Drinking water quality, exposure and health risk assessment for the school-going children at school time in the southwest coastal of bangladesh. J Water Sanitation Hygiene Develop. 2021;11:612–28. https://doi.org/10.2166/washdev.2021.016.
26. Rahman MM, Bodrud-Doza M, Muhib MI, Hossain KFB, Sikder MT, Shammi M, Akter R, Uddin MK. Human health risk assessment of nitrate and trace metals via groundwater in Central Bangladesh. Pollution. 2020;6:253–266. https://doi.org/10.22059/POLL.2019.288090.691
27. Kabir MM, Hossain N, Islam ARMT, Akter S, Fatema KJ, Hilary LN, Hasanuzzaman M, Didar-ul-Alam M, Choudhury TR. Characterization of groundwater hydrogeochemistry, quality, and associated health hazards to the residents of southwestern Bangladesh. Environ Sci Pollut Res. 2021;28:68745–61. https://doi.org/10.1007/s11356-021-15152-2.
28. Islam ARMT, Kabir MM, Faruk S, Al Jahin J, Bodrud-Doza M, Didar-ul-Alam M, Bahadur NM, Mohinuzzaman M, Fatema KJ, Safiur Rahman M, Choudhury TR. Sustainable groundwater quality in southeast coastal Bangladesh: co-dispersions, sources, and probabilistic health risk assessment. Environ Dev Sustain. 2021;23:18394–423. https://doi.org/10.1007/s10668-021-01447-4.
29. Ganguli S, Hosen Rifat MA, Howlader S, Hasan MA, Islam S, Alam MNE, Islam MN. Assessment of Bhatiari Lake water quality: pollution indices, hydrochemical signatures and hydro-statistical analysis. J Indian Chem Soc. 2022;99: 100585. https://doi.org/10.1016/j.jics.2022.100585.
30. Rifat MAH, Howlader S, Alam MA, Islam MN, Afrin A, Ali F, Ganguli S. Hydrogeochemical characteristics, quality assessment and health impact analysis of groundwater for drinking in the University of Chittagong, Bangladesh. Malays J Chem. 2021;23:74–102.
31. Hossain S, Islam A, Zim AFMIU, Bin QF. Assessment of Pb, Cr, Ni contamination in water from different selected sites of the Karnaphuli river, Chattogram, Bangladesh. Int J Nat Soc Sci. 2019;6:8–14.
32. APHA. Standard methods for the examination of water and wastewater, American Public Health Association, 23rd ed. Washington; 2017.
33. Tyagi S, Sharma B, Singh P, Dobhal R. Water quality assessment in terms of water quality index. Am J Water Resourc. 2020;1:34–8. https://doi.org/10.12691/ajwr-1-3-3
34. Venkata Mohan S, Nithila P, Jayarama Reddy S. Estimation of heavy metals in drinking water and development of heavy metal pollution index. J Environ Sci Health Part A Toxic/Hazard Substances Environ Eng. 1996;31:283–9. https://doi.org/10.1080/10934529609376357.

35. Siegel F. Environmental geochemistry of potentially toxic metals. Berlin: Springer; 2002.
36. DoE. The environmental conservation rules, 1997. The department of environment, ministry of environment and forest, government of the People's Republic of Bangladesh, Dhaka; 1997.
37. Ghaderpoori M, kamarehie B, Jafari A, Ghaderpoury A, Karami M. Heavy metals analysis and quality assessment in drinking water—Khorramabad city, Iran. Data Brief; 2018;16:685–92. https://doi.org/10.1016/j.dib.2017.11.078
38. Rahman MATMT, Paul M, Bhoumik N, Hassan M, Alam MK, Aktar Z. Heavy metal pollution assessment in the groundwater of the Meghna Ghat industrial area, Bangladesh, by using water pollution indices approach. Appl Water Sci. 2020;10:1–15. https://doi.org/10.1007/s13201-020-01266-4.
39. Nemerow NL, Sumitomo H. Benefits of water quality enhancement. Report No. 16110 DAJ, prepared for the U.S. environmental protection agency. New York: Syracuse University, Syracuse; 1970.
40. USEPA. Human health evaluation manual, supplemental guidance, standard default exposure factors; US environmental protection agency: Washington. USA: DC; 1991.
41. Kusin FM, Azani NNM, Hasan SNMS, Sulong NA. Distribution of heavy metals and metalloid in surface sediments of heavily-mined area for bauxite ore in Pengerang, Malaysia and associated risk assessment. CATENA. 2018;165:454–64. https://doi.org/10.1016/j.catena.2018.02.029.
42. Maleki A, Jari H. Evaluation of drinking water quality and non-carcinogenic and carcinogenic risk assessment of heavy metals in rural areas of Kurdistan, Iran. Environ Technol Innov. 2021;23: 101668. https://doi.org/10.1016/j.eti.2021.101668.
43. Hasan MF, Nur-E-alam M, Salam MA, Rahman H, Paul SC, Rak AE, Ambade B, Towfiqul Islam ARM. Health risk and water quality assessment of surface water in an urban river of bangladesh. Sustainability. 2021;13:1–22. https://doi.org/10.3390/su13126832.
44. Sarkar T, Alam MM, Parvin N, Fardous Z, Chowdhury AZ, Hossain S, Haque ME, Biswas N. Assessment of heavy metals contamination and human health risk in shrimp collected from different farms and rivers at Khulna-Satkhira region, Bangladesh. Toxicol Rep. 2016;3:346–50. https://doi.org/10.1016/j.toxrep.2016.03.003.
45. Singh SK, Ghosh AK. Health risk assessment due to groundwater arsenic contamination: children are at high risk. Hum Ecol Risk Assess. 2012;18:751–66. https://doi.org/10.1080/10807039.2012.688700.
46. USEPA. Regional screening level (RSL) summary table. Washington, DC, USA: US Environmental Protection Agency; 2022.
47. OEHHA. California office of environmental health hazard assessment (OEHHA). Technical support document for cancer potency factors 2009, Appendix A: hot spots unit risk and cancer potency values; 2019.
48. Molla MH, Chowdhury MAT, Md. Barkat Ali K, Bhuiyan MHR, Mazumdar RM, Das S. Supply water quality in Urban Bangladesh: a case study of Chittagong metropolitan city. Asian J Water, Environ Pollut. 2014;11:27–38.
49. Sarker S, Mahmud S, Sultana R, Biswas R, Sarkar PP. Quality assessment of surface and drinking water of Nakla Paurosova, Sherpur, Bangladesh. Adv Microbiol. 2019;9:703–27. https://doi.org/10.4236/aim.2019.98043.
50. Isa NM, Aris AZ, Sulaiman WNAW. Extent and severity of groundwater contamination based on hydrochemistry mechanism of sandy tropical coastal aquifer. Sci Total Environ. 2012;438:414–25. https://doi.org/10.1016/j.scitotenv.2012.08.069.
51. Mukanyandwi V, Kurban A, Hakorimana E, Nahayo L, Habiyaremye G, Gasirabo A, Sindikubwabo T. Seasonal assessment of drinking water sources in Rwanda using GIS, contamination degree (Cd), and metal index (MI). Environ Monit Assess. 2019;191:1–13. https://doi.org/10.1007/s10661-019-7757-9.
52. WHO. Guidelines for drinking-water quality, 3rd edn. Geneva; 2004.
53. Detay M, Carpenter M. Water wells: implementation, maintenance and restoration. London: Wiley; 1997.
54. Asadullah A, Nisa K, Khan S. Physico-chemical properties of drinking water available in educational institutes of Karachi city. Sci Technol Dev. 2013;32:28–33.
55. Afroze S. Presence of coliform and Fecal coliform and evaluation of the drinking water quality in Chittagong University Campus. Front Environ Microbiol. 2019;5:8. https://doi.org/10.11648/j.fem.20190501.12
56. Sawyer GN, McCarthy DL. Chemistry of sanitary engineers. 2nd ed. New York: McGraw Hill; 1967.
57. Al-ahmadi ME, El-Fiky AA. Hydrogeochemical evaluation of shallow alluvial aquifer of Wadi Marwani, western Saudi Arabia. J King Saud Univ Sci. 2009;21:179–90. https://doi.org/10.1016/j.jksus.2009.10.005.
58. Ameen HA. Spring water quality assessment using water quality index in villages of Barwari Bala, Duhok, Kurdistan Region, Iraq. Appl Water Sci. 2019;9:1–12. https://doi.org/10.1007/s13201-019-1080-z.
59. Mahaqi A, Mehdi M, Marzieh M, Mohmmad M, Moheghy A. Hydrogeochemical characteristics and groundwater quality assessment for drinking and irrigation purposes in the Mazar-i-Sharif city, North Afghanistan. Appl Water Sci. 2018;8:1–10. https://doi.org/10.1007/s13201-018-0768-9.
60. Ebrahimi M, Kazemi H, Ehtashemi M, Rockaway TD. Assessment of groundwater quantity and quality and saltwater intrusion in the Damghan basin, Iran. Chemie der Erde. 2016;76:227–41. https://doi.org/10.1016/j.chemer.2016.04.003.
61. Sridharan M, Nathan DS. Groundwater quality assessment for domestic and agriculture purposes in

Puducherry region. Appl Water Sci. 2017;7:4037–53. https://doi.org/10.1007/s13201-017-0556-y.

62. Islam A, Rakib M, Islam M, Jahan K, Patwary M. Assessment of health hazard of metal concentration in groundwater of Bangladesh. Am Chem Sci J. 2015;5:41–9. https://doi.org/10.9734/acsj/2015/13175.
63. Singh N, Sharma M. Assessment of the quality of drinking water sources and human health in a Rural Area of Solan, North India. Mapan. 2019;35:301–8. https://doi.org/10.1007/s12647-019-00354-4.
64. Rahman A, Hashem A. Arsenic, iron and chloride in drinking water at primary school, Satkhira, Bangladesh. Phys Chem Earth. 2019;109:49–58. https://doi.org/10.1016/j.pce.2018.09.008.
65. UNICEF. Bangladesh national drinking water quality survey, Bangladesh Bureau of Statistics, Dhaka, Bangladesh; 2009.
66. Raju NJ. Iron contamination in groundwater: a case from Tirumala-Tirupati environs, India. The Researcher. 2006;1:28–31.
67. BGS. British geological survey technical report WC/00/19. British Geological Survey: Keyworth; 2001.
68. Wasserman GA, Liu X, Parvez F, Ahsan H, Levy D, Factor-Litvak P, Kline J, van Geen A, Slavkovich V, Lolacono NJ, Cheng Z, Zheng Y, Graziano JH. Water manganese exposure and children's intellectual function in Araihazar, Bangladesh. Environ Health Perspect. 2006;114:124–9. https://doi.org/10.1289/ehp.8030.
69. Powers KM, Smith-Weller T, Franklin GM, Longstreth WT, Swanson PD, Checkoway H. Parkinson's disease risks associated with dietary iron, manganese, and other nutrient intakes. Neurology. 2003;60:1761–6. https://doi.org/10.1212/01.WNL.0000068021.13945.7F.
70. Farina M, Avila DS, Da Rocha JBT, Aschner M. Metals, oxidative stress and neurodegeneration: a focus on iron, manganese and mercury. Neurochem Int. 2013;62:575–94. https://doi.org/10.1016/j.neuint.2012.12.006.
71. Zhang Z, Xiao C, Adeyeye O, Yang W, Liang X. Source and mobilization mechanism of iron, manganese and arsenic in groundwater of Shuangliao City, Northeast China. Water. 2020;12:1–17. https://doi.org/10.3390/w12020534.
72. Ahmad SA, Khan MH, Haque M. Arsenic contamination in groundwater in Bangladesh: implications and challenges for healthcare policy. Risk Management and Healthcare Policy; 2018. p. 251–61.
73. Singh R, Venkatesh AS, Syed TH, Reddy AGS, Kurakalva KM, Mohan R. Assessment of potentially toxic trace elements contamination in groundwater resources of the coal mining area of the Korba Coalfield, Central India. Environ Earth Sci. 2017;76:1–17. https://doi.org/10.1007/s12665-017-6899-8.
74. Edet AE, Offiong OE. Evaluation of water quality pollution indices for heavy metal contamination monitoring. A study case from Akpabuyo-Odukpani area, lower cross River Basin (Southeastern Nigeria). GeoJournal. 2002;57:295–304. https://doi.org/10.1023/B:GEJO.0000007250.92458.de.
75. Backman B, Bodiš D, Lahermo P, Rapant S, Tarvainen T. Application of a groundwater contamination index in Finland and Slovakia. Environ Geol. 1998;36:55–64. https://doi.org/10.1007/s002540050320.
76. Brown RM, McClellan NI, Deininger RA, Tozer RG. A water quality index—do we dare? Water Sew Works. 1972;117:339–43.
77. Shil S, Singh UK, Mehta P. Water quality assessment of a tropical river using water quality index (WQI), multivariate statistical techniques and GIS. Appl Water Sci. 2019;9:1–21. https://doi.org/10.1007/s13201-019-1045-2.
78. Liu CW, Lin KH, Kuo YM. Application of factor analysis in the assessment of groundwater quality in a blackfoot disease area in Taiwan. Sci Total Environ. 2003;313:77–89. https://doi.org/10.1016/S0048-9697(02)00683-6.
79. Bhuiyan MAH, Bodrud-Doza M, Islam ARMT, Rakib MA, Rahman MS, Ramanathan AL. Assessment of groundwater quality of Lakshimpur district of Bangladesh using water quality indices, geostatistical methods, and multivariate analysis. Environ Earth Sci. 2016;75:1–23. https://doi.org/10.1007/s12665-016-5823-y.
80. Barzegar R, Moghaddam AA, Tziritis E, Fakhri MS, Soltani S. Identification of hydrogeochemical processes and pollution sources of groundwater resources in the Marand plain, northwest of Iran. Environ Earth Sci. 2017;76:1–16. https://doi.org/10.1007/s12665-017-6612-y.
81. Gautam RK, Sharma SK, Mahiya S, Chattopadhyaya MC. Contamination of heavy metals in aquatic media: transport, toxicity and technologies for remediation. In: Sharma SK, editors. Royal society of chemistry. 2014. p. 1–24.
82. Bodrud-Doza M, Islam ARMT, Ahmed F, Das S, Saha N, Rahman MS. Characterization of groundwater quality using water evaluation indices, multivariate statistics and geostatistics in central Bangladesh. Water Sci. 2016;30:19–40. https://doi.org/10.1016/j.wsj.2016.05.001.
83. Islam ARMT, Shen S, Bodrud-Doza M, Atiqur Rahman M, Das S. Assessment of trace elements of groundwater and their spatial distribution in Rangpur district, Bangladesh. Arab J Geosci. 2017;10. https://doi.org/10.1007/s12517-017-2886-3
84. Chakraborti D, Mahmudur M, Das B, Murrill M, Dey S, Chandra S, Kumar R, Kumar B, Kumar U, Roy S, Sorif S. Status of groundwater arsenic contamination in Bangladesh: a 14-year study report.

Water Res. 2010;44:5789–802. https://doi.org/10.1016/j.watres.2010.06.051.
85. Mukherjee AB, Bhattacharya P. Arsenic in groundwater in the Bengal delta plain: slow poisoning in Bangladesh. Environ Rev. 2001;9:189–220. https://doi.org/10.1139/er-9-3-189.
86. Juahir H, Zain SM, Kamil M, Hanidza TIT, Armi ASM, Ekhwan M, Mazlin T. Spatial water quality assessment of Langat River Basin (Malaysia) using environmetric techniques. Environ Monit Assess. 2011;173:625–41. https://doi.org/10.1007/s10661-010-1411-x.
87. Gu C, Zhang Y, Peng Y, Leng P, Zhu N, Qiao Y, Li Z, Li F. Spatial distribution and health risk assessment of dissolved trace elements in groundwater in southern China. Sci Rep. 2020;10:1–11. https://doi.org/10.1038/s41598-020-64267-y.
88. Ahmed I, Anisa N. Assessment of water quality and identification of pollution sources of three lakes in Kashmir, India, using multivariate analysis. Environ Earth Sci. 2012;66:2367–78. https://doi.org/10.1007/s12665-011-1458-1.
89. Usman UA, Yusoff I, Raoov M, Alias Y, Hodgkinson J. Natural sources of iron and manganese in groundwater of the lower Kelantan River Basin, North-eastern coast of Peninsula Malaysia: water quality assessment and an adsorption—based method for remediation. Environ Earth Sci. 2021;80:1–17. https://doi.org/10.1007/s12665-021-09717-0.
90. Hossain MG, Selim Reza AHM, Lutfun-Nessa M, Ahmed SS. Factor and cluster analysis of water quality data of the groundwater wells of Kushtia, Bangladesh: implication for arsenic enrichment and mobilization. J Geol Soc India. 2013;81:377–84. https://doi.org/10.1007/s12594-013-0048-0.
91. Omo-Irabor OO, Olobaniyi SB, Oduyemi K, Akunna J. Surface and groundwater water quality assessment using multivariate analytical methods: a case study of the Western Niger Delta, Nigeria. Phys Chem Earth. 2008;33:666–73. https://doi.org/10.1016/j.pce.2008.06.019.
92. Rasouli F, Pouya AK. Hydrogeochemistry and water quality assessment of the Kor–Sivand Basin, Fars province, Iran. Environ Monitor Assess. 2012;184:4861–77. https://doi.org/10.1007/s10661-011-2308-z.
93. Caboi R, Cidu R, Fanfani L, Lattanzi P, Zuddas P. Environmental mineralogy and geochemistry of the abandoned Pb-Zn Montevecchio-Ingurtosu mining district, Sardinia, Italy; 1999. p. 21–8.
94. Herojeet R, Rishi MS, Kishore N. Integrated approach of heavy metal pollution indices and complexity quantification using chemometric models in the Sirsa Basin, Nalagarh valley, Himachal Pradesh, India. Chin J Geochem. 2015;34:620–33. https://doi.org/10.1007/s11631-015-0075-1.
95. Zakir HM, Sharmin S, Akter A, Rahman MS. Assessment of health risk of heavy metals and water quality indices for irrigation and drinking suitability of waters: a case study of Jamalpur Sadar area, Bangladesh. Environ Adv. 2020;2: 100005. https://doi.org/10.1016/j.envadv.2020.100005.
96. Mohammadi AA, Zarei A, Majidi S, Ghaderpoury A, Hashempour Y, Saghi MH, Alinejad A, Yousefi M, Hosseingholizadeh N, Ghaderpoori M. Carcinogenic and non-carcinogenic health risk assessment of heavy metals in drinking water of Khorramabad, Iran. MethodsX. 2019;6:1642–51. https://doi.org/10.1016/j.mex.2019.07.017.
97. Islam ARMT, Siddiqua MT, Zahid A, Sanchita S, Tasnim RMM. Drinking appraisal of coastal groundwater in Bangladesh: an approach of multi-hazards towards water security and health safety. Chemosphere. 2020;255: 126933. https://doi.org/10.1016/j.chemosphere.2020.126933.

Groundwater Pollution in Urban Areas

Fahmida Parvin, Mabia Hossain Shetu,
Shafi M. Tareq, and Syeda Moontaha Ahmed

Abstract

The access to safe drinking water is a basic human right. Groundwater is the largest and relatively easily accessible freshwater supply on earth. In the past, groundwater was assumed to be protected from pollution by layers of rock and soil that act as filters. Recently, however, groundwater has become more and more polluted by chemicals. Drinking water contaminated with microorganisms is the main route of transmission for many waterborne diseases. This paper reviews the current state of research on groundwater contamination, entry pathways and the impact of urban areas on groundwater quality. It discusses the relationship between different parameters that influence water quality.

Keywords

Faecal pollution · Groundwater · Heavy metals · Ions · Urban area

1 Introduction

Groundwater is an important resource for domestic, industrial, agricultural and recreational activizies. Groundwater is also a very important source of water in arid and semi-arid regions, but here too (as in urban areas) the resource is increasingly becoming a rare commodity as a result of climate change and uncontrolled exploitation [1]. Almost one third of the world's population has no or very limited access to clean drinking water [2, 3] and the need is increasing due to growing populations, especially in countries of the global South. Unfortunately, clean groundwater availability is decreasing due to anthropogenic activities and improper management of groundwater [4, 5]. Groundwater especially in most urban areas, especially in developing and low income countries is polluted with heavy metals ions, radioactive substances, pathogenes, and toxic chemicals [1, 6, 7]. Additionally, groundwater in some areas is contaminated by natural processes, which include deposition of salts, gypsum, metals in soil which leach through soil into water, rock weathering and leaching, seawater intrusion [7]. However, in urban areas, groundwater is mainly polluted by anthropogenic activities such as uncontrolled waste dumping, sewage systems, over-exploitation, untreated wastewater discharge from different industrial activities [3, 7, 8].

F. Parvin (✉) · M. H. Shetu · S. M. Tareq · S. M. Ahmed
Hydrobiogeochemistry and Pollution Control Laboratory, Department of Environmental Sciences, Jahangirnagar University, Savar, Dhaka 1342, Bangladesh
e-mail: fahmidaprvn497@juniv.edu

A. M. Bahadir et al. (eds.), *Hydrology and Urban Water Supply*, Water and Wastewater Management,
https://doi.org/10.1007/978-3-031-72589-0_10

Industrial wastewater contains organic pollutants (wastewater of paper industries and food industries) and inorganic pollutants (wastewater from textile, cement or chemical industry, industrial agriculture and hospital waste and wastewater) [6, 9–12]. In such wastewaters often high concentrations of pathogens (virus, bacteria, and anti-biotic resistant bacteria), toxic chemical (pharmaceuticals and drug residues, heavy metals, radioisotopes, etc.) are found [13]. These pollutants through surface water or soil are then seeped into groundwater. Mining activity is another source of groundwater pollution. In mining practices, due to excavation protective layer of the soil is removed and thus pollutants can reach the groundwater faster. Mining activities lead to drainage of acidic water which directly percolates into groundwater [14].

This paper gives an overview on various studies regarding groundwater pollution in urban cities, especially in Southeast Asian countries. The chemical composition of contaminated groundwater, as well as the origin and significance of pollutants from different cities is discussed.

2 Heavy Metals

2.1 Health Impact of Heavy Metals Ingestion

The presence of different toxic metals ions in groundwater has become an urgent and severe issue worldwide. The origin of source contamination like Mercury (Hg) as a by-product of gold mining, non-metal extraction, dental laboratories and hospitals, fish preservation, municipal waste, mercury lamps, batteries etc. are well known. Lead (Pb) is produced in the environment from lead-acid batteries used in vehicles and various industrial applications. Lead (Pb) poisoning in humans can cause anemia, weakness, kidney and brain damage [15, 16]. The standards or maximum limit for occurrence of Pb in drinking water is 0.01 mg/l by WHO 2014 guideline [15, 16]. WHO set a threshold limit of Nickel (Ni) in drinking water (0.02 mg/L) and the long term drinking of Ni polluted water can make problem in lungs and nervous system, dry cough and cancer [17]. Though Cadmium (Cd) and Chromium (Cr) (VI) are both are highly toxic and carcinogenic water contaminants, but Cd has greater toxicity (standard: 0.005 mg/L) than Cr (standard: 0.005 mg/L) [18, 19]. Manganese (Mn) and Iron (Fe) are both essential nutrient for human health and the maximum drinkable concentrations for them are 0.2 mg/L for Mn and 0.3 mg/L for Fe [19]. However, excess uptake of these metals through food/water can cause adverse health risks including, Parkinson disease, hyperkeratosis, diabetes mellitus, cardiovascular disease, pigmentation changes, kidney, liver, and neurological disorders [20, 21]. On the other hand, standard of Zinc (Zn) and Copper (Cu) for drinking water are 3 and 2 mg/l [19]. Consuming excessive concentration of Zn and Cu through ingestion of water/food is also harmful for human health. If any ground water sources contain multiple metals in considerable concentrations, then synergistic or antagonistic effects may appear in the exposed population who drinks the water [6]. The groundwater of different urban cities of South Asian countries are polluted with toxic metals from different sources.

2.2 Heavy Metal Pollution in Groundwater

Tables 1, 2 and 3 shows the concentrations of different metals in groundwater of different urban areas in India, Bangladesh and Pakistan. In India, groundwater of different cities is found to be polluted with different toxic metals (Table 1). Diversified geological and complex tectonic formations and anthropogenic activities have heavily polluted the groundwater system. The concentration of Fe, Pb, Ni, Cr, and Cd in the groundwater of Aligarh city, India is found to exceed WHO standards of maximum occurrence [22]. There are about 5506 industrial units for which the Aligarh city is the center of brassware, electroplating and lock smithy manufacturing [23]. The environmental quality of this city thus deteriorates for these large scale industrial

Table 1 Heavy metal concentration in groundwater in urban cities of India (mg/L) from 2010 to 2022

Location	Cu	Mn	Fe	Pb	Ni	Cr	Zn	Cd	References
WHO drinking water standard	1.5	0.5	0.3	0.05	0.0303	0.05	5	0.005	[19]
Aligarh city pre-moonsoon	0.77	0.09	0.59	0.09	0.61	0.34	BDL	0.03	[22]
Post-moonsoon	0.85	1.14	2.54	1.9	1.07	30.2	BDL	0.64	
Ankleshwar and Panoli in Gujarat, Western India	0.106	0.105	BDL	BDL	0.182	0.079	BDL	0.015	[24]
Goa mining region monsoon	0.003	0.051	0.075	0.0081			0.015	0.0009	[26]
Post-monsoon	0.015	0.02	0.031	0.005			0.022	0.0011	
Chennai pre-moonsoon	0.142	0.478	0.63	0.163	0.082	0.262	0.048	BDL	[27]
Post-moonsoon	0.042	0.518	0.546	0.168	0.025	0.052	0.062	BDL	
Nacharam, Hyderabad	0.25	0.52	0.32	0.006	0.0012	0.013	0.51	0.0001	[28]
Khetri copper mine region, Rajasthan	0.69	0.58	1.65	0.02	0.19	BDL	1.45	BDL	[29]
Bhopal city	0.009	0.054	0.29				0.13		[30]
Vijayawada Andhra Pradesh	0.15	0.48	1.88	0.12	0.02	0.11	0.214	0.024	[28]
Industrial township Virudhunagar, Tamil Nadu, India	0.03	0.02		0.47	0.27	0.08	0.23	0.04	[31]
Obra, southern Sonbhadra, Uttar Pradesh, Premonsoon	0.082	0.23	2.73	0.038	0.026	0.059	0.511	0.023	[32]
Postmonsoon	0.082	0.21	2.3	0.036	0.023	0.09	0.452	0.022	
Renukoot, southern Sonbhadra, Uttar Pradesh	0.15	0.26	1.589	0.036	0.041	0.105	1.95	0.035	[32]
Anpara, southern Sonbhadra, Uttar Pradesh Premonsoon	0.11	0.21	2.4	0.13	1.65	0.066	0.72	0.022	[32]
Postmonsoon	0.124	0.18	2.23	0.1	0.167	0.05	0.61	0.018	
Aligarh, Uttar Pradesh	0.34	0.11		0.06	0.07	0.6	0.77	0.01	[33]
Coastal area, Mumbai		0.41	2	0.13			0.08		[34]
West Tripura district	0.01	0.1	1.36	0.02		0.01	0.02	0.01	[35]

activities, which in turn has deteriorated the groundwater condition. In Gujarat of the western India, the groundwater is found to pollute with Ni and Cd [24]. In Gujarat, there are more than 1700 industries which manufacture chemicals, drugs, pharmaceuticals, pesticides, textile, plastics and dye. The effluents from these industries contain large quantities of heavy metals which percolate down and pollute the groundwater. Agra is one the most polluted states in Uttar Pradesh where domestic, industrial wastewater and sewage waste are directly dumped into river without treatment leading to pollution of the groundwater severely. Almost all heavy metals (Fe, Pb, Ni, Cd) are found in high concentrations in this region [25].

The major reason for the pollution of groundwater with metals in Chennai city and Bangalore is due to improper dumping of solid wastes of industries and sewage wastewater. Leachate percolating through the soil also pollutes ground water of these regions [27, 56]. In Khetri Copper mine region of Rajasthan, heavy metals in groundwater system has increased in large proportions due to the drainage of acid water from mining activities, increase of Cu within the soil and plants of the mining territory [57]. Diversified geological and complex tectonic

Table 2 Heavy metal concentration in groundwater in urban cities of Bangladesh (mg/L) from 2010 to 2021

Location	Cu	Mn	Fe	Pb	Ni	Cr	Zn	Cd	References
WHO drinking water standard	1.5	0.5	0.3	0.05	0.03	0.05	5	0.005	[19]
Singair, Manikgonj	0.022	2.08	7.11	0.02	0.043	0.093	0.06		[36]
Faridpur		0.001	5.95	0.001	0.003		0.011		[37]
Rajshahi city	0.08	1.47	3.1	1.167			0.19	0.014	[38]
Rangpur		0.68	7.72				0.033		[39]
Hazaribagh tannery	0.006				0.004	0.01	0.04		[40]
Dhaka metropolitan city	0.009	0.06	0.089	0.049		0.001	0.005		[41]
Jamalpur Sadar	0.008	1.07	0.363	0.016		0.006	0.02	0.008	[42]
Meghna Ghat industrial area				0.18		0.07		0.006	[43]
South Eastern coastal area	0.057	0.23	1.25	0.004	0.144	0.068	0.173	0.017	[44]
Deltaic flood plain, Middle-West Part of Bangladesh		1.61	5.07	0.043		0.031		0.002	[45]
Setabganj sugar mills limited, Dinajpur	0.037	0.41	0.34	234		0.011	0.02		[46]

Table 3 Heavy metal concentration in groundwater in urban cities of Pakistan (mg/L) from 2010 to 2021

Location	Cu	Mn	Fe	Pb	Ni	Cr	Zn	Cd	References
WHO drinking water standard	1.5	0.5	0.3	0.05	0.03	0.05	5	0.005	[19]
Sheikhupura, Punjab		0.1	0.78		0.25			0.02	[47]
Sahiwal, Punjab			0.06						
Sindh			0.35		0.25				
Jawar Dharki, Sindh	0.055		0.22		0.01				
Taluka Qamber-Shahdadkot, Sindh	0.02	0.022	0.07	0.01	0.067	0.07		0.007	[48]
Various locations of Faisalabad city area	0.038	0.012	0.024	0.024	0.046		0.224	0.02	[49]
Zhob River Basin, Balochistan	0.16	0.16	0.8	0.04		0.2	0.39		[50]
Urban areas of Lahore, Punjab			1.17	0.004					[51]
Deep wells of Palosi drain Peshawar	0.58	0.153	0.73	0.25	0.26		0.041	0.038	[52]
Near tannery effluent, Peshawar			1				0.28		
Kasur, Punjab Province	2.15	0.28	0.47	0.013	0.12	1.06	2.9	0.03	[53]
Sargodha city	0.57		0.25						[54]
Vehari, Vehari, Punjab	0.44	0.2	1.86	0.16	0.04	BDL	0.64	0.02	[55]
Mailsi, Vehari, Punjab	0.1	0.22	1.34	0.18	0.13	BDL	0.55	0.01	
Burewala, Vehari, Punjab	0.21	0.13	1.61	0.11	0.08	BDL	0.6	0.01	

formations and anthropogenic activities have heavily polluted the groundwater with Fe and Cd in Guwahati city of Assam [58, 59].

Bhopal city has experienced the world's worst industrial tragedy where toxic gases were released from a pesticide industry which severely affected the air and water of the city. To escape from the continuous pollution, the government of Madhya Pradesh has buried the deadly waste of the factory. This waste has leaked and polluted the groundwater severely. Till now the groundwater of region is heavily polluted with Mn and

Fe [30]. In the groundwater of Vijayawada Andhra Pradesh, concentrations of heavy metals include Fe, Pb, Cr and Cd has exceeded the drinking water standard [60]. Vijayawada is one of the densely populous cities in India having a population of about 1.5 million where wastewater and waste is not disposed properly, thus leading to increase of heavy metals in the city. In Ludhiana of Punjab, the indiscriminate disposal of domestic, industrial and waste and over exploitation of groundwater in this region increased Mn concentration to an alarming level which is about 25–27 mg/l in different seasons [61].

In Bangladesh about 80% of the population depend on groundwater for daily purposes. The fatal thing is that 74% of the groundwater is used for irrigation. In addition, various industries of the country consume large quantities of groundwater [40]. In Bangladesh, various uncontrolled sources such as effluent from industries, septic tanks and sewage line leakage, improper disposal of waste, hazardous waste improper management, landfill leachate, pesticide and fertilizer use, mining activities, etc. contribute to high levels of groundwater pollution (Table 2). Besides heavy metal ion pollution, arsenic pollution is currently a major problem [62]. Uncontrolled disposal of urban wastewater and untreated industrial wastewater massively pollute the groundwater [49].

3 Groundwater Contamination by Dissolved Ions

There are maximum limits for the occurrence of various dissolved ions in groundwater. Tables 4, 5 and 6 show the concentrations of various dissolved ions in groundwater from urban areas in India, Bangladesh and Pakistan.

3.1 Fluoride Pollution

Fluoride pollution can lead to fluorosis causing severe damage to teeth, bones, etc. Intake of Fluorine more than 1.5 mg/L has adverse impact on human health [63]. Fluoride in groundwater can increase by a number of reasons. In Bhopal city, India due to the presence of large number of industries and also the hazardous waste from the pesticide industry explosion led to increase of fluorides and the concentration rose to ten times the maximum limit [30, 64]. The untreated disposal of domestic, industrial and agricultural wastewater directly and improper management of solid waste led to the increase of fluorides in Agra [65]. Improper dumping of solid waste and percolating leachate from this waste lead to the contamination of groundwater and increase of fluoride within Chennai city and Vijayawada city [66].

3.2 Nitrate Pollution

Consuming water polluted with nitrate ions can lead to blue baby syndrome, thyroid diseases, cancer of different organs also gastric cancer [82]. 45 mg/L nitrate in drinking water is considered as the standard for drinking water [63]. Nitrate ions are released into groundwater through sewage disposal, from dairy, poultry and livestock wastes and using of nitrogen-based fertilizers. The fertilizers leach through the soil as soluble nitrates and pollute the groundwater. All these phenomena lead to the increase of nitrate ions in Maharashtra. Due to similar phenomenon, disposal of untreated domestic, industrial and agricultural wastewater directly and improper management of solid waste led to the increase of nitrates in Agra [65].

4 Faecal Pollution of Groundwater

The presence of *E. coli* bacteria in groundwater resources represents a severe health problem. Very often sewage wastewater, municipal wastes are directly disposed into the environment without any control. According to WHO, in drinking water faecal and total coliform bacteria populations should be completely absent [19]. The presence of coliform bacteria in water in a range of 1–10 CFU/100 mL has possibly low risk; 10–

Table 4 Physicochemical parameters of the groundwater of urban cities of India

Location	pH	EC (mS/cm)	TDS (mg/L)	Alkalinity (mg/L)	PO_4 (mg/L)	F (mg/L)	SO_4 (mg/L)	NO_3 (mg/L)	References
WHO drinking water standard	6.5–8.5	1000	1000	200	0.05	1.5	200	45	[19]
Maharashtra	8.5	2,173	739			0.35	111	49	[67]
Coimbatore city, Tamil	7.44	3.15	635	585			119		[68]
Dindigul Town, Tamil	7.25	1.07	692	143	0.03				[69]
Tirupur and Coimbatore	8.26	1811	1103						
Bhopal city	7.3		576	263.3		10.9	126	18.4	[30]
Khetri Copper Mine Region, Rajasthan	7.08	1974							[57]
Surat city	7.4	2086	1143						[70]
Kolkata	7.8		1326			0.43		0.07	[71]
Obra, Southern Sonbhadra, Uttar Pradesh	7.07	1074							[32]
Renukoot, Southern Sonbhadra, Uttar Pradesh	7.15	849							[32]
Coastal Area, Mumbai	7.23	135.1	161	85			227		[34]
West Tripura district	6.1	120	78	40.72		0.24	7.78	0.9	[35]
Dhapa Uncontrolled Landfill site in Kolkata city	7.3		1148	376			21.18	1.7	[72]

100 CFU/100 mL poses intermediate risk and the concentration of coliform bacteria greater than 100 CFU/100 mL poses the high risk for human health. Table 7 shows the microbial contamination in groundwater of different urban areas of India, Bangladesh and Pakistan.

In Kolkata, effluent from tanneries and other industries, leakage from landfill sites, sewage water disposal lead to increase of fecal and total coliforms within the groundwater (13–21 CFU/100 ml) [88]. Open sanitation in rural and peri-urban regions, pit latrines, improper sewage disposal, etc. led to biological pollution of groundwater of Agra [25]. Presence of open drainage system where disposal of raw sewage, industrial waste and unchecked garbage lead to biological contamination in Meerut, Ludhiana and Chennai city [27, 33, 61]. This indicates Ludhiana and Chennai are least risked by biological contamination.

Table 5 Physicochemical parameters of the groundwater of different urban cities of Bangladesh

Location	pH	EC (mS/cm)	TDS (mg/L)	Alkalinity (mg/L)	PO_4 (mg/L)	F (mg/L)	SO_4 (mg/L)	NO_3 (mg/L)	References
WHO drinking water standard	6.5–8.5	1000	1000	200	0.05	1.5	200	45	[19]
Dhaka city	6.4	600		205.62			10.64	6.24	[73]
Comilla	7.2	1078	676	3.6			0.09		[74]
Southwest Coastal	7.85	1,983	1,089	350					[75]
Rajshahi city	7.1	454.98	241	188			36.6	1.67	[38]
Barishal city	7.58	4264	2191	369			46.86	4.23	[44]
Sylhet city corporation	6.77		240						[76]
Tangail city	7.3	377	299	143					[77]
Jhenaidah city	7.23	619	598	370.5	0.26		0.92		[78]
Khulna city	7.3	1650	1188					0.029	[79]
Koyra, Khulna	7.33	1,400	841	429	0.61			1.43	[80]
Dhaka Metropolitan city	6.54	459	321	184	0.06		11.23		[41]
Jamalpur Sadar	6.87	270.5	162	205	0.27		9.18		[42]
Meghna Ghat Industrial Area	8.06	1350	554						[43]
South Eastern Coastal Area	7.69	804	522						[45]
Rupsha Upazila	7	2425	1365	516	0.272	0.94	1	2.13	[80]

Table 6 Physicochemical properties of the groundwater of urban cities in Pakistan

Location	pH	EC (mS/cm)	TDS (mg/L)	Alkalinity (mg/L)	PO_4 (mg/L)	F (mg/L)	SO_4 (mg/L)	NO_3 (mg/L)	References
WHO Standard	6.5–8.5	1000	1000	200	0.05	1.5	200	45	[19]
Sheikhupura, Punjab	8.29	685		225			24	5	[47]
Sahiwal, Punjab	7.81	887		319	0.25		45		
Sindh	8.03	46,300		233	55		6140	1.18	
Jawar Dharki, Sindh	8.19	276		121	50		44		
Dharki, Sindh	8.44	12,960		583			960		
Talukaqamber-Shahdadkot, Sindh	7.87	3921	2626	327	0.23	0.005	513	4.18	[48]
Islamabad	7.2	354	641	256			40	3.12	[81]
Kasur, Punjab Province	8.1	1102	560	334					[55]
Sargodha city	7.97	2856	1942	450			227.7	25.2	[54]

Table 7 Microbial contamination (CFU/100 ml) in groundwater of different urban areas of India, Bangladesh and Pakistan

Location	Faecal coliform	Total coliform	References
WHO drinking water standard	0	0	[19]
India			
West Tripura District (Hezamara, Lefunga, Mandawi And Jirania)	7.11	10.24	[35]
Tiruchirappalli District, South India		9.9	[83]
Andhra Pradesh	3.12	49	[84]
Bihar	309	789	
Chandigarh	2	23	
Daman Diu and Dadra Nagar Haveli	2	2	
Delhi	15	31	
Bore Well, Cuncolim Industrial Estate, Goa	651	1105	
Gujarat	2	7	
Sonpat, Haryana	1	512	
Jammu & Kashmir	0	0	
Shimla D/S Of MSW dumping site, Himachal	1015	1600	
Vammanjoor, Karnataka	2	6	
Kolkata	13	21	[85]
Bangladesh			
Rajshahi city	0.11	3.7	[38]
Sylhet city corporation	0	0.4	[76]
Upazilas of Satkhira	25.25	3.9	[86]
Kashipur, Narayanganj city	0	0	[87]
Rupsha Upazila, South-Western Coastal area	1058	3495	[80]
Pakistan			
Sheikhupura, Punjab	70	94	[47]
Sindh	49	94	
Dharki, Sindh	39	46	
Islamabad	0	123	[81]
Faisalabad	0	1408	[49]
Bahawalpur	0	29	[81]
Tube well water of Urban Peshawar	0	3	[52]

5 Physical Parameters of Groundwater

5.1 pH Value

The parameter pH is one of the control parameter for water quality. It indicates whether the water is corrosive, determines the solubility and mobility of many dissolved metals ions, and provides information about the types of gases and minerals with which the groundwater has reacted. Soil water in contact with decaying organic matter may have a pH of less than 4.0, and the pH of water that has reacted with iron sulphide minerals in coal or shale may be even lower. In the absence of coal or iron sulphide minerals, the pH value of groundwater is usually between 6.0 and 8.5, depending on the type of soil and rock it comes into contact with. Reactions between groundwater and sandstones result in pH values between about 6.5 and 7.5. Groundwater that perculates through limestone layers can have pH values higher than 8.5 [89]. According to WHO, the acceptable pH range in groundwater is between 7 and 8.5 [19]. Almost in every urban region of south Asia, the pH level is within the permissible limit. In the groundwater of various regions of India, the pH value is between 7.9 and 8.7 in pre-monsoon and between 7.6 and 8.2 in post-monsoon seasons [90].

5.2 Electrical Conductivity (EC) and Total Dissolved Solids (TDS)

Electrical conductivity is the measure of dissolved ions within water. According to WHO, the permissible limit for EC is 1000 μS/cm and for TDS 1000 mg/L [19]. EC and TDS are used to decide whether the water is suitable for domestic use or for irrigation purposes [91].

5.3 Alkalinity

The WHO permissible limit for alkalinity is 200 mg/L [19]. Alkalinity can increase if water perculates through rocks. In Coimbatore, the high alkalinity levels of the groundwater are mainly due to percolation through the rock [8]. Bhopal city is surrounded by several hills. Here, weathering of rocks of nearby hills leads to the increase of the groundwater alkalinity. Weathering and dissolution of rocks also led to increase of alkalinity in areas like Meerut, Lucknow, Ludhiana, Chennai, Madurai and Vijayawada too [27, 60, 91, 92].

6 Conclusion

This chapter shows how the groundwater systems in urban areas are loaded with different pollutants. Previously, groundwater was thought to be protected from pollution by layers of rock and soil that act as filters. Unfortunately, in recent times, groundwater in urban areas has been polluted by synthetic organic and inorganic chemicals and microorganisms due to increased anthropogenic activities and leaching from the soil. The pollution extent in the groundwater of urban areas is high due to anthropogenic activities and urbanization than natural processes. In the cities of South Asian countries, with a high industrial density and high population numbers, clean/uncontaminated groundwater can hardly be found. Only in the post-monsoon season does the situation improve a little due to dilution effects. Among the pollutants, Fe is the most prevalent heavy metal, pH is moderate, TDS and EC are higher, and fluoride ions are higher than sulfate ions.

Groundwater is vital for human health, and degradation of groundwater is a global issue at present. Thus, it is necessary to take proper remedial actions to control groundwater pollution because it is worsening with time.

References

1. Li P, Tian R, Xue C, Wu J. Progress, opportunities, and key fields for groundwater quality research under the impacts of human activities in China with a special focus on western China. Environ Sci Pollut Res. 2017;24(15):13224–34. https://doi.org/10.1007/s11356-017-8753-7.
2. Grönwall J, Danert K. Regarding groundwater and drinking water access through a human rights lens: self-Supply as a norm. Water (Switzerland). 2020;12 (2). https://doi.org/10.3390/w12020419
3. Li P, Karunanidhi D, Subramani T, Srinivasamoorthy K. Sources and consequences of groundwater contamination. Arch Environ Contam Toxicol. 2021;80(1):1–10. https://doi.org/10.1007/s00244-020-00805-z.
4. Lioubimtseva E. A multi-scale assessment of human vulnerability to climate change in the Aral Sea basin. Environ Earth Sci. 2015;73(2):719–29. https://doi.org/10.1007/s12665-014-3104-1.
5. Parvin F, Haque MM, Tareq SM. Recent status of water quality in Bangladesh: a systematic review, meta-analysis and health risk assessment. Environ Challenges. 2022;6: 100416. https://doi.org/10.1016/j.envc.2021.100416.
6. Wu J, Sun Z. Evaluation of shallow groundwater contamination and associated human health risk in an alluvial plain impacted by agricultural and industrial activities, Mid-west China. Expo Health. 2016;8 (3):311–29. https://doi.org/10.1007/s12403-015-0170-x.
7. Stefanakis AI, Zouzias D, Marsellos A. Groundwater pollution: human and natural sources and risks. Environ Sci Eng. 2015;4:82–102.
8. Selvakumar S, Chandrasekar N, Kumar G. Hydrogeochemical characteristics and groundwater contamination in the rapid urban development areas of Coimbatore, India. Water Resourc Indus. 2017;17:26–33. https://doi.org/10.1016/j.wri.2017.02.002.
9. Amin N, Rahman M, Raj S, Ali S, Green J, Das S, Doza S, Mondol MH, Wang Y, Islam MA, Alam M-U, Huda TMN, Haque S, Unicomb L, Joseph G, Moe CL. Quantitative assessment of fecal contamination in multiple environmental sample types in urban communities in Dhaka, Bangladesh using SaniPath microbial approach. PLoS ONE. 2019;14 (12): e0221193. https://doi.org/10.1371/journal.pone.0221193.
10. Parvin F, Ferdaus Z, Tareq SM, Choudhury TR, Islam JMM, Khan MA. Effect of gamma-irradiated textile effluent on plant growth. Int J Recycl Org Waste Agric. 2015;4(1):23–30. https://doi.org/10.1007/s40093-014-0081-z.
11. Parvin F, Nayna OK, Tareq SM, Rikta SY, Kamal AK. Facile synthesis of iron oxide nanoparticle and synergistic effect of iron nanoparticle in the presence of sunlight for the degradation of DOM from textile wastewater. Appl Water Sci. 2018;8 (2):73. https://doi.org/10.1007/s13201-018-0719-5.
12. Panouillères M, Boillot C, Perrodin Y. Study of the combined effects of a peracetic acid-based disinfectant and surfactants contained in hospital effluents on Daphnia magna. Ecotoxicology. 2007;16(3):327–40. https://doi.org/10.1007/s10646-007-0136-2.
13. Kümmerer K. Erratum: drugs in the environment: Emission of drugs, diagnostic aids and disinfectants into wastewater by hospitals in relation to other sources—A review (Chemosphere (2001) 45 (957–969)(PII:S0045653501001448)). Chemosphere. 2002;48(3):383. https://doi.org/10.1016/S0045-6535(02)00088-7.
14. Giri S, Singh AK. Human health risk assessment via drinking water pathway due to metal contamination in the groundwater of Subarnarekha River Basin, India. Environ Monitor Assess. 2015;187(3). https://doi.org/10.1007/s10661-015-4265-4
15. Anawar HM, Akai J, Mostofa KMG, Safiullah S, Tareq SM. Arsenic poisoning in groundwater: health risk and geochemical sources in Bangladesh. Environ Int. 2001;27(7):597–604. https://doi.org/10.1016/S0160-4120(01)00116-7.
16. Smith AH, Lingas EO, Rahman M. Contamination of drinking-water by arsenic in Bangladesh: a public health emergency. Bull World Health Organ. 2000. https://doi.org/10.1590/S0042-96862000000900005.
17. Genchi G, Carocci A, Lauria G, Sinicropi MS, Catalano A. Nickel: human health and environmental toxicology. Int J Environ Res Public Health. 2020;17 (3):679. https://doi.org/10.3390/ijerph17030679.
18. Jaishankar M, Tseten T, Anbalagan N, Mathew BB, Beeregowda KN. Toxicity, mechanism and health effects of some heavy metals. Interdiscip Toxicol. 2014. https://doi.org/10.2478/intox-2014-0009.
19. WHO. Guidelines for drinking-water quality: fourth edition incorporating the first addendum. World Health Organization; 2014.
20. Ghosh GC, Khan MJH, Chakraborty TK, Zaman S, Kabir AHME, Tanaka H. Human health risk assessment of elevated and variable iron and manganese intake with arsenic-safe groundwater in Jashore, Bangladesh. Sci Rep. 2020;10(1):5206. https://doi.org/10.1038/s41598-020-62187-5.
21. O'Neal SL, Zheng W. Manganese toxicity upon overexposure: a decade in review. Curr Environ Health Rep. 2015. https://doi.org/10.1007/s40572-015-0056-x.
22. Irfan A, Khurshid S. Heavy metal pollution in groundwater of Kali River, Sub-Basin, Aligarh City, U.P. (India). Int J Adv Res. 2018;6(6):520–29. https://doi.org/10.21474/ijar01/7241
23. Anwar K, Aggarwal V. Analysis of groundwater quality of Aligarh City, (India): using water quality index. Curr World Environ. 2014;9(3):851–57. https://doi.org/10.12944/cwe.9.3.36
24. Chinchmalatpure A, Vibhute S, Kumar S. Assessing groundwater contamination due to heavy metals and

their spatial distribution in an industrial area of Western India; 2019.
25. Yadav H, Keshari AK. Assessment of groundwater pollution from industries in Agra region. Int J Renew Energy Res. 2017;7(4):2068–74. https://doi.org/10.20508/ijrer.v7i4.6417.g7254
26. Singh G, Kamal RK. Heavy metal contamination and its indexing approach for groundwater of Goa mining region, India. Appl Water Sci. 2017;7(3):1479–85. https://doi.org/10.1007/s13201-016-0430-3.
27. Mahapatra SR, Nimmy PM. Groundwater contamination with heavy metals in Chennai city, India—A threat to the human population. IOP Conf Ser: Earth Environ Sci. 2021;889(1). https://doi.org/10.1088/1755-1315/889/1/012037
28. Rao BV, Murthy NN. Heavy metal contamination of groundwater in Nacharam Industrial Area. India: Hyderabad; 2016.
29. Punia A, Siddaiah NS. Assessment of heavy metal contamination in groundwater of Khetri copper mine Region, India and health risk assessment. Asian J Water Environ Pollut. 2017;14(4):9–19. https://doi.org/10.3233/AJW-170032.
30. Kori R, Saxena A, Upadhyay N, Shrivastava K, Kulshreshtha A, Mishra S, Sen S, Dubey P. Study of ground water quality of Bhopal City, Madhya Pradesh, India; 2019. https://doi.org/10.24941/ijcr.36871.10.2019
31. Raja V, Lakshmi RV, Sekar CP, Chidambaram S, Neelakantan MA. Health risk assessment of heavy metals in groundwater of industrial township Virudhunagar, Tamil Nadu, India. Arch Environ Contam Toxicol. 2021;80(1):144–63. https://doi.org/10.1007/s00244-020-00795-y.
32. Ahamad A, Raju NJ, Madhav S, Khan AH. Trace elements contamination in groundwater and associated human health risk in the industrial region of southern Sonbhadra, Uttar Pradesh, India. Environ Geochem Health. 2020;42(10):3373–91. https://doi.org/10.1007/s10653-020-00582-7.
33. Ghasera KM, Rashid SA, Gupta K. Heavy metals abundance and distribution in soil, groundwater and vegetables in parts of Aligarh, Uttar Pradesh, India: implication for human health risk assessment. Curr Sci. 2021;121(8):1056–63. https://doi.org/10.18520/cs/v121/i8/1056-1063
34. Nimje JJ, Wayal AS. Determination of ground water quality index in coastal area of Mumbai. ARPN J Eng Appl Sci. 2020;15(8):1016–22.
35. Brindha K, Paul R, Walter J, Tan ML, Singh MK. Trace metals contamination in groundwater and implications on human health: comprehensive assessment using hydrogeochemical and geostatistical methods. Environ Geochem Health. 2020;42(11):3819–39. https://doi.org/10.1007/s10653-020-00637-9.
36. Halim MA, Majumder RK, Rasul G, Hirosiro Y, Sasaki K, Shimada J, Jinno K. Geochemical evaluation of arsenic and manganese in shallow groundwater and core sediment in Singair Upazila, Central Bangladesh. Arab J Sci Eng. 2014. https://doi.org/10.1007/s13369-014-1111-x
37. Bodrud-Doza M, Islam ARMT, Ahmed F, Das S, Saha N, Rahman MS. Characterization of groundwater quality using water evaluation indices, multivariate statistics and geostatistics in central Bangladesh. Water Sci. 2016;30(1):19–40. https://doi.org/10.1016/j.wsj.2016.05.001.
38. Mostafa MG, Uddin SMH, Haque ABMH. Assessment of hydro-geochemistry and groundwater quality of Rajshahi City in Bangladesh. Appl Water Sci. 2017. https://doi.org/10.1007/s13201-017-0629-y.
39. Islam ARMT, Shen S, Bodrud-Doza M, Atiqur Rahman M, Das S. Assessment of trace elements of groundwater and their spatial distribution in Rangpur district, Bangladesh. Arab J Geosci. 2017;10(4):95. https://doi.org/10.1007/s12517-017-2886-3.
40. Alam MS, Han B, Al-Mizan, Pichtel J. Assessment of soil and groundwater contamination at a former Tannery district in Dhaka, Bangladesh. Environ Geochem Health. 2020;42(7):1905–20. https://doi.org/10.1007/s10653-019-00457-6
41. Sharmin S, Mia J, Miah MS, Zakir HM. Hydrogeochemistry and heavy metal contamination in groundwaters of Dhaka metropolitan city, Bangladesh: assessment of human health impact. HydroResearch. 2020;3:106–17. https://doi.org/10.1016/j.hydres.2020.10.003.
42. Zakir HM, Sharmin S, Akter A, Rahman MS. Assessment of health risk of heavy metals and water quality indices for irrigation and drinking suitability of waters: a case study of Jamalpur Sadar area, Bangladesh. Environ Adv. 2020;2: 100005. https://doi.org/10.1016/j.envadv.2020.100005.
43. Rahman MATMT, Paul M, Bhoumik N, Hassan M, Alam MK, Aktar Z. Heavy metal pollution assessment in the groundwater of the Meghna Ghat industrial area, Bangladesh, by using water pollution indices approach. Appl Water Sci. 2020;10(8):1–15. https://doi.org/10.1007/s13201-020-01266-4.
44. Goswami S, Rahman AKMM, Fuad MIM, Roy DK, Rahim A. Groundwater quality assessment of Barisal city corporation in Bangladesh. Barisal Univ J Part. 2017;1(2).
45. Deeba F, Rahman SH, Kabir MZ, Dina NT, Das SC, Sultana S, Ahmed FT. Heavy metals distribution and contamination in groundwater of the south eastern coastal Area of Bangladesh. J Water Environ Technol. 2021;19(5):267–82. https://doi.org/10.2965/jwet.20-169.
46. Hossain SMS, Haque ME, Pramanik MAH, Uddin MJ, Al Harun MAY. Assessing the groundwater quality and health risk: a case study on Setabganj sugar mills limited, Dinajpur, Bangladesh. Water Sci. 2020;34(1):110–23. https://doi.org/10.1080/11104929.2020.1790184.

47. Deeba F, Abbas N, Butt M, Irfan, M. (2020). Ground Water Quality of Selected Areas of Punjab and Sind Provinces, Pakistan: Chemical and Microbiological Aspects. *SSRN Electronic Journal*. https://doi.org/10.2139/ssrn.3407494
48. Lanjwani MF, Khuhawar MY, Khuhawar TMJ. Groundwater quality assessment of Shahdadkot, Qubo Saeed Khan and Sijawal Junejo Talukas of District Qambar Shahdadkot, Sindh. Appl Water Sci. 2020;10 (1). https://doi.org/10.1007/s13201-019-1098-2
49. Nasir MS, Nasir A, Rashid H, Shah SHH. Spatial variability and long-term analysis of groundwater quality of Faisalabad industrial zone. Appl Water Sci. 2017. https://doi.org/10.1007/s13201-016-0467-3.
50. Umar M, Waseem A, Sabir MA, Kassi AM, Khan AS. The impact of geology of recharge areas on groundwater quality: a case study of Zhob River Basin, Pakistan. Clean: Soil, Air, Water. 2013;41 (2):119–27. https://doi.org/10.1002/clen.201100581.
51. Ashraf S, Abbas F, Ibrahim M, Rashid U, Khalid S, Ahmad HR, Hakeem KR, Majeed T. Application of GIS for the identification and demarcation of selective heavy metal concentrations in the urban groundwater. J Geog Sci. 2015;25(2):225–35. https://doi.org/10.1007/s11442-015-1164-0.
52. Waseem A, Arshad J, Iqbal F, Sajjad A, Mehmood Z, Murtaza G. Pollution status of Pakistan: a retrospective review on heavy metal contamination of water, soil, and vegetables. BioMed Res Int. 2014. https://doi.org/10.1155/2014/813206
53. Ashraf I, Ahmad F, Sharif A, Altaf AR, Teng H. Heavy metals assessment in water, soil, vegetables and their associated health risks via consumption of vegetables, District Kasur, Pakistan. SN Appl Sci. 2021;3(5). https://doi.org/10.1007/s42452-021-04547-y
54. Riaz O. Assessment of ground water quality: a case study in Sargodha City; 2017.
55. Afzal M, Shabir G, Iqbal S, Mustafa T, Khan QM, Khalid ZM. Assessment of heavy metal contamination in soil and groundwater at leather industrial area of Kasur, Pakistan. Clean 0Soil, Air, Water. 2014;42 (8):1133–9. https://doi.org/10.1002/clen.201100715.
56. Natesan U, Deepthi K. Groundwater pollution modelling—a case study for Chennai (India). Pollut Res. 2012;31(4):513–8.
57. Punia A, Siddaiah NS, Singh SK. Source and assessment of metal pollution at Khetri copper mine tailings and neighboring soils, Rajasthan, India. Bull Environ Contam Toxicol; 2017. https://doi.org/10.1007/s00128-017-2175-6
58. Borah M, Das PK, Borthakur P. An assessment of surface and ground water quality of some selected locations in Guwahati. 2020;15(1):93–108.
59. Lahkar M, Bhattacharyya KG. Heavy metal contamination of groundwater in Guwahati City, Assam, India. Int Res J Eng Technol. 2008;1520–25.
60. Tadiboyina R, Ptsrk PR. Trace analysis of heavy metals in ground waters of Vijayawada industrial area. Int J Environ Sci Educ. 2016;11(10):3215–29.
61. Singh DD, Thind PS, Sharma M, Sahoo S, John S. Environmentally sensitive elements in groundwater of an industrial town in India: spatial distribution and human health risk. Water (Switzerland). 2019;11(11). https://doi.org/10.3390/w11112350
62. Harvey CF, Ashfaque KN, Yu W, Badruzzaman ABM, Ali MA, Oates PM, Michael HA, Neumann RB, Beckie R, Islam S, Ahmed MF. Groundwater dynamics and arsenic contamination in Bangladesh. Chem Geol. 2006;228(1–3 SPEC. ISS.):112–36. https://doi.org/10.1016/j.chemgeo.2005.11.025
63. WHO. Potable water specification, 2nd edn. Rwanda Bureau of standard, Kigali; 2012.
64. Kataria HC, Bux S, Ambhore S, Shrivastava SC, Pathak G, Namdeo M. Analysis of fluoride concentration in groundwater in and around Bhopal city. M. P. India. Biosci Biotechnol Res Asia. 2008;5(2):699–700.
65. Ali S, Gupta SK, Sinha A, Khan SU, Ali H. Health risk assessment due to fluoride contamination in groundwater of Bichpuri, Agra, India: a case study. Model Earth Syst Environ. 2022;8(1):299–307. https://doi.org/10.1007/s40808-021-01105-8.
66. Sujatha M, Rao NR, Satyanarayana T, Asadi SS, Babu KSK. Assessment of groundwater pollution along Thippagunta drain in Prasadampadu, Vijayawada Rural, Krishna District, Andhra Pradesh. Int J Civil Eng Technol. 2016;7(6):738–46.
67. Wagh V, Mukate S, Muley A, Kadam A, Panaskar D, Varade A. Study of groundwater contamination and drinking suitability in basaltic terrain of Maharashtra, India through pig and multivariate statistical techniques. J Water Supply Res Technol AQUA. 2020;69(4):398–414. https://doi.org/10.2166/aqua.2020.108.
68. Jayarajan SK, Kuriachan L. Exposure and health risk assessment of nitrate contamination in groundwater in Coimbatore and Tirupur districts in Tamil Nadu, South India. Environ Sci Pollut Res. 2021;28 (8):10248–61. https://doi.org/10.1007/s11356-020-11552-y.
69. Kuttimani R, Raviraj A, Pandian BJ, Kar G. An overview of groundwater quality in Tamil Nadu. Environ Conserv J. 2017:18(3):27–37. https://doi.org/10.36953/ecj.2017.18304
70. Saraswat C, Kumar P, Dasgupta R, Avtar R, Bhalani P. Sustainability assessment of the groundwater quality in the Western India to achieve urban water security. Appl Water Sci. 2019;9(4). https://doi.org/10.1007/s13201-019-0956-2
71. Mukherjee S, Sikdar PK, Pal S, Schütt B. Assessment of environmental water security of an Asian deltaic megacity and its Peri-urban wetland areas.

Sustainability (Switzerland). 2021;13(5):1–33. https://doi.org/10.3390/su13052772.
72. De S, Maiti SK, Hazra T, Debsarkar A, Dutta A. Appraisal of seasonal variation of groundwater quality near an uncontrolled municipal solid waste landfill in Kolkata, India. Glob Nest J. 2017;19(3). https://doi.org/10.30955/gnj.002172
73. Nahar MS, Zhang J, Ueda A, Yoshihisa F. Investigation of severe water problem in urban areas of a developing country: the case of Dhaka, Bangladesh. Environ Geochem Health. 2014;36(6):1079–94. https://doi.org/10.1007/s10653-014-9616-5.
74. Roy PK, Roy B, Roy BC. Assessment of groundwater quality for drinking and irrigation purposes in Comilla District of Bangladesh. Int J Sci Res Publ. 2016;6(7).
75. Rahman MA, Islam MR, Kumar S, Al-Reza SM. Drinking water quality, exposure and health risk assessment for the school-going children at school time in the southwest coastal of Bangladesh. J Water Sanitation Hygiene Develop. 2021;11(4):612–28. https://doi.org/10.2166/washdev.2021.016.
76. Md Munna G. A study on water quality parameters of water supply in Sylhet city corporation area. Hydrology. 2015;3(6):66. https://doi.org/10.11648/j.hyd.20150306.12
77. Chandra Bhowmick A, Ashrafuzzaman MD. Assessment of drinking water quality in Mbstu campus. J Sci Technol. 2016;6(2).
78. Kundu A, Kumar Datta D, Tonu NT, Haque MR. Engineering and technology: a high impact factor. Int J Innov Res Sci. 2018;7. https://doi.org/10.15680/IJIRSET.2018.0710063
79. Mahmud A, Sikder S, Joardar JC. Assessment of groundwater quality in Khulna city of Bangladesh in terms of water quality index for drinking purpose. Appl Water Sci. 2020;10(11). https://doi.org/10.1007/s13201-020-01314-z
80. Das TK, Shaibur MR, Rahman MM. Groundwater chemistry at deep aquifer in Koyra: Khulna, Bangladesh. Curr World Environ. 2021;16(2):460–71. https://doi.org/10.12944/cwe.16.2.12
81. Mahboob MA, Atif I, Iqbal J. Geospatial modelling for groundwater quality with contamination potential assessment of three urban areas in Pakistan. Int J Water Resourc Environ. 2017;6(2):216–26.
82. Adelana SMA. Nitrate health effects. Water Encyclopedia. 2005. https://doi.org/10.1002/047147844x.dw21.
83. Pavendan P, Anbu Selvan S, Sebastian Rajasekaran C. Physico chemical and microbial assessment of drinking water from different water sources of Tiruchirappalli District, South India. Eur J Exp Biol. 2011;1(1):183–89.
84. Max M, Min MM. Table 1: water quality data of ground water under Nwmp in Andhra Pradesh; 2020.
85. Banerji S, Mitra D. Geographical information system-based groundwater quality index assessment of northern part of Kolkata, India for drinking purpose. Geocarto Int. 2019;34(9):943–58. https://doi.org/10.1080/10106049.2018.1451922.
86. Saha R, Dey NC, Rahman M, Bhattacharya P, Rabbani GH. Geogenic arsenic and microbial contamination in drinking water sources: exposure risks to the coastal population in Bangladesh. Front Environ Sci. 2019;7:1–12. https://doi.org/10.3389/fenvs.2019.00057.
87. Islam T, Acharjee M, Tabassum N, Acharjee MR. Bacterial propagation in municipal water and deep tube-well water in Kashipur locality of Narayanganj City, Bangladesh. J Water Environ Technol. 2020;18(5):327–37. https://doi.org/10.2965/JWET.20-049.
88. Singh RB, Haque MS, Grover A. Drinking water, sanitation and health in kolkata metropolitan city: contribution towards urban sustainability. Geogr Environ Sustain. 2015;8(4):64–81. https://doi.org/10.24057/2071-9388-2015-8-4-64-81
89. Zhou X, Shen Y, Zhang H, Song C, Li J, Liu Y. Hydrochemistry of the natural low pH groundwater in the coastal aquifers near Beihai, China. J Ocean Univ China; 2015. https://doi.org/10.1007/s11802-015-2631-z
90. Jacintha TGA, Rawat KS, Mishra A, Singh SK. Hydrogeochemical characterization of groundwater of peninsular Indian region using multivariate statistical techniques. Appl Water Sci. 2017;7(6):3001–13. https://doi.org/10.1007/s13201-016-0400-9.
91. Singh AV. District ground water brochure Agra District, Uttar Pradesh District at a glance Agra District, Uttar Pradesh; 2013.
92. Sajjad H, Mohammad Rashid S, Prasad S. Assessment of groundwater quality in Meerut City, India. Int J Environ Prot. 2013;3(2):20–6.

Sustainable Ways to Manage Soil and Groundwater Resources

Pankaj Kumar Gupta

Abstract

Water as a renewable resource is presumed to be the prime concern and focal point of attention in forthcoming years and groundwater will remain an important component of it. Nonetheless, the management of groundwater is very costly and a herculean task for water resource managers even after interventions of the state-of-the-art technology. In most of the cases, the knowledge gaps and thus effort gaps may increase the groundwater management cost. This chapter aims to present the basic and advance knowledge about soil and groundwater systems, sources of contaminants, its fate and transport under varying subsurface conditions. Given that, contaminants can exist in a soil matrix as free-colloidal dissolved or gas phase affecting soil-groundwater quality. Therefore, this chapter also provide valuable information about expertise approach for the contaminated landscape, which highlights local and climatic solutions. In addition, it also talks about the field-scale hydrological practices to improve groundwater quality and bring back it's pristinely at a low cost. This chapter will provide in-depth analysis of the cost related to the remediation and management of soil-groundwater systems. This chapter will help the resource managers, policy makers, and hydrogeologists to frame designs, methods, policies, and actions plans to use very low-cost natural climatic solutions.

Keywords

Aquifer · Groundwater · Contamination · Restoration · Soil Management

1 Introduction

Soil is a living system, thus an inseparable part of our life. Soil can be defined as a natural body that is from either geogenic or organic materials, sustaining biodiversity and biogeochemical cycle. Soil quality refers to the soil's capacity to perform its functions. The soil can be viewed as a mixture of mineral and organic particles of varying size and composition. Space among soil particles, referred to as pore, contain water and air and serve as channels for the transportation of water, solute, and air in the subsurface. Rainwater that gets infiltrated into the soil of upper layer and accumulate in the underlined soil layers, thereafter such water is referred to as groundwater.

P. K. Gupta (✉)
Indian Institute of Technology (I.I.T.) Delhi, New Delhi 110 016, India
e-mail: pk3gupta@uwaterloo.ca

University of Waterloo, 200 University Ave W, Waterloo, ON, Canada

A. M. Bahadir et al. (eds.), *Hydrology and Urban Water Supply*, Water and Wastewater Management,
https://doi.org/10.1007/978-3-031-72589-0_11

Vadose zone is the layer of soil present between the earth's surface and the water table, and it is of utmost physical, chemical and biological importance (Fig. 1). Vadose zone is associated with the recharge of groundwater, degradation and uptake of contaminants thus preventing or accelerating these chemicals to leach to the groundwater. Preferential flow and transport hastens the movement of agricultural contaminants like fertilizers, pesticides, pathogens, trace elements to the groundwater through vadose zone. Bottom layer of the vadose zone is under capillarity, which is the difference of pressure that exist across the interface of two (immiscible fluid) water–air phases. The relative permeability, ability of fluid to flow in the presence of other phase, and capillarity controls the flow of water, air, and any contaminants with the connected pores. Saturation of water increases with increasing depth of the vadose zone and finally got fully saturated at the groundwater table.

Groundwater table dynamics play a significant role in the subsurface structure and functional services, for example, in maintaining geochemical cycling, oxygen supply to deep zone, vadose zone temperature regulations, microbial actions, etc. Groundwater table can also after the thickness of the vadose zone, thus more or less heat and gas flux to the atmosphere, thus global warming, and climate change implications. Groundwater table dynamics also enhance the spreading of residual contaminants from vadose zone to the underlaying groundwater and accelerate its dissolution and transportation to the downstream locations. Rise and fall of the groundwater table also improve oxygen supply to the capillary and deeper zones, which ultimately enhance the growth/actions of soil microbial communities. Therefore, in general, a high degradation of contaminants has been reported in the groundwater fluctuating zone as compared to deep saturated zone.

Aquitard occurs below saturated layer, in general, it is a formation which has low to medium permeability which is not sufficient to be a source of water supply but allows certain quantity of water to flow on a regional scale due to leakage. Formations having predominance of silt, loam, and clay along with kankar form typical aquitards. Aquitards behave as semi-confining layers. One can refer the final layer as aquiclude that is an impermeable formation that does not transmit water, or aquifuge. Aquitards are becoming increasingly important layers due to their use as chemical waste disposal, CO^2 storage, and nuclear waste disposal sites. We will explore more about issues related to soil and groundwater resources in the next section.

This chapter aims to present the basic and advanced knowledge about soil and groundwater systems, sources of contaminants, their fate, and transport under varying subsurface conditions. Given that, contaminants can exist in a soil matrix as free-colloidal dissolved or gas phase affecting soil water quality. Therefore, this chapter also provides valuable information about expertise approaches for the contaminated landscape, which highlights local and climatic solutions. In addition, it also discusses the field-scale hydrological practices to improve groundwater quality and restore its hygiene at a low cost. This chapter will offer an in-depth analysis of the cost related to the remediation and management of soil and groundwater systems. This chapter will help the resource managers, policymakers, and hydrogeologists to frame designs, methods, policies, and action plans to use very low-cost natural climatic solutions (Fig. 1).

2 Issues Related to Soil and Groundwater Resources

One cannot ignore implications of human activities on the soil–water resources. Major challenges are associated with the soil mining, loss of soil biodiversity, desertification, water scarcity, soil–water quality depletion, and finally disappearances of hydrogeological resources. For example, globally, our annual sand and gravel consumption is estimated to be around 53 billion tons. That's equal to around 7.5 tons/a (20 kg/d) for every person on Earth. It turns out that the rivers in the world collectively supply around 16 billion tons of sand per year to the coast. Humans are currently using sand at a rate that is more than

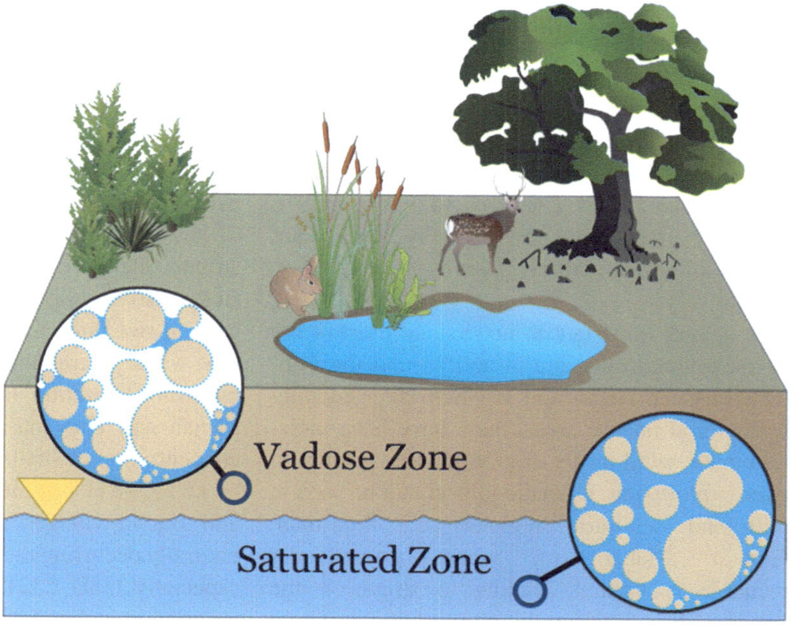

Fig. 1 Groundwater system representing vadose and saturated zones

three times faster than it is being produced by nature. Likewise, other examples are from water scarcity side. Drained groundwater left the reservoirs in Chennai completely dry, similar stories are from Cape Town, California, Istambul, Shimla (India), etc. Groundwater contamination is of equal concern. Gupta [1] reported highly contaminated sites in India, very surprisingly most of the aquifers are contaminated by multiple chemicals either geogenic or industrials. Initial conditions of groundwater contamination have been magnified due to emerging contaminants, for example microplastics, health care chemicals, nanoparticles, microbial agents, etc. At the same time, human has started exploring deep aquifers for chemical, nuclear waste disposals and for CO_2 storage for long term to cap global warming and climate change. Sea level has risen about 19 cm in the last 100 years, and the rate is getting faster; it was rising about 1.7 mm/a at the end of the twentieth century and is now rising at around 3.4 mm/a. Indian National Center for Ocean Information Services, say the sea level along India is rising at 0.33–5.16 mm/a. Overuse of coastal groundwater and sea level rise cumulatively accelerate sea water intrusion into the coastal aquifers. We will explore more specific details on the issues related to soil–water resources in sub-sections.

2.1 Overuse of Groundwater

Overuse of groundwater can cause subsurface drought, thus very low ground productivity, and ultimately may turn as ecological crisis. Groundwater continues to over exploitation to meet agriculture/irrigation water demands around the globe. In India, several parts of the landscape are experiencing rapid declining groundwater, thus reduction in supply, saline water intrusion, drying of the spring, etc. Northern Gujarat, southern Rajasthan, Maharashtra, several districts of Tamil Nadu, Karnataka, almost each district of Andhra Pradesh, Punjab, Haryana, and Uttar Pradesh are undergoing the subsurface drought just due to speedy decline (1–2 m/a) in groundwater. It has been reported that declining water levels could reduce India's harvest by 25% or more. Further dangerous consequences of falling water levels are an increase in arsenic levels in several regions of the country, Fluoride contamination of the drinking water supply and many other forms of groundwater contamination, making tube well water unit for human consumption and farming [2]. According to national compilation on dynamic groundwater resources of India, 2020 report published by the Central Groundwater Board (CGWB), 4427, 1057, 270, 1114, 97 assessment units, mainly blocks, are safe, semi-critical, critical, over-exploited, saline, respectively.

It is very important to understand most of the over-exploited units are in and around the Punjab, Haryana, Delhi, and Western UP, generally agricultural intense areas, there have been indiscriminate withdrawals of groundwater leading to over-exploitation. Climatic impact can be seen in arid and semi-arid regions i.e., Rajasthan, Gujarat, Maharashtra, where groundwater recharge is limited, leading to stress on the aquifers. In crystalline aquifers of the southern part of peninsular India (Karnataka, Andhra Pradesh, Telangana, and Tamil Nadu), the groundwater availability is low. Also, very important to note that the over-exploited assessment units are increasing from 839 in 2004, 1071 in 2010, to 1114 in 2020 as per CPCB.

Groundwater extraction (>40%) for domestic uses is also major concern in many states like Arunachal Pradesh, Delhi, Goa, Kerala, Manipur, Meghalaya, Mizoram, Nagaland, Sikkim, Tripura, Andaman and Nicobar Island, Chandigarh, Dadra and Nagar Haveli, Jammu and Kashmir, Ladakh and Lakshadweep. The groundwater report by the CGWB [3] will be very useful for the field practitioners, district administrators, and policy makers for the management of soil–water resources. Extraction for industrial uses is also important in Delhi, Himachal Pradesh, Jharkhand, Sikkim, etc. Key massage is to improve groundwater storage by implementing sustainable recharge techniques.

2.2 Groundwater Contamination

Recharging groundwater can compensate for the overuse, but aquifer contamination is a threatening condition, this is because restoration of contaminated aquifers is a complex and costly task. Soil–groundwater system, simply aquifers, of India is home of major contaminants, Arsenic, Fluoride, Nitrate, Selenium, Uranium, Hydrocarbons and Salinity. Dhillon and Dhillon [4] reported high selenium concentration, to be ranged between 0.25 and 69.5 μg/L with an average value of 4.7 μg/L, in groundwater samples in and around seleniferous region of Panjab, India. This study also highlight that the shallow tube wells contain two/three time more selenium than deep tube wells. Thereafter, Bajaj et al. [5] found selenium concentration in soil and groundwater in Panjab and Haryana State and reported a high concentration in shallow (73 m deep) groundwater i.e., 45–341 μg/L in two village named Jainpur and Barwa villages in Punjab. Like selenium, naturally occurring radioactive material (NORM), especially 238U, 232Th, 40K and 137Cs has been reported in soil and groundwater in Panjab State, India [6]. Recently, Uranium contaminations were also reported in aquifers of Bihar [7]; Karnataka [8, 9]. Balaram et al. [10] has listed Uranium hotspots in parts of India and world.

Nitrate is emerging as one of the potential groundwater contaminants due to its intensive use as chemical fertilizers required for increasing the agricultural production to cater the demand of growing population, and its high mobility through variably saturated soils [1]. Nitrate contamination of the world's groundwater poses potentially serious health hazard to the mankind at large [11]. Most of the intense agriculture region of India is experiencing high nitrate load in the subsurface [1]. A high concentration of F—(i.e., >1.5 mg/L) are found in groundwater in China, Syria, Jordan, Ethiopia, Sudan, Tanzania, Kenya, and Uganda [12]. Groundwater contamination with F—is well reported in India at numerous places of Andhra Pradesh, Gujarat, Karnataka, Madhya Pradesh, Rajasthan, Chhattisgarh, Haryana, Orissa, Punjab, Haryana, Uttar Pradesh West Bengal, Bihar, Delhi, Jharkhand, Maharashtra, and Assam. Likewise, 97 assessment units are affected by the salinity in India. One can read a review by Gupta [1] for details on geogenic and Mahajan et al. [13] for industrial contaminants.

Microplastic, nanoparticles, microbial contaminations are of emerging concern for soil-groundwater resources. Microplastics have been reported in soil–water systems by several researchers [14]. Likewise, microbial contaminants (pathogens) are frequently reported in groundwater [15]. Leakage from septic tanks is main contributor to the microbial contaminants in groundwater. Similarly, leakage from deep CO_2 storage is a major threat to the above laying groundwater as the dissolution of CO_2 may accelerate metal mobilisations [16]. In the next section, we will learn more about how to manage our most vital resources in a sustainable way.

3 Soil Management

To manage soil in a sustainable way, we must explore cases in which the energy supplied by human must be relatively small compared to the energy from natural resources to maintain the soil processes and functions. Thus, input and output rules by [17] can be take care while managing soils, where inputs should be within the regenerative capacity of the natural system and outputs should be within the assimilative capacity of local environment (In case of soil, without unacceptable losses of its future waste absorptive capacity). Theory has not yet connected with the practice, especially in case of soil management in post-green revolution years. Chemical farming, introduced by green revolution, is an ongoing practice to meet the global food demands at the cost of soil quality losses/contamination. To restore soil quality, in the term of ecological services, several concepts have been emerged, for example organic farming, natural farming, agroforestry, sustainable agroecological practices, etc. Concept of natural farming is in a close approximation of ecological/ecotechnological practices.

Natural farming is a collective practice which moves centripetally toward nature, means it is based on the self-designing capacity of ecosystems. Natural farming in practical is all about no cultivation, no fertilizer, no weeding, and no pesticides. Instant of plowing, one can left soil as it is, soil can swell and grow more porous with each passing years to maintain an abundance of air and water. Which is absolutely true for microorganism to grow, for plant uptake, for roots to penetrate deep zones. By taking full advantage of the inherent powers of nature, one can obtain yields equal to those that can be had with heavy fertilization. The soil is teeming with life, each life (millions of bacteria, yeasts, molds, diatoms, and other lives) is there for a purpose, struggling, cooperating, and carrying on the cycles of nature i.e., biogeochemical cycles. Natural farming is a way to protect soil microbiome from the harmful chemical fertilizers. Next principle of natural farming is "no weeding". Not having to weed or plow might sound like asking for too much, but if one stops to think about what repeatedly weeding and running a plow through a field means, it becomes clear that weeding is not as indispensable as we have been led to believe. Weed roots penetrate deep into the soil and allow water and air to infiltrate, soil microbes to proliferate on humus due to root materials, so weeds and grasses are essential for a soil to remain organic and alive, just referred as weed utility. There are no non-contaminating pesticides, thus natural farming is based on "no pesticides". The main truth of natural farming is that nothing need be done to grow crops, simply means no energy input from human side.

Sustainable soil management in non-agricultural landscape can be achieved by reclamation or restoration techniques. For example, a mining site can be restore using local natural climatic resources, like wetland. Likewise, soils at the contaminated sites, industrial dump sites, can be reclaimed by ecotechniques like bioremediation. Where one can apply locally available plants, algae, moss, etc. and manipulate hydrological conditions to accelerate the microbial actions to degrade target contaminants from soils. In the next section, we will learn more about sustainable ways to manage groundwater resources.

4 Sustainable Ways to Recharge Groundwater

Decentralised rainwater harvesting systems in the structural method of check dams in various parts of India were an operational solution to the water crisis through their capability to channel runoff into shallow aquifers. One good example is "Chauka" rainwater harvesting system of semi-arid region of Rajasthan. Water conservationist, Shri Laxman Singh, who invented the Chauka system in Lapodiya village, Rajasthan, is now able to facilitate water demands even in summer days to not only own village but also to the nearby habitats. Similar solutions were also in operation in various parts of India, Paar system in the western Rajasthan region, Bandhis in the Bundelkhand region, Pat in Madhya Pradesh, Ahars and Pyanes in Bihar, Ery system in Andhra Pradesh, Karnataka, and Tamil Nadu, etc. (see more on www.indiawaterportal.org). Disappearance of traditional rainwater harvesting systems, aquifer recharge systems, is also a major cause of water stress conditions in India.

In contemporary time, our land surface is home of waste disposal sites, and our subsurface environment is heavily loaded with municipal, industrial, and geogenic contaminants. The residual contaminants in the vadose zone due to the seepage from landfill, open dump sites, farmlands, underground tanks, septic tanks, and surface discharged wastewater may act as long-term source of contamination [18]. Considering most of the rainwater harvesting systems are surface structure, there is high chance of aquifer contamination by the recharging water, which is moving through vadose zone along with residual contaminants [19]. Very little, if any, research has been done yet on the relationship between recharging water and contaminant mobility in recharge systems. Thus, there is urgent need to implement more sustainable methods to recharge aquifers with clean water, simply means we must re-think our rainwater harvesting systems and come up with solution to concurrently treat recharging water.

"Managed Aquifer Recharge (MAR)" can be a good tool to recharge aquifers by incorporating concurrent treatment of contaminations. MAR is widely accepted approach to manage groundwater recharge safely at a low cost. MAR includes infiltration, direct injection, and filtration methods. Soil aquifer treatment (SAT), aquifer storage and recovery (ASR) and aquifer storage transfer and recovery (ASTR) are more attractive MAR techniques [20, 21]. Very important to note that MAR is an artificial method, neither a natural and nor an ecologically sound method. This is because it is not based on the self-designing capacity of soil–water systems.

Generally, people forget that the MAR system largely depends on local hydrogeological conditions. Thus, the decentralised natural recharge methods/structure, like traditional rainwater harvesting systems, can be a sustainable solution to manage local aquifers. Here, natural recharge structure means an integration of natural farming with localised water conservation practices. For example, the restoration of Ahars and Pyanes along with natural farming in Bihar can be a sustainable way to manage soil-groundwater resources.

5 Sustainable Ways to Improve Groundwater Quality

Ecological approaches are used to solve a contamination problem, for example phytoremediation for terrestrial sites, treatment wetlands for aquatic sites. More complexity appears in case of the subsurface pollution management. This is due to pollutants can exist in a subsurface matrix as free, colloidal, dissolved or gas phase affecting soil–water quality. Pollutant source zone treatment can be physical, chemical, or biological methods of remediation depending on the initial and hydrogeological conditions at the target site [16]. For sure, physical, and chemical treatment may cost a lot (>\$100/m^3 of contaminated soils), and or destructive, may takes years to recover ecological functions of the site [13]. Therefore, bioremediation is an attractive technology, which may cost less ~$10/m^3 of contaminated soils [13].

Plants and microbes can solve the problem on the contaminated sites. Basu et al. [22] recently reported a high degradation of target contaminants in soil column integrated with treatment wetlands. Supply of root zone water, simply root microbes, to the polluted zone were found very effective. Treatment wetland of 1 ha construction size would cost almost $200,000, a 10 ha wetland would cost $60,000 per hectare as suggested by Mitsch and Gosselink [23]. Thus, technology like treatment wetland is still a costly affair to implement in low economics. To keep remediation cost almost zero, Gupta et al. [24] designed a unique bioremediation system by incorporating the practical experiments, process-based models, and optimization simulators to remediate NAPL contaminated sediments under varying soil moisture and temperature conditions. The remediation time was also found to be significantly less for a high moisture—high temperature combination, at this combination remediation cost is relatively low. Gupta et al. [24] suggests that the application of in-situ bioremediation methods in the region of high temperature, with moisture content near field capacity, is most efficient and cost-effective. Although, their design is laboratory scale and hypothetical works, but their study provides us very important idea on hydrological restoration of contaminated sites.

Groundwater table manipulation can be a sustainable hydrogeological restoration technique, which can be design based on self-designing capacity of soil–water systems [25]. Groundwater table rise and fall first provides sufficient air to the deep saturated zone to accelerate microbial growth [25]. Further, groundwater fluctuations may re-mobilise residual contaminants and help to decontaminate source zone by dissolving residual contaminants [26]. Furthermore, rise of groundwater table can help release of soil gas phase through the unsaturated zone into the atmosphere, and help us to clean vadose zone [25]. Groundwater manipulation may not be a good solution for many target pollutants, for example pathogens. One can try soil moisture alternation to regulate microbial communities, pathogens, by moderating groundwater table. To design self-operating remediation system, one can modify decentralised rainwater harvesting system to optimise groundwater table dynamics for effective treatment of pollution at a target site.

6 Remediation Cost

Since the prime goal of divergent remediations attempts for contaminated sites is to reinstate the pristineness and flawlessness of soil water system, a paradigm shift in technological advancement and vision has been incorporated in remedial actions for cost effective treatment of the contaminated sites., apropos this intent bioremediation has been ascertained an economical biological method to administer the contaminants, for instance in In situ bioremediation (US$ 50–150), and phytoremediation cost for metal and metalloids is evaluated around US $10–35 [13]. The global market for remediation was quantified around US$30–35 billion in 2001 and computation also show positive feedbacks in forthcoming future with favour of 1.5 billion per annum [27] but on the other hand physicochemical treatments like solidification, soil venting, solvent extraction and incineration shows much costlier figure as US$240–340, US$ 20–220, US$360–440, and US$200–1500 respectively. Now moving towards biological remediation, for bioremediation, the economic cost is assessed by different valuation approaches including 'willingness to pay, substitution cost and hedonic price analysis, hedonic price analysis deliver much promising value as compared to substitution cost method, the hedonic price is approximately 14,600 and 14,850€h^{-1}, However there are diverse implicit factors like site conditions, climatic condition, types of contaminants, production costs for the remediation crop and after all the suitability should be taken into consideration prior to the selection of particular method and simulation-optimization tools for target contaminated sites must be employed in order to estimate the cost and moreover for multicriteria decision making.

7 Conclusion

We are now able to make an extensive contribution to the remediation, restoration, and management of soil–water resources through natural farming and ecologically sound restoration techniques. There are many examples of how traditional methods help us to maintain the sustainability of soil–water resources. Considering the emerging contamination issues, the ultimate goal is to return site/soil–water systems to original conditions at acceptable cost at local level. Natural farming can be a sustainable solution for soil management in farmlands, while just groundwater table manipulation can be a good solution for underlaying aquifers. Most importantly, theory has not yet connected with the practice. Future studies must take care of linkage of different components of ecosystem, most importantly how changes in one ecosystem (soil) can produce changes in neighbouring ecosystems (aquifer). This chapter, I hope, help resource managers, policy makers, and hydrogeologists to frame designs, methods, policies, and actions plans.

References

1. Gupta PK. Pollution load on Indian soil-water systems and associated health hazards: a review. J Environ Eng. 2020;146(5):03120004.
2. Singh DK, Singh AK. Groundwater situation in India: problems and perspective. Int J Water Resour Dev. 2002;18(4):563–80.
3. CGWB. Report-national compilation on dynamic ground water resources of India; 2020. http://cgwb.gov.in/documents/2021-08-02-GWRA_India_2020.pdf. Dated 15 Aug 22.
4. Dhillon KS, Dhillon SK. Quality of underground water and its contribution towards selenium enrichment of the soil–plant system for a seleniferous region of northwest India. J Hydrol. 2003;272(1–4):120–30.
5. Bajaj M, Eiche E, Neumann T, Winter J, Gallert C. Hazardous concentrations of selenium in soil and groundwater in North–West India. J Hazard Mater. 2011;189(3):640–6.
6. Srivastava A, Lahiri S, Maiti M, Knolle F, Hoyler F, Scherer UW, Schnug EW. Study of naturally occurring radioactive material (NORM) in top soil of Punjab State from the North Western part of India. J Radioanal Nucl Chem. 2014;302(2):1049–52.
7. Kumar D, Singh A, Jha RK, Sahoo SK, Jha V. Using spatial statistics to identify the uranium hotspot in groundwater in the mid-eastern Gangetic plain, India. Environ Earth Sci. 2018;77(19):1–12.
8. Srinivasan R, Pandit SA, Karunakara N, Salim D, Kumara KS, Kumar MR, Ramkumar KD. High uranium concentration in groundwater used for drinking in parts of eastern Karnataka, India. Curr Sci. 2021;121(11):1459–69.
9. Lapworth DJ, Brauns B, Chattopadhyay S, Gooddy DC, Loveless SE, MacDonald AM, Nara SN. Elevated uranium in drinking water sources in basement aquifers of southern India. Appl Geochem. 2021;133: 105092.
10. Balaram V, Rani A, Rathore DPS. Uranium in groundwater in parts of India and world: a comprehensive review of sources, impact to the environment and human health, analytical techniques, and mitigation technologies. Geosyst Geoenviron. 2022;100043.
11. Suthar S, Bishnoi P, Singh S, Mutiyar PK, Nema AK, Patil NS. Nitrate contamination in groundwater of some rural areas of Rajasthan, India. J Hazard Mater. 2009;171(1–3):189–99.
12. Ando M, Tadano M, Yamamoto S, Tamura K, Asanuma S, Watanabe T, Chen X. Health effects of fluoride pollution caused by coal burning. Sci Total Environ. 2001;271(1–3):107–16.
13. Mahajan M, Gupta PK, Singh A, Vaish B, Singh P, Kothari R, Singh RP. A comprehensive study on aquatic chemistry, health risk and remediation techniques of cadmium in groundwater. Sci Total Environ. 2021;151784.
14. Kumar R, Manna C, Padha S, Verma A, Sharma P, Dhar A, Ghosh A, Bhattacharya P. Micro (nano) plastics pollution and human health: how plastics can induce carcinogenesis to humans? Chemosphere. 2022;298: 134267.
15. Dey U, Sarkar S, Duttagupta S, Bhattacharya A, Das K, Saha S, Mukherjee A. Influence of hydrology and sanitation on groundwater coliform contamination in parts of Western Bengal Basin: Implication to safe drinking water. Front Water 84.
16. Gupta PK, Yadav B. Leakage of CO_2 from geological storage and its impacts on fresh soil–water systems: a review. Environ Sci Pollut Res. 2020;27(12):12995–3018.
17. Lemons J, Westra L, Goodland R. Ecological sustainability and integrity: concepts ad approaches. Boston: Kluwer Academic; 1998.

18. Gupta PK, Yadav BK. Remediation and management of petrochemical-polluted sites under climate change conditions. In: Environmental contaminants: ecological implications and management. Singapore: Springer; 2019. p. 25–47.
19. Gupta PK, Yadav BK. Subsurface processes controlling reuse potential of treated wastewater under climate change conditions. In: Water conservation, recycling and reuse: issues and challenges. Singapore: Springer; 2019. p. 147–70.
20. Dillon P. Future management of aquifer recharge. Hydrogeol J. 2005;13(1):313–6.
21. Dillon P, Stuyfzand P, Grischek T, Lluria M, Pyne RDG, Jain RC, Bear J, Schwarz J, Wang W, Fernandez E, Stefan C. Sixty years of global progress in managed aquifer recharge. Hydrogeol J. 2019;27(1):1–30.
22. Basu S, Yadav BK, Mathur S, Gupta PK. In situ bioremediation of toluene-polluted vadose zone: integrated column and wetland study. CLEAN–Soil, Air, Water. 2020;48(5–6):2000118.
23. Mitsch WJ, Gosselink JG. Wetlands. 3rd ed. New York: Wiley; 2000. p. 920.
24. Gupta PK, Yadav B, Yadav BK, Sushkova S, Basu S. Engineered bioremediation of NAPL polluted sites: experimental and simulation-optimization approach under heterogeneous moisture and temperature conditions. J Environ Eng. 2021;147(8):04021023.
25. Gupta PK, Yadav B, Yadav BK. Assessment of LNAPL in subsurface under fluctuating groundwater table using 2d sand tank experiments. J Environ Eng. 2019;145(9):04019048.
26. Cavelan A, Golfier F, Colombano S, Davarzani H, Deparis J, Faure P. A critical review of the influence of groundwater level fluctuations and temperature on LNAPL contaminations in the context of climate change. Sci Total Environ. 2022;806: 150412.
27. Khalid S, Shahid M, Niazi NK, Murtaza B, Bibi I, Dumat C. A comparison of technologies for remediation of heavy metal contaminated soils. J Geochem Explor. 2017;182:247–68.

Insight into the Coastal Brackish Water Intrusion: A State-of-the-Art Review

Paulami Sahu and Swayam Siddha

Abstract

Brackish water intrusion (BWI) is the major concerning processes that degrades the groundwater quality and jeopardises the availability of freshwater in coastal aquifers. Massive extraction of groundwater resources as a result of population hike, urban sprawl, and sea-level rise have resulted in salinization of many coastal aquifer systems around the globe. BWI results in major fluctuations in the biogeochemistry of the disturbed aquifer system and influences the drinking, irrigation, and industrial water qualities and as a consequence, on wellbeing of the people in coastal areas in terms of their economic and socio-cultural conditions. BWI leads to drastic change in their economic and socio-cultural activities which have impelled an extensive array of studies and research. Therefore, the purpose of the this chapter is to demonstrate a complete overview of the methods that regulate BWI, the present worldwide distribution scenario of brackish water, numerous investigative techniques to detect BWI and finally advanced mitigation methods, in brief. BWI generally occurs in coastal aquifer systems known as subterranean estuaries, which are dynamic in nature. These estuaries get both freshwater from the terrestrial surface and saline water from the sea, exposing this fluid combination to intensive biogeochemical dynamics since they are loaded with the aquifer and aquicludes materials. Therefore, to preserve valuable resources of coastal ecosystem from degradation, BWI and other pollutants, it is necessary to take practical steps. Recently, experts' interest in brackish water desalination technologies has expanded worldwide. Due to its low salt concentration compared to that of saline water, brackish water, may be served as an excellent substitute for freshwater conversion and is commonly employed in several water-stressed nations. However, it is essential that freshwater generated by partial or entire desalination of brackish water should meet the standard of productivity and living. Since the last decades, usage of green technologies, particularly use of renewable energy (RE) in desalination has turned the unusable brackish water into a valuable and usable natural resource.

Keywords

Brackish water intrusion · Desalination · Renewable energy · Subterranean estuaries · Water quality

P. Sahu (✉) · S. Siddha
School of Environment and Sustainable Development, Central University of Gujarat, Sector-30, Gandhinagar, Gujarat 382030, India
e-mail: paulamisahu@gmail.com

A. M. Bahadir et al. (eds.), *Hydrology and Urban Water Supply*, Water and Wastewater Management,
https://doi.org/10.1007/978-3-031-72589-0_12

1 Introduction

Availability to drinking water quality has become constrained because of massive usage of freshwater resources in comparison to their recyclability, the continual degradation of groundwater and surface water quality, as well as latest climate change, and also in new regions where water availability was not a severe issue earlier [1]. Coastal zones feature a few of the densest populations, with a mean density of population of roughly 80 people per square km, which really is double the global average [2]. As a result of the increased demand for freshwater, substantial overuse of groundwater resources in the aquifer system is observed. Brackish water intrusion (BWI) caused by massive pumping is among the most serious risks to groundwater in the coastal locations. Many nations are concerned about the prevalence of BWI in coastal areas.

Seawater rests beneath freshwater in the aquifer system, which is much less thick, pushing the seawater interface seaward. The region of interaction with fresh and salt water is referred to as brackish water [3]. The BWI crisis is among the most serious environmental challenges affecting groundwater resources because groundwater salinity reduces the freshwater's accessibility as well as degrades the groundwater quality [4]. The phenomena of sea and river water mingling is the mechanism of BWI, and estuarine layering is the conjunction of comparatively tiny turbulence and experienced significant growth of the area of island forming velocities which are variable with time, place, or directions [5]. As a result, research on seawater intrusion into the coastal aquifers is critical to identify impacted zones where issues may be avoided or remedied quickly.

Excessive saltwater intrusion affects these systems' ionic strength and oxidative capability, leading to higher levels of specific chemical entities in groundwater, which runs from underground estuaries into to the oceans as submarine groundwater discharge (SGD) [6]. Nutrients, carbon, trace gases, sulphur, metals, even radionuclides are abundant in these substantially changed fluids. The underground estuary grows due to seawater intrusion. Climate change magnifies changes in sea level in short and seasonal time frames. These alterations may result in greater SGD fluxes, speeding nutrient release as well as improving biological productivity in nutrient-depleted environments [7]. However, this procedure may have a negative impact on the ecosystem, and it may disrupt the local ecology too. As additional seawater enters the coastal aquifers due to saltwater intrusion, the underground estuary extends inland, drastically affecting biogeochemical dynamics across large areas [8].

2 Brackish Water Intrusion and Its Triggering Factors

Coastal aquifers are very vulnerable to both local and worldwide events such as rise in sea levels, tidal waves, climate change, coastline erosion, as well as local flooding in coastal regions [9]. Furthermore, anthropogenic events are hastening the development of salinization in coastline. Aside from these aquifers, surface water resources are impacted too by their association with seawater. Waterways and estuaries enable brackish water to enter on a regular basis on account of backwaters commencing from the ocean, leading the freshwater to turn brackish [10].

Lithological, geomorphological, and structural characteristics, and other geological elements all contribute essential responsibilities in regulating the seawater intrusion into groundwater sources. The rock formation or lithology of any aquifer governs the inland circulation of freshwater in the aquifers of coastal regions [11]. In addition to regeneration, age of the aquifer structure, groundwater flow direction, amount of groundwater abstraction etc. have a substantial impact on the level of BWI [12]. Geographical structures for instance fissures, entrenched faults, including lineaments, as per the study, could enabling brackish water to permeate the coastline groundwater and mingle with the available freshwater [13].

Furthermore, coastal topography as well as local land use practices are important in limiting surface and subterranean runoff. They influence seawater movement inland and can account saline accumulation in coastal soils [14]. Paleochannels, which are loaded with highly permeable sediments, are favoured conduits for groundwater as well as seawater migration in coastal areas [15]. The flow of submarine groundwater from paleochannels limits the saline front and the leeward migration. Nevertheless, when groundwater extraction on land increases, such paleochannels might enable the ocean water incursion [16].

Numerous scientists have already carried out scientific experiments, fieldworks, as well as computational analyses to investigate the effects of tidal action upon coastline aquifer. Employing infographic data, the quality of aquifer in the coastlines of Kimje, Korea, suggests cyclical fluctuations in tidal movement was demonstrated [17]. Some other analysis scrutinizes the implications of tidal actions on the shoreline aquifers of Jeju Island, Korea, up to 3 km inward from the coastline [18]. Numerical simulations have shown that tidal changes as being the prime origins of freshwater-brackish water intercourse within the Atoll Islands [19]. Tidal action, according to [20] increases the BWI process and causes the freshwater-brackish water interface to expand. The research also discovered that tidal cycles have a significant impact on the groundwater level within the shallow aquifer, and that the coastline gradient has a massive influence on flow direction of groundwater into the ocean [21].

Both the changing climate as well as rising sea level are the two most significant climatic factors governing BWI in coastlines. The largest cause of groundwater recharge is meteorological precipitation, although spatio-temporal variations of its volume is discovered. Additionally, there is substantial inter-annual fluctuation in atmospheric precipitation. Groundwater table improves when precipitation accumulates during the monsoon season, but in the summertime, evapotranspiration amplified by a loss in rainfall causing groundwater table to decline [22]. As a result, even the little groundwater extraction during dry seasons might lead to BWI. Consequently, the likelihood of BWI appeared to be increased during dry seasons than rainy seasons. Several climate change-related elements that are explicitly connected to BWI comprise meteorological drought, coastal flooding, and storm surges [23]. "Storm surges" damage marine groundwater system through pouring seawater into sea-level coastal regions, salinizing littoral aquifers and thus producing salt deposits on land [24]. Both the groundwater as well as soil become unsuitable for agrarian uses and numerous other applications consequently.

Even though they are intrinsically related to certain other factors, anthropogenic acts are the primary important element affecting brackish water invasion. Overconsumption of groundwater resources remains the biggest prominent anthropogenic course which causes intrusion of brackish water in coastlines. Groundwater is mostly employed for domestic, agricultural, and industrial applications in regions near the coastline. When water requirement grows, groundwater supplies are being over-exploited, leading in BWI. Excessive groundwater removal has been identified as the primary cause of BWI in several places of the world, including Africa, Australia, China, Europe, India, the United States, and Vietnam. The aquifer sediments are alternately exposed to freshwater and saltwater due to periodic fluctuations in the freshwater–seawater interface [2].

3 Recent Assessment Methods

Evaluating the geographical pattern of subterranean physicochemical parameters like electrical conductivity (EC), salinity, quality of water, total dissolved solids (TDS), and seawater mixing models, amongst many other factors, is part of the examination of seawater intrusion. Additionally, remote sensing and geographic information systems (GIS) are employed in amalgamation with directly or indirectly to detect the coastline groundwater in a variety of conditions (Table 1).

Table 1 Brackish water intrusion assessment methods

Methods	Descriptions
Hydrogeochemical techniques	• 4 processes of geochemistry: (I) freshwater–brackish water mixing, (II) precipitation of carbonate, (III) ion exchange, (IV) redox reactions • Use of tracers to detect seawater intrusion have included Na/Cl, Cl/Br, Ca/Mg, Cl/HCO_3, $Ca/(HCO_3 + SO_4)$, and other ionic ratios [25]
Geophysical techniques	• Resistivity and electromagnetic techniques such as "vertical electrical sounding" (VES) and "electrical resistivity tomography" (ERT) [26], GPR: both seismic reflection and electromagnetic approaches
Remote sensing and GIS	• Rapidly mapping aquifer, particularly in big and inaccessible zones • Generally utilised in combination with various research like VES and geochemical investigations [27]

4 Present Status of Brackish Water Intrusion Worldwide

The most cited global drivers for BWI are sea level rise, hydropower dams, and land subsidence. The threat of SWI has resulted in the emergence of new adaptive livelihood models throughout the Vietnamese Mekong Delta (VMD), the most prominent of which is the prawn rice rotational crop (PRRC). Even though research on SI-driving factors is rapidly expanding, little synthesis has occurred. Similarly, several studies have investigated the economic benefits of PRRC; however, less emphasis has been given on environmental and societal aspects, resulting in questionable sustainability [28].

Many investigations were carried out in the Cutler Ridge region near Deering Estate, Florida, USA, with Henry's challenge and the Biscayne aquifer serving as the global baseline dilemma. In computational models, the finite-difference algorithm SEAWAT was employed. As per the simulation solution, raising the groundwater recharge reservoir volume decreases the SWI in moist and damp locations with significant floods [29]. Mathematical research was undertaken to analyse and manage SWI in the Gaza aquifer, Palestine, using the modeling of three-dimensional variable-density groundwater flow (SEAWAT) algorithm, taking into consideration excessively pumping owing to growing population and rising sea level [30]. Egypt's Nile Delta aquifer (NDA) holds a significant volume of water. SWI has grown because of risen aquifer extraction and rising sea levels, extending up to 100 km inland. Consequently, real steps must be taken to avoid additional SWI. In this scenario, an ideal well network for managing salty water incursion in NDA is established employing a range of management strategies like brackish water pumping, aquifer replenishment and freshwater abstraction.

SWI in the aquifer is simulated by the SEAWAT algorithm across several pumping and sea level rise scenarios [31]. Conversely, in northeast Brazil, the Electrodialysis Reversal (EDR) approach is chosen to purify saline water from the Almada River Basin in southern Bahia, recognized as the Cocoa Region. During the time of emergencies, saltwater intrusion produces excessive saline concentrations in Brazil's rivers and freshwater wells [32].

5 Management and Mitigation Measures

Brackish water delivers a boundless supply of water for desalination procedures. Hybrid structures that associate with several desalination processes along with energy sources seem to be the utmost viable alternatives [33]. In places with developing water shortage and strong solar radiation, innovative hybrid solar and wind energy directed structures paired with incredibly effective desalination technologies display possibilities. Renewable energy and desalination are separate solutions that can be coupled in diverse manners. The integration of professionals from two diverse sectors—renewable energy and desalination—are required for efficient embedded solutions. On-site generation of electricity from locally reachable renewable energy resources could support the process of desalination. This energy may be produced through several means, like heat, electricity, and mechanical energy [34].

Reduction in groundwater withdrawal volumes particularly in basins with groundwater overdraft is a highly significant and low-cost mitigating technique for reducing preventable brackish water infiltration into the littoral groundwater system [25]. Shifting cultivation practices on a routine basis as well as eliminating excessive water-intensive crops such as rice, soya and wheat can substantially minimize the groundwater extraction in coastlines [35]. Additionally, effluent and water consumed in industrial operations can be recovered after suitable treatments for other applications. Water delivery from distant interior places to coastal regions can dramatically reduce pumping activities.

Pumping activities along the coast may be the origin of fast inland migration of brackish water. In particular, the freshwater—brackish water interface develops at low depths near the sea and at deeper depths inland. Shifting sea-side pumping wells inland might thereby limit brackish water incursion [25]. Decreasing groundwater withdrawal operations in saltwater-affected locations helps to restore the natural groundwater flow gradient among the brackish water and freshwater, preventing seawater from flowing landward [35]. This approach is ideal for littoral groundwater sources with substantial horizontal variability. Figure 1 presents the overall management and mitigative measures controlling BWI management.

Fig. 1 Management and mitigative measures for BWI

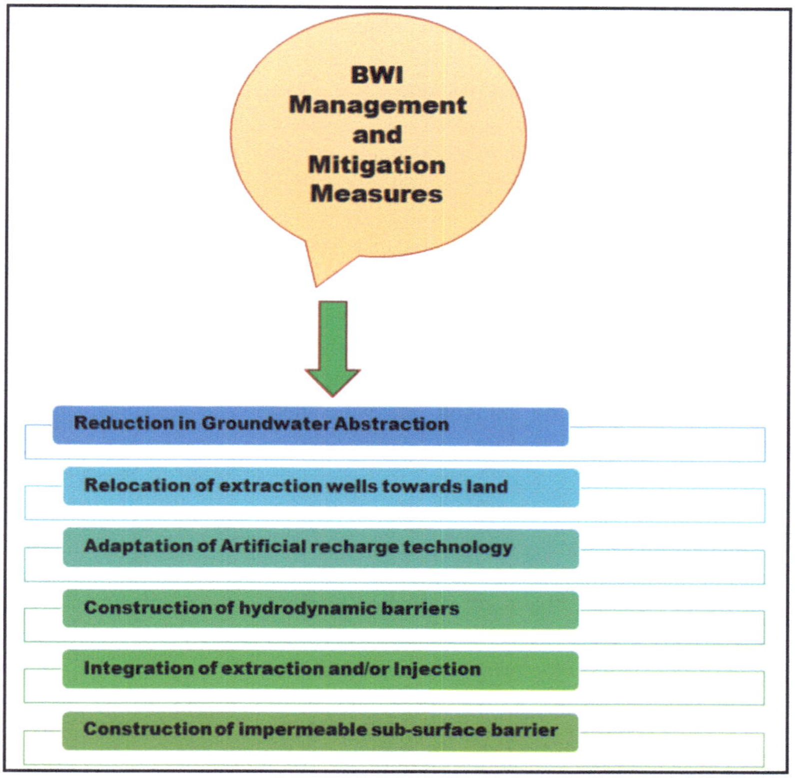

If stored and used appropriately, rainwater which is a renewable natural resource, may be highly effective in coping with groundwater shortages and associated difficulties. Storm-water runoff hazards can be mitigated by storing rainwater. The stored rainwater can subsequently be utilized for the recharge of aquifers, home and irrigational usage, and industrial tenacities.

The installation of a strings of injection wells derived from freshwater near the coasts may support to maintain the freshwater—brackish water balance. The freshwater-seawater interaction is pushed back into the sea by the high-pressure freshwater injection [25]. Another method for relocating the freshwater-seawater interface is to construct several brackish water extraction wells near the coastline. The collected brackish water might be utilised in "desalination plants" [35]. Physical subsurface barriers can be constructed to reduce inland seawater intrusion into coastal aquifers. To prevent seepage from underneath, these barriers are frequently constructed atop impermeable strata with concrete, grout, slurry walls, and other impermeable materials. They live longer and require less attention [35].

Numerous water regulations and laws have indeed been implemented globally at the state and national sectors to enhance the management and conservation of coastline water resources. According to others, high resolution groundwater quality data is required for analysing the present condition of brackish water intrusion, which could only be obtained with a very well surveillance system. Frequent water resources assessment also aids in detecting dwindling surface water produced by siltation and watershed removal in catchment regions.

6 Conclusion

High level treated effluent or reclaimed surface water could have been used to replenish groundwater resources in coastal aquifers for further brackish water management and planning. As per the current research, using these low-cost waterways will help to reduce the risks of both brackish water intrusion as well as local floods. In coastal areas, complex detection methods such as "observation wells" at varied distances, sensors, and programmed catalogue storage systems are required to manage the saltwater inland growth. One more potential area for subsequent BWI management is the creation of a rigorous decision analysis system that incorporates a huge spectrum of hydraulic features of surface and groundwater acquaintances for example artificial recharge, evaporation, groundwater withdrawal and natural recharge. This evaluation techniques should be used with computational methods to simulate aquifer reactions to various management conditions. This comprehensive tool may be developed to determine the best management framework for any circumstance. Prior implementing any BWI preventive actions, the implications of parameter variation on calibration process and the optimal outcomes of each BWI control method should be investigated.

References

1. UNESCO. The united nations world water development report 2020, water and climate change. Paris: UNESCO—United Nations Educational, Scientific and Cultural Organization; 2020.
2. Prusty P, Farooq SH. Seawater intrusion in the coastal aquifers of India—a review. HydroResearch. 2020;3:61–74.
3. Chithra S, Joseph S, Kannan N. A study of saltwater intrusion in the Kallada River, southwest coast of Kerala, India. Water Supply. 2022;22(2):1–18.
4. Dipu S, Josna RP. Saline water intrusion in urban coastal area: a case study of Kuttiyadi River, Kerala, India. Am J Water Resourc. 2020;8(3):128–33.
5. Antony MM, Ignatius J. A hydrological study of Ashtamudi Lake, Kerala, India with special reference to its ecological difference. Int J Sci Res (IJSR). 2016;5(8):1841–6.
6. Moore WS, Joye SB. Saltwater intrusion and submarine groundwater discharge: acceleration of biogeochemical reactions in changing coastal aquifers. Front Earth Sci. 2021;9:1–14.
7. Duque C, Michael HA, Wilson AM. The subterranean estuary: technical term, simple analogy, or source of confusion? Water Resour Res. 2020;56:1–14.
8. Adyasari D, Hassenrück G, Montiel D, Dimova N. Microbial community composition across a coastal

hydrological system affected by submarine groundwater discharge (SGD). PLoS ONE. 2020;29:1–26.
9. Barlow PM. Ground water in freshwater-saltwater environments of the Atlantic Coast. Virginia: USGS; 2003
10. Vijay R, Khobragade P, Mohapatra PK. Assessment of groundwater quality in Puri City, India: an impact of anthropogenic activities. Environ Monit Assess. 2011;177:409–18.
11. Michael HA, Russoniello CJ, Byron LA. Global assessment of vulnerability to sea-level rise in topography-limited and recharge-limited coastal groundwater systems. Water Resour Res. 2013;49:2228–40.
12. Ammar SB, Taupin JD, Zouari K, Khouatmia M. Identifying recharge and salinization sources of groundwater in the Oussja Ghar el Melah plain (northeast Tunisia) using geochemical tools and environmental isotopes. Environ Earth Sci. 2016;75 (606):1–16.
13. Gupta C, et al. Geoelectrical studies for delineating seawater intrusion in parts of Konkan coast, Western Maharashtra. Int J Environ Earth Sci. 2010;1:62–79.
14. Yu J, et al. The spatial distribution characteristics of soil salinity in coastal zone of the Yellow River Delta. Environ Earth Sci. 2014;72:589–99.
15. Falls W, et al. Hydrogeology, water quality, and saltwater intrusion in the Upper Floridan Aquifer in the offshore area near. Sci Invest Rep. 2005;2005-5134:1999–2002.
16. Mulligan A, Evans R, Lizarralde D. The role of paleochannels in groundwater/seawater exchange. J Hydrol. 2007;335:313–29.
17. Kim J, et al. Use of time series analysis for the identification of tidal effect on groundwater in the coastal area of Kimje, Korea. J Hydrol. 2005;300(1–4):188–98.
18. Kim K, et al. Tidal effects on variations of fresh–saltwater interface and groundwater flow in a multilayered coastal aquifer on a volcanic island (Jeju Island, Korea). J Hydrol. 2006;330:525–42.
19. Underwood M, Peterson F, Voss C. Groundwater lens dynamics of Atoll Islands. Water Resour Res. 1992;28:2889–902.
20. Ataie-Ashtiani B, Volker R, Lockington D. Tidal effects on sea water intrusion in unconfined aquifers. J Hydrol. 1999;216:17–31.
21. Li L, Dong P, Barry D. Tide-induced water table fluctuations in coastal aquifers bounded by rhythmic shorelines. J Hydraul Eng. 2002;128:925–33.
22. Rapti-Caputo D. Influence of climatic changes and human activities on the salinization process of coastal aquifer systems. Ital J Agron. 2010;5:67–79.
23. Terry J, Falkland A. Responses of atoll freshwater lenses to storm-surge overwash in the Northern Cook Islands. Hydrogeol J. 2010;18:749–59.
24. Rezaie A, Ferreira C, Rahman M. Storm surge and sea level rise: threat to the coastal areas of Bangladesh. In: Extreme hydro climatic events and multivariate hazards in a changing environment. Elsevier; 2019. p. 317–42.
25. Manivannan V, Elango L. Management of coastal groundwater resources. In: Coastal Management. s. n.: Elsevier; 2018. p. 383–97.
26. Kumar V, Dhakate R, Amarender B, Sankaran S. Application of ERT and GPR for demarcating the saline water intrusion in coastal aquifers of Southern India. Environ Earth Sci. 2016;75:393.
27. Dhakate R, et al. Demarcating saline water intrusion pathways using remote sensing, GIS and geophysical techniques in structurally controlled coastal aquifers in Southern India. Environ Earth Sci. 2016;75:363.
28. Loc H, et al. How the saline water intrusion has reshaped the agricultural landscape of the Vietnamese Mekong Delta, a review. Sci Total Environ. 2021;794(10): 148651.
29. Abd-Elaty I, Javadi A, Abd-Elhamid H. Management of saltwater intrusion in coastal aquifers using different wells systems: a case study of the Nile Delta aquifer in Egypt. Hydrogeol J. 2021;29:1767–83.
30. Abd-Elaty I, Abd-Elhamid H, Qahman K. Coastal aquifer protection from saltwater intrusion using abstraction of brackish water and recharge of treated wastewater: case study of the Gaza Aquifer. J Hydrol Eng. 2020;25(7):5020012.
31. Abd-Elaty E, Straface S, Kuriqi A. Sustainable saltwater intrusion management in coastal aquifers under climatic changes for humid and hyper-arid regions. Ecol Eng. 2021;171: 106382.
32. Kunrath C, et al. Electrodialysis reversal as an alternative treatment for producing drinking water from brackish river water: a case study in the dry season, northeastern Brazil. J Environ Chem Eng. 2020;8(2): 103719.
33. Ghaffour N, et al. Membrane distillation hybrids for water production and energy efficiency enhancement: a critical review. Appl Energy. 2019;254: 113698.
34. Yang Q et al. Research status of Brackish water desalination technology. E3S Web Conf. 2021;329 (01039):1–5.
35. Hussain M, Abd-Elhamid H, Javadi A, Sherif M. Management of seawater intrusion in coastal aquifers: a review. Water. 2019;11(12):246.

Groundwater Quality Assessment for Bursa—A Multi-catchment Region

Babak Vaheddoost, Soroosh Sharifi, Soroush Abolfathi, and Egemen Aras

Abstract

This study represents a comprehensive analysis of the groundwater quality in a data poor catchment located in Bursa region of Turkey. The dataset includes three years of groundwater monitoring data from 12 observations wells. The data were recorded between 2015 and 2018 at a biannual temporal scale. The spatiotemporal variations in groundwater indices were investigated using an advanced analytical method to determine the role of anthropogenic activities on the quality of groundwater for irrigation and drinking purposes. The primary properties of the groundwater quality are determined using sodium percentage, sodium absorption ratio (SAR), magnesium hardness (MH), permeability index (PI), residual sodium bicarbonate (RSBC), soluble sodium percentage (SSP), Kelly's ratio (KR), base exchange index (BEI), meteoric genesis index (MGI), and the water quality index (WQI). Following the primary analysis of groundwater characteristics, a holistic data analysis is proposed using Piper-, Durov-, and Wilcox-diagrams to provide a detailed overview of the groundwater quality in the region. The results indicate that, in general, groundwater quality in the region is suitable for irrigation purposes. However, it was also observed that there are discrepancies among the samples, indicating variations in the groundwater quality within the study area. The data analyses revealed that the groundwater in the Bursa region is predominantly of the alkaline type, while the prevailing anions and cations in most of the samples are HCO_3^- and Ca^{2+}. These findings suggest a correlation with rock dissolution and geomorphological processes in the study region. The disturbances in EC and variance also indicate possible pollution in the groundwater that could be attributed to a combination of anthropogenic activities and climate change.

Keywords

Groundwater · Quality index · Groundwater hydrology · Water quality

B. Vaheddoost (✉) · E. Aras
Department of Civil Engineering (G Blk-RN 131), Bursa Technical University, Mimar Sinan Mah. Mimar Sinan Blv. Eflak St. No:177, 16310 Bursa, Turkey
e-mail: babak.vaheddoost@btu.edu.tr

S. Sharifi
School of Engineering, Department of Civil Engineering, University of Birmingham, Edgbaston Birmingham B15 2TT, UK

S. Abolfathi
School of Engineering, University of Warwick, Coventry CV4 7AL, UK

A. M. Bahadir et al. (eds.), *Hydrology and Urban Water Supply*, Water and Wastewater Management,
https://doi.org/10.1007/978-3-031-72589-0_13

1 Introduction

Water is an abundant substance in the biosphere, but the spatiotemporal variations in its properties urges decision makers to develop alternative realization for the worst scenarios that could exacerbate the ongoing climate change and failure in supply–demand chain. Groundwater is one of the key water resources that are extensively used for agricultural, urban and industrial purposes. Hence, the quality of the groundwater is of great importance in preserving biota, and maintaining a sustainable surface–groundwater interaction. Given the significance of understanding groundwater quality, several studies in recent years have been dedicated to understand and quantify the quality and characteristics of groundwater [1–8].

Likewise, some aspects of groundwater including its characteristics, quality and geomorphological dynamics at Bursa region have partially been investigated by past studies. Balderer et al. [9] investigated the thermal groundwater sources in Bursa and Kuzuluk. Two thermal spring namely Kukurtlu and Cekirge located at the northern slope of Mount Uludag (150–200 m above sea level) were studies and it was found that the thermal waters includes Ca, Mg, Na, HCO_3, and SO4 minerals (total dissolved solution $\sim$0.5 g/l) resolved in. The existing data also showed that the tritium content <1.3–6.3, corresponds to the local meteoric water line. Additionally, the study of δ18O–δ2H diagram of thermal and cold groundwater also showed that recharge of the groundwater is concentrated by Mount Uludag and dominated by the regional hydrological processes including heavy precipitation in winter and snow melt during spring and summer. Sen [10] studied Nilufer and Ayvali basins in Bursa region using data collected from 35 wells and springs. It was concluded that nitrate concentration is dominant in the samples. Imbach [11] investigated the outflows from Kukurtlu and Cekirge thermal springs in Bursa and compared the water quality by means of physical, chemical, and geographical differences in groundwater samples. It was concluded that, the dominant processes in the groundwater circulation at both sites are mixture with shallow groundwater, energy loss through heat conduction, and interaction of crust or atmosphere with gases raised from the mantle. Baba and Tayfur also investigated the groundwater contaminations across Turkey including Iznik region, Bursa [12]. Analysis showed that the marble and alluvium units in Iznik region and the associated fractures cause wastewater intrusion through ancient quarries which adversely affect the groundwater. Simsek et al. [13] investigated the hydrogeological, hydrogeochemical, and waste disposal in a karstic system related to ancient marble pits near Iznik in Bursa region, and suggested that the contamination of the karstic systems is related to Na (3260 mg/l), K (903 mg/l) and Cl (10,396 mg/l). Akar Tarim et al. [14] also measured the Radon concentrations (222Rn) in 27 wells and 19 tabs with consideration of several geologic formations in Bursa region. It was concluded that the Rn concentration was ranged from 1.46–53.64 Bq/l and 0.91–12.58 Bq/l, for well and tab samples, respectively. Fidan et al. [15] investigated the properties of the groundwater level time-series at Adakoy, Aticilar, Cayirkoy, Delicay, Karasil, Kursunlu, Narlidere, Yenice and Yenicabat. Then, the groundwater flow direction was determined through analysis of hydraulic gradient between groundwater levels, correlation, and skewness, kurtosis, and variation coefficients. The analysis of data also indicated the connectivity of the aquifers and that most of the aquifers are recharged by the snow melts originated from the Mount Uludag.

The review of the existing studies on the groundwater properties in the Bursa region reveals that the groundwater quality in this region has not been fully understood and addressed, mainly due to the lack of information, poor spatial and temporal data records, as well as high uncertainties associated with the existing data. However, in recent years more comprehensive groundwater data has been collected that enables further investigations on the groundwater. Therefore, with consideration to the concerns

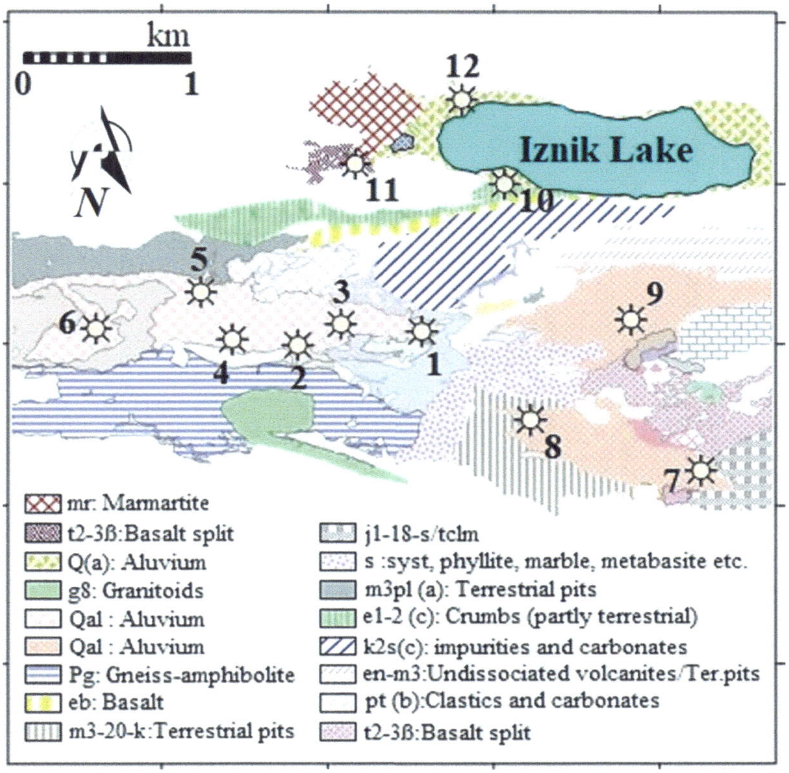

Fig. 1 Study area and related lithology in the observation wells' vicinity

against resilience against climate change and the allocation of the groundwater as an active water resource in the region, it is important to acquire more information about the situation and dynamics of groundwater in the area. In this respect, this study aims to outline methodological approach to conduct a comprehensive analysis of groundwater properties and water quality assessment in a data-poor multi-catchment region. To do this, the historical data were addressed and the current state of groundwater in Bursa for different uses are investigated.

2 Materials and Methods

2.1 Study Area

Bursa is the 4th major city in Turkey and important hub for agro-food security and industrial activity in the western Anatolia and Marmara region. The Bursa district covers 1,036 km^2 area with average 100 m above mean sea level. Figure 1 shows the geographical position of the Bursa, together with the morphological features and the location of the 12 observation wells, used in the analysis. According to Koeppen–Geiger classification map [16, 17] the Bursa region can be classified under Csa (i.e. hot summer Mediterranean) and Csb (i.e. warm summer Mediterranean) climates. The total annual precipitation in the region is 704.9 mm, while the average annual temperature is about 14 °C [18].

2.2 Data

In this study we have selected 12 observation wells across Bursa region (see Fig. 1) with the water sampling data measured at bi-annual scale (i.e. every 6 month). The data collection campaign was conducted from 2015 to 2018 by State Hydraulic Works (DGSHW). Table 1 summarizes the details of the data collected from the

Table 1 The average values, measured or calculated in the observation wells during 2015–2018 (*Allu., Bas., Eoc., Org., Kuv., and Quat., Ter.* respectively indicates to Alluvium deposits, Eocene age, mixed organic soil, Kuvars, Quaternary age, and Teras age)

Station	Narlidere	Delicay	Adakoy	Aticilar	Yeniceabat	Cayirkoy	Kursunlu	Yenice	Karasil	Soloz	Bolge Trafik	Yenikoy
Station number	1	2	3	4	5	6	7	8	9	10	11	12
Basin	Susurluk	Susurluk	Susurluk	Susurluk	Susurluk	Susurluk	Sakarya	Sakarya	Sakarya	Marmara	Marmara	Marmara
Sub Basin	Bursa	Bursa	Bursa	Cayirkoy	Bursa	Cayirkoy	Inegol	Inegol	Yenisehir	Orhangazi	Gemlik	Orhangazi
Latitude (N)	40.220	40.190	40.230	40.210	40.264	40.219	40.043	40.106	40.230	40.399	40.425	40.504
Longitude (E)	29.314	29.164	29.216	29.085	29.047	28.921	29.650	29.445	29.566	29.413	29.233	29.361
Lithology	*Allu*	*Allu*	*Allu*	*Allu*	*Allu*	*Allu*	*Allu*	*Allu*	*Allu*	*Allu*	*Allu*	*Allu*
Soil type	*Kuv*	*Kuv*	*Kuv*	*Kuv*	*Kuv*	*Kuv*	*Kuv*	*Kuv*	*Kuv*	*Org*	*Bas*	*Kuv*
Formation age	*Quat*	*Quat*	*Quat*	*Quat*	*Quat*	*Quat*	*Quat*	*Quat*	*Quat*	*Eoc*	*Ter*	*Quat*
pH	7.68	6.75	7.36	7.15	7.29	7.88	7.12	7.34	7.55	7.28	8.09	7.52
EC (25 °C mS/m)	40.17	12.52	64.20	67.23	77.64	41.92	77.64	66.40	41.98	29.38	29.10	69.57
$Na^{(+)}$ (meq/l)	0.54	0.24	0.36	0.93	0.68	1.07	1.88	1.47	0.68	1.17	1.12	2.12
$K^{(+)}$ (meq/l)	0.09	0.07	0.1	0.06	0.06	0.08	0.14	0.21	0.068	0.081	0.0663	0.11
$Ca^{(2+)}$ (meq/l)	2.48	1.70	4.64	5.38	5.05	1.84	4.93	2.47	2.75	1.02	1.35	3.42
$Mg^{(2+)}$ (meq/l)	1.92	0.74	1.95	1.22	3.53	2.58	2.9	1.86	2.44	1.22	0.68	1.95
$HCO_3^{(-)}$ (meq/l)	4.44	6.48	3.55	7.52	5.95	4.5	1.53	4.44	6.48	3.55	1.26	5.31
$Cl^{(-)}$ (meq/l)	0.19	0.29	0.64	0.76	0.49	0.64	0.41	0.51	0.65	1.24	1.90	1.08
$SO_4^{(2-)}$ (meq/l)	0.19	0.19	0.5	0.96	1.49	0.82	0.71	0.2	0.33	0.02	0.03	0.59
NO_2 (mg/l)	0.01	0.03	0.01	0.03	0.01	0.02	0.02	0.02	0.03	0.02	0.04	0.01
NO_3 (mg/l)	4.85	2.76	0.56	51.01	6.40	0.67	26.14	4.71	0.74	0.43	0.80	4.69
$CaCO_3$ (mg/l)	22.01	12.21	32.99	33	40.61	22.44	43.02	21.6	25.93	11.13	10.13	25.32
TDS (mg/l)	251.25	130.29	414.16	396.3	491.43	288.30	530.40	363.65	307.01	224.15	182.29	383.79

observational wells and the lithology of the studied region. The samples are related to three different basins of Marmara, Sakarya, and Susustluk. The lithology of the soils is mostly related to the alluvium deposits (alluv.) shaped in Eocene (Eoc.), Quaternary (Quat.), and Teras (Ter.) ages. The analysis of the data collected shows the wells No. 1–9 and 12 are of the Kuvars (Kuv.) type while the remaining wells No. 10 and 11 are mixed and Basaltic lithology, respectively (see Fig. 1 and Table 1). Further data indices including cations, anions, potential of hydrogen (pH), electrical conductivity (EC), hardness ($CaCO_3$), total dissolved solids (TDS) and water class are collected to conduct a comprehensive analysis of groundwater quality in Bursa region.

The pH of the groundwater samples collected in the region shows that the hydrogen ion concentration in the samples results in an alkaline solution (i.e. basic) with an average value of 7.41, while the pH in the Delicay samples is more acidic (6.75). The highest pH recorded belongs to Bolge Trafik observation well with 8.09 which is considerable for daily usage. Therefore, a general conclusion of alkaline type groundwater for this region is valid. EC, related to the electrical conductance of the solution varies, on average, between 12.52–77.64 mS/m (25 °C). Cations, which are measured based on Table 1 are, Sodium (Na^+), Potassium (K^+), Calcium (Ca^{2+}), Magnesium (Mg^{2+}); while the anions are Bicarbonate (HCO_3^-), Chlorine (Cl^-), sulfate (SO_4^{2-}), all of which are measured with meq/l scale. The Nitrate (NO_3^-) and Nitrite (NO_2^-) are also measured using mg/l scale which are contaminants in the groundwater samples. Similarly, the hardness of the Calcium Carbonate ($CaCO_3$) in the samples is given that is a common substance of rocks with calcite and aragonite origin. The scaling used in measurement of the $CaCO_3$ is in French unit (°fH or °f) which is 10 mg/L $CaCO_3$, equivalent to 10 ppm. The total dissolved solids (TDS) is determined to identify the solution class. The measured TDS, on average, is ranging from 130.29 to 491.43 mg/l, and the Delicay and Yeniceabat regions have the lowest and highest recorded TDS in the samples, respectively.

The data addressed and discussed in Table 1 are then employed to develop data analysis framework to investigate the groundwater quality in the data-poor multi-catchment regions. To ensure high precision results and avoid the data skewness due to the data loss, this study adopts the average values over 3 year monitoring period (Table 1). Preliminary analyses are conducted to obtain standard metrics necessary for evaluation of the water quality. This section outlines the proposed methodological approach for investigating the groundwater quality from a holistic and case-based analytical approach suitable for data-poor regions.

2.3 Preliminary Analysis

The preliminary analysis is carried out to gain a deeper understanding of the groundwater samples based on established and standardised indices. Hence, the sodium percentage, sodium absorption ratio, magnesium hardness, permeability index, residual sodium bicarbonate, dissolved sodium percentage, Kelly's ratio, base exchange index, meteoric genesis index, and water quality index are used for the preliminary analysis of groundwater data in the 12 monitoring wells.

2.3.1 Sodium Percentage

The sodium cation is important for the quality of water for irrigation. Sodium content reduces the permeability of the soil and is an indication of point and non-point pollution or salt intrusion. Together with chloride or carbonate, sodium can be a source of saline soil or alkalinity in the region. Equation 1 based on the methodology proposed by Sadashivaiah et al. [19] and Hagras [20] is used to assess the sodium content in groundwater samples:

$$Na(\%) = \left(\frac{Na^+}{Na^+ + Ca^{2+} + Mg^{2+} + K^+}\right) \times 100 \quad (1)$$

Thus, the groundwater quality can be categorized based on the percentage of *Na* into < 20, 20–40, 40–60, 60–80, and > 80, representing excellent, good, permissible, doubtful, and unsuitable water quality, respectively.

2.3.2 Sodium Absorption Ratio (SAR)

Along with the *Na* ratio, the *SAR* value defines the relative activity of sodium ion in exchange with soil particles. The SAR is calculated according to Todd and Mays [21] as follows:

$$SAR = \frac{Na^+}{\sqrt{\frac{Ca^{2+} + Mg^{2+}}{2}}} \quad (2)$$

Based on SAR values, groundwater properties can be classified as S1 (0–10), S2 (10–18), S3 (18–26) and S4 (>26), representing low, medium, high and very high sodicity, respectively. The SAR value is used together with other indices for the case study to provide a robust and comprehensive analysis of water quality.

2.3.3 Magnesium Hardness (MH)

In general, Ca^{2+} and Mg^{2+} in water are the two main factors that define the water hardness. According to the World Health Organization (WHO), drinkable water should have magnesium hardness less than 40 mg/l. In this paper, the methodology of Szabolcs and Darab [22] is used to determine the MH level at the 12 case study sites:

$$MH = \frac{Mg^{2+}}{Ca^{2+} + Mg^{2+}} \times 100 \quad (3)$$

In cases MH $\geq$ 50%, the water is unsuitable for irrigation purposes [23, 24].

2.3.4 Permeability Index (PI)

The permeability index is a qualitative measure to classify the estimated rates of the vertical movement of water from the surface of the ground through the unsaturated zone, between the land surface and the water table. The *PI* is calculated according to Kacmaz and Nakoman [25] as:

$$PI = \frac{Na^+ + \sqrt{HCO_3^-}}{Ca^{2+} + Mg^{2+} + Na^+} \times 100 \quad (4)$$

PI together with the total concentration of ions in Doneen-classification [26], can be categorized into Class-I ($\geq$ 75), Class-II (25–75), and Class-III (<25), indicating excellent, good, and unsuitable groundwater quality for irrigation use, respectively.

2.3.5 Residual Sodium Bicarbonate (RSBC)

RSBC defines the excess concentration of HCO_3^- compared to Ca^{2+} and is used to evaluate the irrigation water quality [27]. Gupta method is adopted in this study to determine the *RSBC* as described by Eq. 5 [28]:

$$RSBC = HCO_3^- - Ca^{2+} \quad (5)$$

With respect to RSBC, the irrigation water quality is classed as safe when RSBC < 5, marginal for RSBC ranging from 5 to 10, and unsatisfactory for RSBC > 10.

2.3.6 Soluble Sodium Percentage (SSP)

SSP index describes the excess of Na^+ in the groundwater absorbed by the Clay and eventually disperses the Ca^{2+} and Mg^{2+}. During this phenomena Na^+ with Ca^{2+} and Mg^{2+} result in the reduction of permeability rate in the soil sample. SSP is determined as:

$$SSP = \frac{Na^+ + K^+}{Na^+ + K^+ + Ca^2 + Mg^{2+}} \times 100 \quad (6)$$

The results of *SSP* evaluation are classified as excellent (<20), good (20–40), permissible (40–60), doubtful (60–80), and unsuitable (>80) water samples.

2.3.7 Kelly's Ratio (KR)

The Kelly's ratio (KR) proposed by Kelly [29] is determined for the data in order to evaluate the excessive Na^{2+} ion in the groundwater samples:

$$KR = \frac{Na^+}{Ca^{2+} + Mg^{2+}} \quad (7)$$

The water quality with *KR*s lower that one is considered suitable for irrigation and those *KR*s larger than one are deemed unsuitable for irrigation.

2.3.8 Base Exchange Index (BEI) and Meteoric Genesis Index (MGI)

The joint BEI and MGI indices, also known by Soltan-classification indices are previously used by Matthess [30] and Soltan [31, 32]. This study evaluates the *BEI* and *MG* as follows:

$$BEI = \frac{Na^+ - Cl^-}{SO_4^{2-}} \quad (8)$$

$$MG = \frac{(Na^+ + K^+) - Cl^-}{SO_4^{2-}} \quad (9)$$

The results from *BEI* analysis can be categorized as $BEI < 1$ (Na^+–SO_4^{2-}) or $BEI > 1$ (Na^+–HCO_3) type. The *MGI* shows if the groundwater samples are deep meteoric ($MGI < 1$), or shallow meteoric percolation ($MGI > 1$) types.

2.3.9 Water Quality Index (WQI)

The *WQI* is proposed by Horton [33], to provide an overall quality measurement for drinking and irrigation water. *WQI* is determined using Eq. 10–12:

$$W_i = \frac{w_i}{\sum w_i} \quad (10)$$

$$q_i = \frac{C_i}{S_i} \times 100 \quad (11)$$

$$WQI = \sum_{i=1}^{n} (W_i \times q_i) \quad (12)$$

where q_i and w_i are the rating scale, and unit weight for the respective water quality parameter, respectively. In present study, the pH, EC, Na^+, K, Ca, Mg, HCO_3, Cl, SO_4, NO_2, NO_3, $CaCO_3$, and TDS values together with their acceptable range provided by World Health Organisation (WHO) and The Food and Agriculture Organization for drinking water was used to define the rating scales [34, 35]. Finally, the results of *WQI* can be categorized into excellent ($WQI < 50$), good (WQI = 50–100), poor (WQI = 100–200), very-poor (WQI = 200–300), and unsuitable ($WQI > 300$) for drinking or general use.

2.4 Assessment of Regional Groundwater Quality

This study utilized all the data collected from the 12 observation wells across our case study region to undertake a holistic assessment of the state of groundwater quality in Bursa as well as providing in depth analysis of data anomalies and potential uncertainties associated with the data-set. The holistic assessment was performed by conducting data visualization and statistical analysis using piper-, Durov-, Gibbs-, and Wilcox-diagrams are used [39]. The ensemble of these analyses is then used to evaluate and determine the overall condition of the groundwater quality in the region. The analytical approach for the proposed holistic assessment are described in the following sections.

2.4.1 Piper-Diagram

The Piper diagram, suggested by Piper [36], is used for presenting water chemistry data to help in understanding the sources of the dissolved constituent salts in water. This procedure is based on the premise that cations and anions in water are in such amounts to assure the electroneutrality of the dissolved salts, in other words the algebraic sum of the electric charges of cations and anions is zero [4, 40, 41]. For this, a trilinear-diagram is used to segregate relevant statistics in understanding the sources of the dissolved constituents in groundwater. Since the cations and anions in the groundwater samples are usually in equilibrium, most of the abundant cations are Ca^{2+} and Mg^{2+} (alkaline earths) and Na^+ (Alkali). While, the most frequent anions are HCO_3^- (weak acid), SO_4^{2-} (strong acid), Cl^- and CO_3^{2-}.

2.4.2 Durov Diagram

The Durov diagram [37] provides extended information as it depicts the hydrochemical processes in solution. It combines three rectangles and sometimes two additional triangle figures for joint evaluation of cations, anions, pH, ion exchange, and TDS in the water samples.

2.4.3 Gibbs Diagram

Gibbs diagram [38] generates information about the origin of the variation in chemistry and dominance of ions in the groundwater with respect to the precipitation, or evaporation events.

2.4.4 Wilcox Diagram

Wilcox diagram [39] assesses the quality of groundwater for irrigation, using the Na% versus electrical conductivity EC, across the 12 case study sites. Additionally, the variations of EC versus SAR is used to evaluate the absorption rate and the relation between permeability and EC in the groundwater samples. The SAR and EC are providing a classification of S1–S4 and C1–C4 for data samples, respectively.

2.5 Assessment of Local Groundwater Quality

This study also adopts a case by case analysis to conduct a comprehensive and systematic analysis of the groundwater quality across all the collected samples. Hence, the Stiff- and Schuller-diagrams are also used to recognize the disturbance and the total variance across all the samples at the local scale.

2.5.1 Stiff Diagram

The Stiff diagram [42] depicts the major ion composition in the groundwater in the Bursa region. To develop and derive Stiff diagram, ions are plotted in meq/l on the horizontal axis. Therefore, it is very usual in rapid visualization of groundwater quality from different sources.

2.5.2 Schuller Diagram

The semi-logarithmic diagram Suggested by Schoeller [43] is adopted to depict the relative concentrations of ions in the groundwater samples. Then, taken samples can be evaluated using a single diagram to reveal the similarities between the allocated ions. For instance, the projected similarities might be used in defining the common or disparate source of ions in the groundwater.

3 Results

First, provided data in the Table 1 are evaluated based on their simple characteristics and therefore, Na%, SAR, MH%, PI, RSBC, SSP, KR, BE, MG, and WQI indices are used for interpretation and preliminary analyses. Afterward, the Piper-, Durov-, Gibbs-, and Willcox-diagrams are used to project a general overview through the study area. Finally, Stiff- and Schuller-diagrams are applied to conduct a case based analysis.

3.1 Preliminary Analysis

As detailed previously Na%, SAR, MH%, PI, RSBC, SSP, KR, BE, MG, and WQI indices are used for this aim. Given in Table 2, the conducted analysis provides useful insight about the groundwater samples. The Na% ranges between 6.52% (at Adakoy) and 36.89% (at Bolge Trafik). Since, the Na% of <20% and 20%-40% provides excellent and good groundwater quality respectively, it is suitable for irrigation. However, Delicay with 0.24 meq/l (5.52 gr/l) and Yenikoy with 2.12 meq/l (48.74 g/l) showed potential for drinking purpose (i.e. ions more than 50 mg/l is unsuitable for drinking [44]). The SAR index ranging between 0.19 (at Adakoy) and −1.29 (at Yenikoy) shows that all the samples can be considered as S1-type with low Sodicity also indicating to a good water quality

Table 2 Summary of the preliminary analysis

Stations	Na (%)	SAR	MH %	PI	RSBC	SSP	KR	BEI	MGI	WQI
Narlidere	10.80	0.37	43.66	50.86	1.41	12.52	0.12	1.87	2.33	42.69
Delicay	8.67	0.22	30.29	60.19	0.19	11.26	0.09	−0.26	0.11	22.26
Adakoy	5.14	0.19	29.6	42.09	1.94	6.52	0.05	−0.54	−0.35	69.72
Aticilar	12.19	0.51	18.5	40.30	−0.94	13.02	0.14	0.17	0.23	60.17
Yeniceabat	7.30	0.33	41.12	34.85	1.43	7.94	0.08	0.13	0.17	75.67
Cayirkoy	19.12	0.72	58.4	53.74	1.71	20.62	0.24	0.52	0.62	43.48
Kursunlu	19.13	0.95	37.01	47.63	2.59	20.50	0.24	2.08	2.27	83.31
Yenice	24.46	0.99	42.93	67.44	3.48	27.88	0.34	4.79	5.81	61.18
Karasil	11.46	0.42	47.03	47.75	1.75	12.60	0.13	0.08	0.29	50.07
Soloz	33.49	1.1	54.5	70.69	0.52	35.82	0.52	−3.75	0.31	31.03
Bolge Trafik	34.84	1.11	33.64	71.22	−0.09	36.90	0.55	−23.88	−21.85	26.67
Yenikoy	27.89	1.29	36.27	59.06	1.88	29.34	0.39	1.75	1.94	57.98

for irrigation uses. On the other hand, based on the MH of the samples, the Mg rate (18.5–58.4) in all wells except Cayirkoy and Soloz have low hardness (MH $\geq$ 50% unsuitable for irrigation). Therefore, the groundwater quality in Cayirkoy and Soloz are not suitable in this respect. The PI ranging between 34.85 (at Yeniceabat) to 71.22 (at Bolge Trafik) also is an index for permeability; while all of the stations are accepted as Class-II and recommended for irrigation uses. The RSBC, depicts the excess of HCO_3 concentration compared to Ca^{2+} while all of the samples considered to be safe (<5), for irrigation use. The SSP ranging between 6.52 (at Adakoy) to 36.90 (at Bolge Trafik), indicates to excellent (<20; in Narlidere, Delicay, Adakoy, Aticilar, Yeniceabat, and Karasil) and good (20–40; in Cayirkoy, Kursunlu, Yenice, Soloz, Bolge Trafik, and Yenikoy) permeability in irrigation. Similarly, and based on the KR, ranging from 0.09 (at Delicay) to 0.55 (at Bolge Trafik) are suitable ($\leq$ 1) for irrigation. According to the BEI and MGI, the groundwater samples in Narlidere, Kursunlu, Yenice, and Yenikoy are shallow meteoric water type; whilst the remaining samples can be considered as deep meteoric water type. Finally, based on the WQI ranging between 22.26 (at Delicay) to 75.67 (at Yeniceabat) all of the groundwater samples can be considered as either excellent or good water types for drinking use.

3.2 Regional Situation

In Fig. 2, Piper-diagram of the samples shows that the dominant cations and anion in the samples are of the Ca^{2+} and HCO_3^- type respectively. Bicarbonate is an inorganic carbon based

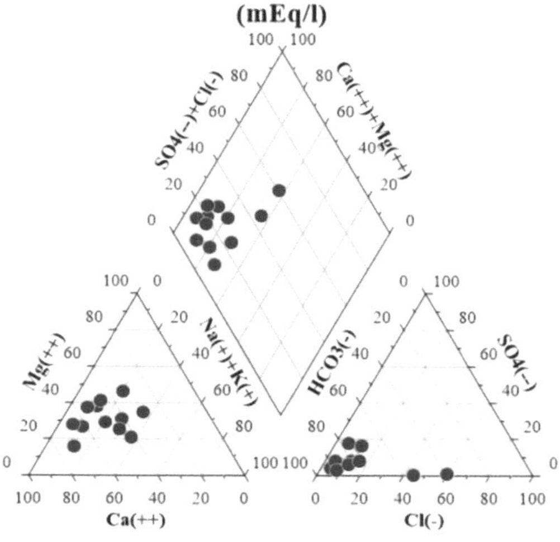

Fig. 2 Piper-diagram of the samples

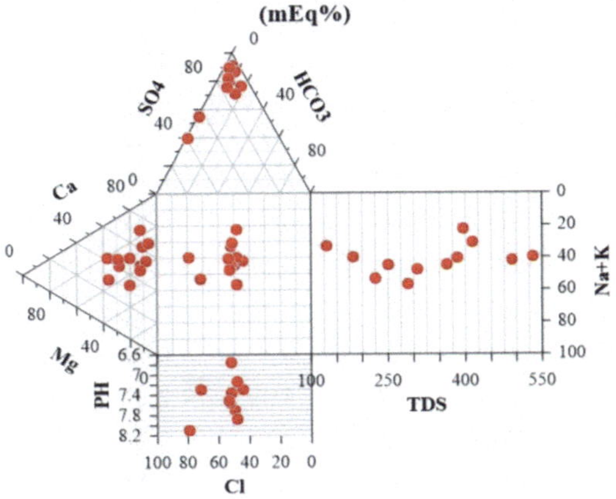

Fig. 3 Durov-diagram of the samples

substance that can be found in sea and fresh water, while the Calcium is one of the main sources of hardness in the water. It is also concluded that, the water is of the Magnesium bicarbonate type. Similarly, and based on Durov-diagram (Fig. 3), groundwater samples are of the Mg-HCO3 type with simple dissolution or linear mixture. The Cl^- concentration is about 40–80 mg/l which is a source for major EC ranging between 6.6 and 8.2 mS/m (25 °C) and TDS ranging between 100 and 250 mg/l with the dominance of Na + K ranging between 20 and 60 mg/l.

Therefore, it is clear that the Na^+, K^+ and Cl^- ions have dominant effect on the EC, TDS and SAR in the solution. For this, Gibbs- (Fig. 4) and Wilcox-diagrams (Fig. 5) are also used. Based on the results of Gibbs-diagram, in all samples the dominance of either Na^+ or the Cl^- ions in the solution are linked to the rocks in the region. Accordingly, due to the alluvium deposits and soluble rock contents, the source of such ion dominance is probably the natural rock decomposition and chemical reactions. In this respect, the dominance of Na^+ (Fig. 4a) depicts more variation in TDS compared to the Cl^- (Fig. 4b) and therefore it might be the main reason for the EC in the solutions. Given in Fig. 5a and b, respectively the SAR and Na^+ are evaluated against EC. According to the Wilcox-diagram (Fig. 5a), all the groundwater samples are categorized as C1-S1 type (low salinity-low sodium) which are excellent or good (Fig. 5b) for irrigation uses.

3.3 Local Situations

Since a holistic approach provides a unilateral result on the samples, each observation well was also evaluated based on specific ions and exclusive condition in each observation well. For this, Fig. 6 shows the Stiff-diagram of each observation well. Based on the concentration of ions (meq/l), results show the dominance of cations or anions in the solution. In this respect, and in the samples provided for Narlidere, Adakoy, Cayirkoy, Kursunlu, Yenice, Karasil, Soloz, and Yenikoy the dominance of HCO_3^- as an anion is clear. It might be due to the Calcite, Dolomite, and/or Gypsum dissolution or even caused by Silicate weathering. In Bolge Trafik however, the

Fig. 4 Gibbs-diagram considering the dominance of **a** Na + K and **b** Cl ions

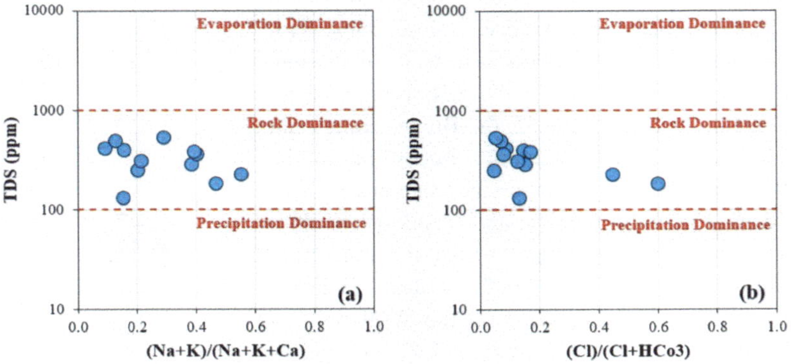

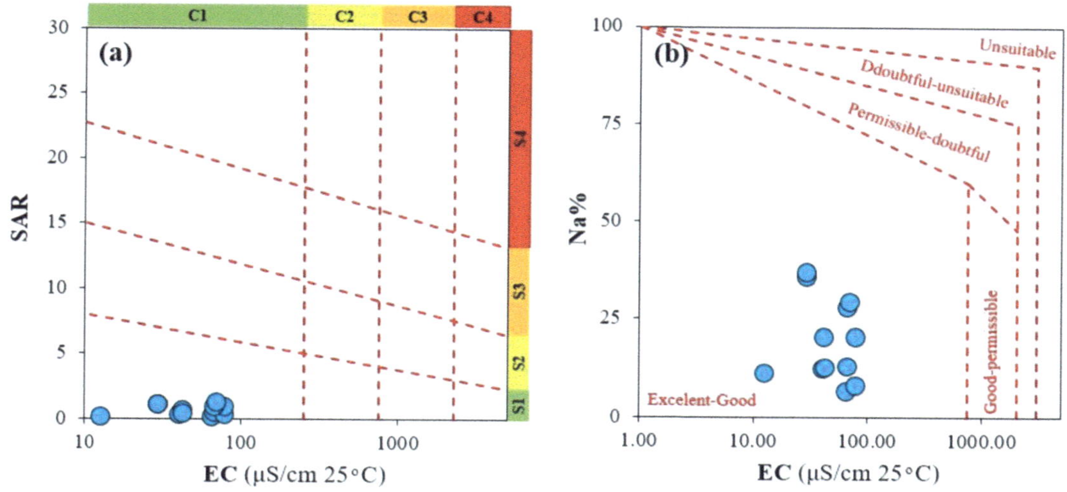

Fig. 5 Wilcox-diagram of the samples based on the relationship between **a** *SAR*-EC and **b** Na%-EC

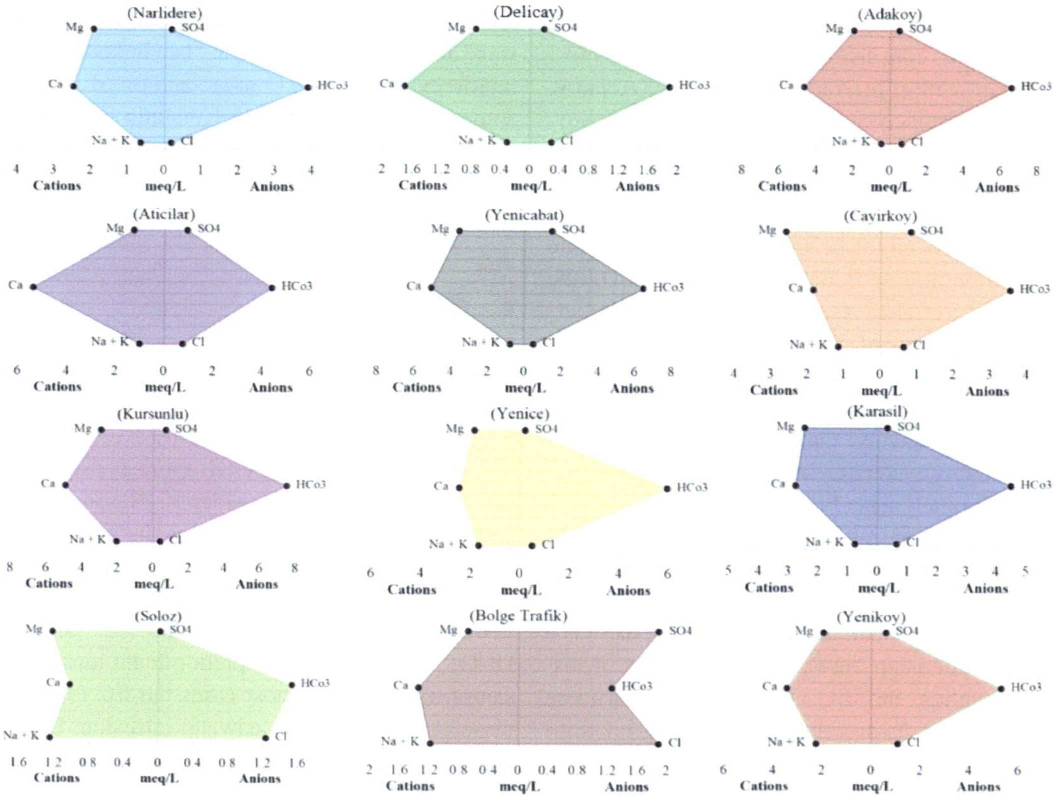

Fig. 6 Stiff-diagram of the samples

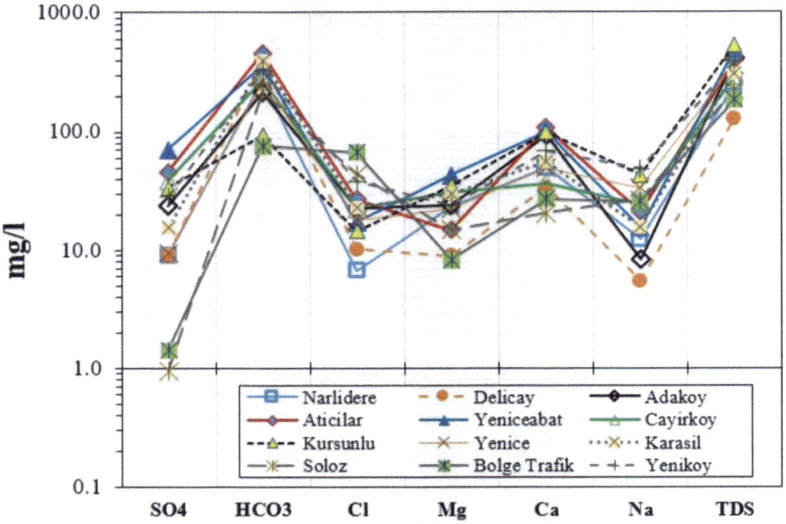

Fig. 7 Schuller classification diagram

most dominant anions are SO_4^{2-}, and Cl^- which may indicate to dispersion of pollution or Gypsum dissolution. On the other hand, and when cations are concerned, the Ca^{2+} is the most dominant cation in Narlidere, Delicay, Adakoy, Aticilar, Yeniceabat, Kursunlu, and Yeni koy. This might have linked to the dissolution of Calcite, Dolomite, and/or Gypsum. In Soloz however, the dominance of Mg^{2+}, Na^+, K^+ is higher which might be the product of Dolomite dissolution. In Fig. 7, the Schuller-diagram of the samples are also given. Based on Schuller-diagram amount of SO_4^{2-}, HCO_3^-, Cl^-, Mg^{2+}, Ca^{2+}, Na^+, and TDS are lower than 200 mg/l, 180 mg/l, 250 mg/l, 70 g/l, 100 g/l, 100 g/l, and 400 g/l respectively, that can be recognized as a good quality water for irrigation and drinking.

Additionally, in Fig. 8, the source of EC based on the ionization of Na^+ and Cl^- is given. In Delicay (Fig. 8b), Adakoy (Fig. 8c), Aticilar (Fig. 8d), Yeniceabat (Fig. 8e), Yenice (Fig. 8h), Karasil (Fig. 8i), and Yenikoy (Fig. 8l), there is a direct relationship between concentration of dissolved solids and EC. However, in Narlidere (Fig. 8a) and Yeniceabat (Fig. 8f) the relationship between Na and Cl concentration is reverse and that can be linked to the presence of inorganic materials such as alkalis, sulfides and carbonate compounds. Surprisingly, in Kursunlu (Fig. 8g), Soloz (Fig. 8j) and Bolge Trafik (Fig. 8k) anomalies were observed, and diverse relationship in Na-EC and Cl-EC relationship was depicted which might be linked to the interference of other ions and/or disturbing anthropogenic activities. In Kursunlu however, it might be linked to the presence of geothermal springs and/or ionization of other salts in the environment.

4 Final Discussion

Based on the results, it can be concluded that the groundwater quality in the region is considerably good for irrigation purposes. However, if drinking water quality is affected, more conservative measures should be considered. Based on the chemical analysis, the source of the cations and anions in the area could be related to the dissolution or weathering of the rocks in the region, with evidence of anomalies in the samples. HCO_3^- and Ca^{2+} are the predominant ions in the samples, and yet in most cases the EC is due to dissolution of NaCl or seawater intrusion. Due to the recent concerns regarding the emerging water deficit and wide spread droughts, the groundwater consumption became an active option for the authorities. But the analysis showed that the

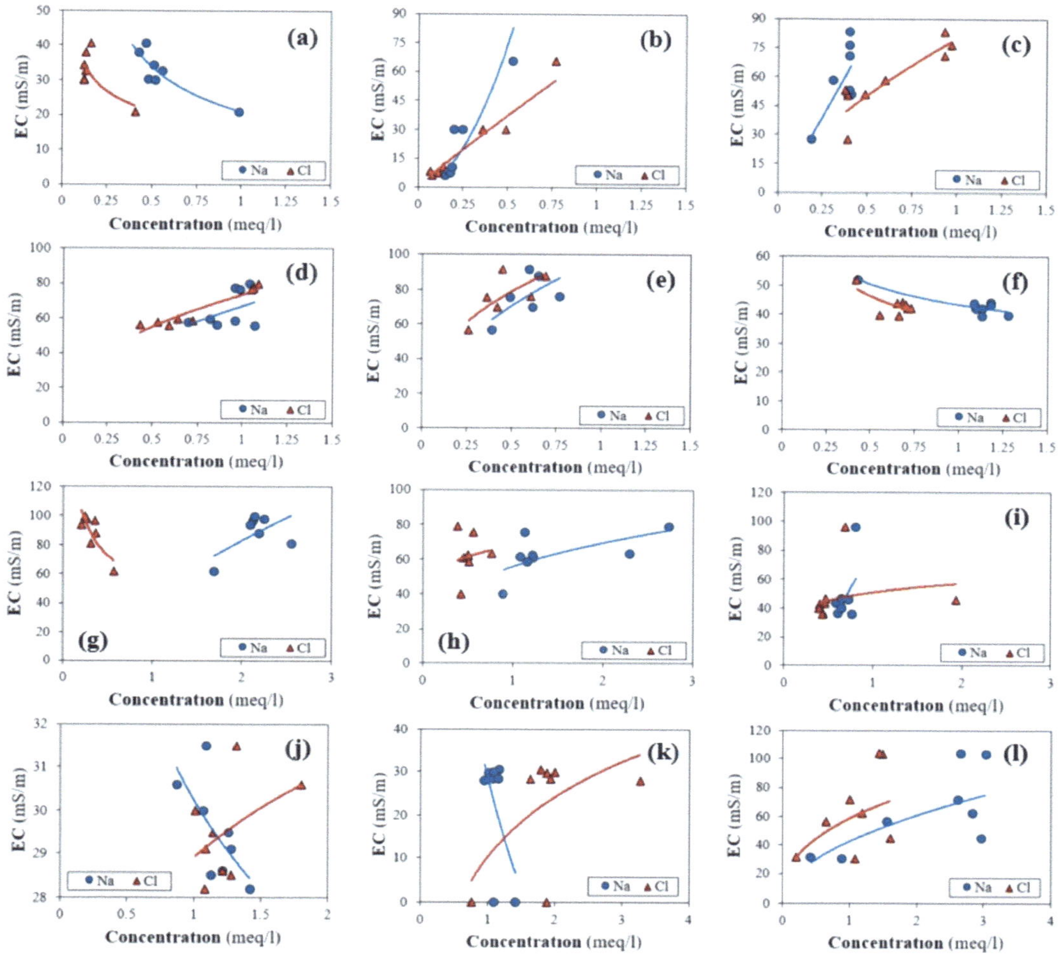

Fig. 8 Relationship between concentration of Na and Cl with EC at **a** Narlidere, **b** Delicay, **c** Adakoy, **d** Aticilar, **e** Yeniceabat, **f** Cayirkoy, **g** Kursunlu, **h** Yenice, **i** Karasil, **j** Soloz, **k** Bolge Trafik, and **l** Yenikoy

sources of some ions in the samples are rocks. Keeping in mind that the formation of the geological layers in the region is of the alluvium type, it is important to be aware of the formation of sinkholes and sudden ground collapses. The rapid groundwater consumption may also trigger groundwater pollution originated from more than 17 industrial zone located at the catchment. Extended drought periods as direct consequences of climate change may also amplfy the problem by lowering the groundwater table and sea water intrusion in coastal areas. It is recommended that future studies include an assessment of the hydrogeology and geohydrology of the region. Since groundwater use is via deep wells, additional studies are needed to determine the age of the groundwater using carbon dating and isotope hydrology, as well as a precise definition of the groundwater potential through drought analyses and hydrological-ecological modelling.

Groundwater data acquired from 12 observation wells in the Bursa region were used for analysis. The main characteristics of the groundwater was determined and evaluated using field-based observational dataset. Subsequently, a holistic and case-based approach was employed to assess the groundwater quality based on the collected samples. The proposed approach described in this study adopts piper-, Durov-, Gibbs-, and Wilcox- diagrams to provide holistic overview of the groundwater quality. Further, a case-based approach was conducted using Stiff-diagram, Schuller classification, and EC. It was concluded that the groundwater quality in Bursa region is generally good for agricultural purposes, and partially for drinking use. However, the results highlighted some anomalies in the water samples which signifies the need for more comprehensive spatiotemporal analysis of the samples, to further investigate the source of these anomalies. This study also conclude that a robust groundwater monitoring strategy would be necessary to ensure sustainable abstraction of groundwater in the region and understand the effects of climate change on the quantity and quality of groundwater. Such monitoring can help to prevent the formation of sinkholes due to excessive use of groundwater.

Acknowledgements This study was part-funded and facilitated through British Council NEWTON Fund grant —Project No. 528205534, "Knowledge Transfer and Capacity Building for Sustainable Water Pollution Management in Bursa, Turkey". Authors would like to thank Bursa Regional Directorate of the State Hydraulic Work (DSİ) for providing the data used in this study.

References

1. Saeedi M, Abessi O, Sharifi F, Meraji H. Development of groundwater quality index. Environ Monit Assess. 2010;163(1–4):327–35.
2. Derby NE, Korom SF, Casey FX. Field-scale relationships among soil properties and shallow groundwater quality. Groundwater. 2013;51(3):373–84.
3. Ghesquiere O, Walter J, Chesnaux R, Rouleau A. Scenarios of groundwater chemical evolution in a region of the Canadian Shield based on multivariate statistical analysis. J Hydrol: Reg Stud. 2015;4:246–66.
4. Akpah FA, Onwuka SO, Oha IA. Hydrogeochemical investigation and characterization of shallow groundwater within Ankpa Town, North Central Nigeria. Int J Phys Sci. 2017;12(5):60–73.
5. Karandish F, Hoekstra AY, Hogeboom RJ. Groundwater saving and quality improvement by reducing water footprints of crops to benchmarks levels. Adv Water Resour. 2018;121:480–91.
6. Pacheco Castro R, Pacheco Avila J, Ye M, Cabrera Sansores A. Groundwater quality: analysis of its temporal and spatial variability in a karst aquifer. Groundwater. 2018;56(1):62–72.
7. Hillel N, Wine ML, Laronne JB, Licha T, Be'eri-Shlevin Y, Siebert C. Identifying spatiotemporal variations in groundwater-surface water interactions using shallow pore water chemistry in the lower Jordan river. Adv Water Resour. 2019;131: 103388.
8. Chegbeleh LP, Akurugu BA, Yidana SM. Assessment of groundwater quality in the Talensi District, Northern Ghana. Sci World J. 2020.
9. Balderer W, Greber E, Imbach T, Rauert W, Trimborn P, Guler S. Environmental isotope study of thermal, mineral and normal groundwater within the Bursa and Kuzuluk/Adapazari areas of northwestern Turkey. In: Proceedings isotope techniques in water resources development 1991, 11–15 March 1991. IAEA and UNESCO. Vienna, Austria; 1992. p. 720–24.
10. Sen E. Nitrate concentration with UV technic on spring water in Nilüfer and Ayvalı Basin of Bursa. In: Uludag University, I. Uludag environmental engineering symposium, June; 1996. p. 24–6.
11. Imbach T. Deep groundwater circulation in the tectonically active area of Bursa, Northwest Anatolia, Turkey. Geothermics. 1997;26(2):251–78.
12. Baba A, Tayfur G. Groundwater contamination and its effect on health in Turkey. Environ Monit Assess. 2011;183(1–4):77–94.
13. Simsek C, Yavuz AB, Elci H, Elci A, Gunduz O. Waste disposal on karstic terrain: a case study from the ancient marble quarries in Iznik (Nicaea), Turkey. Geosci J. 2011;15(3):339.
14. Akar Tarim U, Gurler ORHAN, Akkaya G, Kilic N, Yalcin S, Kaynak G, Gundogdu O. Evaluation of radon concentration in well and tap waters in Bursa, Turkey. Radiat Protect Dosimetry. 2012;150(2):207–12.
15. Fidan R, Aras E, Vaheddoost B. Evaluation of monthly groundwater level at Bursa, Turkey. J Innov Sci Eng (JISE). 2019;3(1):38–46. http://jise.btu.edu.tr/tr/download/article-file/742418
16. Kottek M, Grieser J, Beck C, Rudolf B, Rubel F. World map of the Köppen-Geiger climate classification updated. Meteorol Z. 2006;15(3):259–63.
17. Singla R, Soni S, Kulurkar PM, Kumari A, Mahesh S, Patial V, Padwad SP, Yadav SK. In situ functionalized nanobiocomposites dressings of bamboo cellulose nanocrystals and silver nanoparticles for accelerated wound healing. Carbohyd Polym. 2017;155:152–62.
18. Turkish State Meteorological Service (TSMS). Accessed 11 Dec 2020. https://www.mgm.gov.tr/veridegerlendirme/il-ve-ilceler-istatistik.aspx?m=BURSA
19. Sadashivaiah C, Ramakrishnaiah CR, Ranganna G. Hydrochemical analysis and evaluation of groundwater

quality in Tumkur Taluk, Karnataka state, India. Int J Res Public Health. 2008;5(3):158–64.
20. Hagras MA. Water quality assessment and hydrochemical characteristics of groundwater in Punjab, Pakistan. Int J Res Rev Appl Sci. 2013;16(2):254–62.
21. Todd DK, Mays LM. Groundwater hydrology, 3rd edn. John Wiley & Sons, Inc., New York; 2005. p. 636.
22. Szabolcs I, Darab C. The influence of irrigation water of high sodium carbonate content of soils. Proc Int Congress Trans 2:803–812. Or proceedings of 8th ISSS, Trans. 1964;2:802–12.
23. Singh AK, Mondal GC, Tewary BK, Sinha A. Major ion chemistry, solute acquisition processes and quality assessment of mine water in Damodar valley coalfields, India. Abstract of the Mine Water Conf., Pretoria, South Africa, Proceedings ISBN Number 978-0-9802623-5-3, produced by Document Transformation Technologies cc; 2009. p. 267–76.
24. Brindha K, Neena Vaman KV, Srinivasan K, Sathis Babu M, Elango L. Identification of surface water-groundwater interaction by hydrogeochemical indicators and assessing its suitability for drinking and irrigation purposes in Chennai, south India. Appl Water Sci. 2014;4:159–74.
25. Kaçmaz H, Eran Nakoman M. Shallow groundwater and cultivated soil suitability assessments with respect to heavy metal content in the Köprübaşı U mineralization area (Manisa, Turkey). Bull Environ Contam Toxicol. 2010;85:37–41.
26. Domenico PA, Schwartz FW. Physical and chemical hydro-geology. New York: Wiley; 1990.
27. Hussain I, Hussain J. Evaluation of irrigation water quality of the village situated near river Kothari, Rajasthan (India). Pollut Res. 2004;23:561–4.
28. Gupta SK. Variations of water table in Yamuna Grainage basin of Haryana—implications and management strategies. Paper presented at the Seminar on Strategies for Irrigation Water Management, Patna; 1983.
29. Kelly WP. Permissible composition and concentration of irrigated waters. In: Proceedings of the A.S.C. F, 607; 1940.
30. Matthess G. The properties of groundwater. New York: Wiley; 1982.
31. Soltan ME. Characterization, classification, and evaluation of some groundwater samples in upper Egypt. Chemosphere. 1998;37:735–45.
32. Soltan ME. Evaluation of groundwater quality in Dakhla Oasis (Egyptian Western Desert). Environ Monit Assess. 1999;57:157–68.
33. Horton RK. An index number system for rating water quality. J Water Pollut Control Feder. 1965;37 (3):300–6.
34. World Health Organization (WHO). Guidelines for drinking-water quality: first addendum to the fourth edition; 2017. p. 307–442
35. Ayers RS, Westcot DW. Water quality for agriculture, vol. 29. Rome: Food and Agriculture Organization of the United Nations; 1985. p. 174. https://www.fao.org/3/t0234e/t0234e00.htm
36. Piper A. A graphic procedure in the geochemical interpretation of water-analyses. Trans Am Geophys Union. 1944;25(6):914–28.
37. Durov SA. Classification of natural waters and graphical representation of their composition. Doklady Akademii Nauk, USSR. 1948;59(1):87–90.
38. Gibbs RJ. Mechanisms controlling world water chemistry. Science. 1970;170(3962):1088–90.
39. Wilcox L. Classification and use of irrigation waters (No. 969). US Department of Agriculture; 1955.
40. Freeze RA, Cherry JA. Groundwater. Englewood Cliffs, New Jersey: Prentice-Hall. Inc.; 1979. p. 604.
41. Hounslow A. Water quality data: analysis and interpretation. CRC Press; 1995.
42. Stiff HA Jr. The interpretation of chemical water analysis by means of patterns. J Petrol Technol. 1951;3(10):15–23.
43. Schoeller H. La classification géochimique des eaux. IASH publication. 1964;64:16–24.
44. Patnaik KN, Satyanarayana SV, Rout SP. Water pollution from major industries in Paradip area–a case study. Indian J Environ Health. 2002;44(3):203–11.

Arsenic Problem in Groundwater Sources and Treatment Technologies, Indian Experiences

Modhurima Misra and Amitava Bandyopadhyay

Abstract

Groundwater arsenic (As) contamination is prevalent in different parts of the world to varying extents; however, its severity in India is a major concern, which has garnered significant global attention. The nature of the problem is considered to be geogenic. The identification of this crisis took place not too long ago; still it has managed to acquire the form of an endemic in the country. Millions of individual are getting vulnerable to arsenicosis on getting exposed to As polluted water. The concentration of the metalloid in the Indian groundwater reserves is often found to be as high as few thousand ppm, which is way beyond10 µg/L (10 ppb), the permissible limit of the metalloid in drinking water, set by WHO. Viable options for mitigating this problem and supply clean water to the affected populace include either identification of alternative As-free aquifers or developing new technologies for the treatment of As contaminated water. The major arsenic treatment units (ATUs) or arsenic removal plants (ARPs), installed in different As affected areas all over the country, are based on several conventional principles namely adsorption, oxidation, precipitation-filtration, coagulation-flocculation, electro-coagulation and membrane separation. Implementation of a single technology or devising an end-of-pipe solution for the remediation of As contamination seem far-fetched due to nature of the aquifers, socio-economic conditions and geo-political issues. The review aims at understanding the current scenario of the problems associated with As contaminated groundwater with respect to the Indian peninsula. It also emphasizes on the different technologies implemented for As removal, in terms of sustainability and economic feasibility.

Keywords

Arsenic · Groundwater contamination · Indian scenario · Removal technologies · Sludge disposal

1 Introduction

Ground water accounts for about 30.1% of the entire fresh water resources on the earth. It is an indispensable natural resource supporting the domestic and agricultural needs of the ever increasing global population. Arsenic (As) poisoning of ground water is threatening the well-

M. Misra
Department of Biotechnology, Heritage Institute of Technology, Chowbaga Road, Anandapur, PO, India

A. Bandyopadhyay (✉)
Department of Chemical Engineering, University of Calcutta, 92, APC Road, Kolkata 700 009, India
e-mail: abchemengg@caluniv.ac.in

A. M. Bahadir et al. (eds.), *Hydrology and Urban Water Supply*, Water and Wastewater Management,
https://doi.org/10.1007/978-3-031-72589-0_14

being of mankind with its deleterious effects for the past four decades [1, 2]. Different parts of the world have already been exposed to this menace; however, in Bangladesh and in India, it has manifested itself with unprecedented severity. The catastrophic effects and magnitude of the problem have rendered the highly affected Indian states to become hotspots of the greatest groundwater calamity. In different regions of India, the concentration of As in groundwater sources has been found to be as high as few thousand ppm, which is quite exorbitant with respect to 10 μg/L (10 ppb), the permissible limit of the metalloid in drinking water, set by WHO [3, 4].

Arsenic is categorised as a metalloid, belonging to Group VB and Period 4 of the periodic table. It is universally distributed on the earth's surface and has been exploited both as a medicine and poison since ancient times. This metalloid finds use in metallurgical science, therapeutics, agriculture and technology; however, its notoriety as one of the most toxic chemicals is quite prominent, hence it has been categorized as non-threshold Group 1 carcinogen to humans by the International Agency for Research on Cancer (IARC) [5]. It is a mutagen educing cytotoxic and genotoxic effects through the generation of excess reactive oxygen species (ROS). It affects DNA synthesis, damage and repair, alters signal transduction pathways and interferes with cell cycle and apoptosis. Long term exposure of humans to inorganic As through water and food can cause arsenicosis, a collective term used to describe As related health issues which include skin pigmentation, cancers (skin, bladder, lung, liver and kidney), hypertension, respiratory diseases, cardiovascular problems, neurological and reproductive disorders [5–7]. Arsenicosis have far reaching socio-economic consequences as well.

The introduction of As into the sub-surface environment can occur by natural and anthropogenic means. In the nature, As occurs mainly in association with the sulphide ores of different elements such as iron (Fe), lead (Pb), nickel (Ni), cobalt (Co), copper (Cu), antimony (Sb) and silver (Ag). Volcanic eruptions contribute to the major As influx in the nature while metal smelting, combustion of low-grade brown coal and rampant use of arsenical pesticides are the principal human interventions behind the addition of As in the environment [8]. The technologies for As removal are mostly based on some conventional principles namely adsorption, coagulation-flocculation, electro-coagulation, co-precipitation, oxidation and membrane separation; however none could satisfactorily serve as an ultimate solution in mitigating the excessive As load from the groundwater sources [2]. The efficacy of new innovative technologies and combined techniques are being reviewed with great interest, but their implementation in field applications remain a far-fetched reality.

The review aims at understanding the current scenario of the problems of contaminated groundwater with As and treatment technologies with respect to the Indian peninsula. The first half of the chapter mainly deals with the geochemistry of the metalloid, sources, toxic effects and extent of the problem in the subcontinent. The second half emphasizes on the different technologies implemented for As removal, in terms of sustainability and economic feasibility. Technological solutions for switching over to alternative aquifers for the supply of clean water are also discussed briefly. Lastly, ongoing practices and recent innovations for the safe disposal of As containing sludge from various treatment plants are mentioned with concluding remarks to provide a holistic idea about the remediation plans.

2 Permissible Limit of Arsenic in Drinking Water—A Pragmatic Solution or a Debatable Choice

Over the years, different regulatory bodies have faced several challenges in setting up the maximum permissible limit for As in drinking water, as well as the criteria governing the selection. However, after quite a few changes, the current acceptable value of As in potable water (based on an estimated daily 2L direct water consumption), according to the WHO guidelines, stands at 10 μg/L. It is a provisional value which can be further modified based on the nutritional status, health concerns, and As removal capabilities of a particular region [9]. The United States Environmental Protection Agency has set two different standards for As; a non-enforceable maximum contaminated level goal (MCLG) of zero and enforceable maximum contaminant level (MCL) of 10 µg/L in drinking water [10]. Considering the conditions of the Indian sites seriously affected with As, the Bureau of Indian Standards (BIS) in its recent revision has fixed the standard for As in drinking water as 10 µg/L in line with the value recommended by WHO. However, the permissible limit is elevated to a value of 50 µg/L as prescribed by the BIS, in the absence of alternative As-free drinking water sources, due to resource constraints, lack of awareness, infrastructures and treatment techniques to drastically reduce such an obnoxious concentration [11].

3 Arsenic—Geochemistry and Occurrence

Arsenic is a redox sensitive trace element which ranks 20th in terms of abundance on the surface of the earth. It is prevalent in the environmental matrices in various chemical forms, in which it exists in 4 different oxidation states (Table 1). Metallic or elemental As [As(0)] is found to occur rarely [12]. In groundwater conditions, it mainly occurs as oxy-anions: As(III) (arsenite) and As(V) (arsenate), within the pH range 6–9. Both the species can co-exist in either oxidizing or reducing aquatic environments due to slow redox transformations, however As(III) is found to be predominant under anaerobic or anoxic conditions such as groundwater or deep lake sediments; whereas As(V) is thermodynamically stable in aerobic or oxic environments [13]. Geology, pH, Eh (redox potential) and groundwater environment are the factors responsible for the speciation and mobility of As. Over 200 minerals on the earth have been identified, with arsenic as one of their primary component. They include metal arsenates (60%), sulphides and sulphosalts (20%), arsenides and oxides (20%). Arsenopyrite (FeAsS) is the most abundant one, usually occurring under anaerobic conditions [14]. Different rock forming mineral deposits such as sulfide, carbonate, oxide, silicate and phosphate contain arsenopyrite in varying concentrations.

In India, As pollution in groundwater, is prominent mostly in the shallow alluvial aquifers of the Ganga–Brahmaputra basin (90%), lying 15–70 m below the surface, with few exceptional occurrences (10%) in the hard rock terrain [2, 8]. These aquifers contain late Quaternary sediments of Holocene age, bearing organic matter, clay, slits and micaceous sand, eroded from the Himalayan belt by the Ganga and its tributaries. The As content in the sedimentary aquifers

Table 1 Speciation of arsenic [8]

Sl. no	Properties	Details
1	Allotropes	Black, gray, and yellow
2	Oxidation states	−III, +III, 0, and +V
3	predominant As(III) and As(V) species	As (+III) − $H_3AsO_3, H_2AsO_3^-, HAsO_3^{2-}, AsO_3^{3-}$ As (V) − $H_3AsO_4, H_2AsO_4^-, HAsO_4^{2-}, AsO_4^{3-}$
4	Chemical compounds	Organic—monomethylarsonous acid [MMA(III); CH_5AsO_3] monomethylarsonic acid [MMA(V); $CH_3AsO(OH)_2$], dimethylarsinous acid [DMA(III);$(CH_3)_2AsO_2H$] dimethylarsinic acid [DMA(V); $(CH_3)_2AsO(OH)$], arsenobetaine [AsB; $(CH_3)_3AsCH_2COOH$], arsenocholine [AsC; $(CH_3)_3AsCH_2CH_2OH$], arsenosugars (AsS), arsenolipids Inorganic—arsine (−III),arsenite (+III), arsenic (0), and arsenate (+V)
5	Minerals	Arsenopyrite (FeAsS), cobaltite (CoAsS), löllingite ($FeAs_2$), orpiment (As_2S_3), proustite (Ag_3AsS_3), realgar (As_4S_4), seligmannite ($PbCuAsS_3$)

surrounding ore deposits, with elevated levels of pyrites, iron oxides and hydrous metal oxides and mountains with convergent plate boundaries, is found to be quite high [15, 16]. The level of As in different aquifers varies due to its total sedimentary concentration, redox condition of aquifers, the sorption capacity and size of sediment grains, the duration of groundwater flushing and its extraction. The low As concentration in deeper Pleistocene aquifer can be attributed to elongated duration of effective flushing under larger hydraulic heads with respect to the young Holocene sediments. The extent of As contamination in dug wells is less owing to the aerobic conditions, high flushing and inflow of groundwater [17].

4 Sources and Mobilization of Arsenic in Groundwater

The metalloid occurs as a natural constituent of soil and different rock forming minerals. It is also found frequently in the deposits of various elements such as cadmium (Cd), copper (Cu), lead (Pb), iron (Fe), mercury (Hg), nickel (Ni), phosphorus (K), sulphur (S), gold (Au), silver (Ag), zinc (Zn), etc. [18]. Several anthropogenic and biogeochemical activities are responsible for the introduction and propagation of As into the groundwater environments. The natural causes include volcanic eruptions, atmospheric emissions, geothermal activities, organic deposition, forest fires, leaching of weathered minerals, exudates from wind-blown dusts and vegetation covers and biological processes [2, 8, 19].

The actual source of As in groundwater in the Ganga–Brahmaputra Basin and Bengal Delta Plains is yet to be fully elucidated. The recent and most accepted theory advocates the release and subsequent transportation of As from the Himalayan fluvial sediments [20]. Majority of the state of West Bengal is exposed to severe arsenicism due to alluvial deposits from the Ganga river [21]. Downstream water sources are found to get contaminated by As-enriched water, originating from the hydrothermal springs. Excessive contamination in some parts of the Bengal Delta Plain is caused by the breakdown of As traps consisting of clastic granular iron with precipitates of As-rich secondary oxides and hydroxides [22]. The reducing environment caused by the rise in sea level in the Holocene epoch contributed greatly to the As contamination of the Bengal Basin. Different isolated geological sources of As have been identified such as Bihar mica belt, Darjeeling Himalayan belt, Gondwana coal seams of Rajmahal basin, Son valley gold belt and pyrite-bearing shale of Proterozoic Vindhyan [23, 24].

Among anthropogenic activities, coal combustion is one of the major factors, particularly in north-east India, that exacerbate the As contamination in groundwater aquifers, through leaching of the metalloid from coal-ash tailings. During coal combustion, As is emitted when arsenic trioxide (As_4O_6) is volatilized. It gets incorporated in the water reservoirs after condensing in the flue system [13, 24]. Mining of sulphide ores and mineralization of FeAsS are mainly responsible for high As concentration in various parts of Rajasthan and Karnataka respectively. Digging of wells and indiscriminate exploitation of groundwater make the prevalent condition aerobic which oxidizes As-sulphides to highly water soluble compounds and further increases the As concentration. Boro irrigation is responsible for As contamination in crops (rice) and potable water in West Bengal [25]. Other anthropogenic interferences which contribute to the addition of As in the water courses include unbridled use of arsenical pesticides and insecticides in agriculture, electroplating, smelting, manufacturing of electronic goods, cement manufacturing, wood preservation, paper production, burning of fossil fuels, industrial pollution, waste processing and disposal etc. [2, 13, 18].

Extensive research is being carried out to obtain details about the nature of aquifers and understand the mechanisms behind the mobilization of As in the groundwater systems. Rock-water interactions and biochemical reactions triggers As propagation from the solid phase to the groundwater aquifer systems. The transport of As into the aquifers is governed by certain factors such as, pH, Red-ox potential (Eh), speciation of As, concentration of competing ions, adsorption-desorption, precipitation and dissolution, bio-transformation, etc. [14]. Numerous theories and hypotheses have been proposed to elucidate the physico-chemical processes behind generation of As rich groundwater, however none could satisfactorily explain the phenomena owing to some inherent limitations and discrepancies. The possible mechanisms of As mobilization suggested so far include reductive dissolution of pyrites, Fe oxides or other metal oxyhydroxides, oxidation of As-sulphides, alkali desorption under alkaline and aerobic conditions, desorption from oxides and hydroxides, and microbial activities [26, 27].

5 Metabolism, Toxic Effects and Mechanism of Arsenic Poisoning

The variation in physico-chemical properties of different arsenicals is primarily responsible for their extent of toxicities [28]. The mobility and toxicity of arsenite is about 60 times more than arsenate. As(V) is poorly soluble in water, thus its bioavailability is low. Inorganic forms of As are reported to be 100 times more toxic than the organic ones. However, it has been recently reported that the trivalent methylated As species namely, MMA(III) and DMA(III) are highly reactive and more toxic than As(III). The degree of toxicity of different species of As follows the order −As(V) < MMA(V) < DMA(V) < As (III) < MMA(III)≈ DMA(III) [29–31].

Human beings become victims of As poisoning through occupational exposure or intake of contaminated water and food. Majority of the inorganic As uptake by the body occurs in the form of As(III), which is similar in structure to glycerol and diffuse into the cells through aquaglyceroporins [8]. As(V) is a phosphate analog, which is transported inside the cells in small quantity via phosphate transporters, through an energy-driven process and is instantly reduced to As(III). Water soluble As(III) is metabolized in the liver after getting readily absorbed from the gastro-intestinal tract [32, 33]. Inside the cells, protein bound As(III) is further metabolised into methylated intermediates [MMA(III) and DMA(III)] and end products [MMA(V) and DMA(V)] through successive reductive methylation with the help of the enzyme arsenicmethylatransferase (As3mt). Glutathione (GSH) and S-adenosylmethionine (SAM) act as the cofactor and methyl donor for the reaction respectively. The metabolized end products such as MMA(V) and DMA(V) are less than other arsenicals and are eliminated out of the body mainly through urine [33–35].

The degree of As poisoning is determined by duration of exposure, concentration of the metalloid, proneness of an individual, age, genetic constitution, nutritional status and pregnancy [33, 36]. A person may become susceptible to acute or chronic toxicity when exposed to high As levels. The symptoms of acute toxicity include weakness, loss of appetite, dehydration, vertigo, nausea, vomiting and diarrhoea [35]. Long term exposure to inorganic arsenicals may result in severe chronic health issues, affecting almost all the organ systems of the body (Fig. 1). It can induce blood vascular dysfunction, haematological abnormalities (anaemia, leukopenia and extravascular haemolysis), endocrine disorders like diabetes mellitus (Type I and II), hyperthyroidism, various developmental abnormalities and cancers of bladder, colon, kidney, liver, lung and skin [33, 34]. Arsenic is a potent carcinogen, teratogen and mutagen and its actions as a toxicant or cancerous agent depend on the distribution within the tissues and retention, which is further regulated by the bioconversion of its methylated intermediates [34].

Arsenicosis can be categorised into preclinical, clinical, internal complications and malignancy stage. In the preclinical stage, methylated As metabolites can be detected in the nails, hair and urine of infected individuals, after ingestion of contaminated water and food. However, the individuals do not show any clinical symptoms. The clinical stage is characterized by the appearance of dermatological symptoms such as melanosis, keratosis, non-pitting edema, conjunctival congestion etc. [2]. The toxicity of As can be confirmed by analysing nails, hairs and

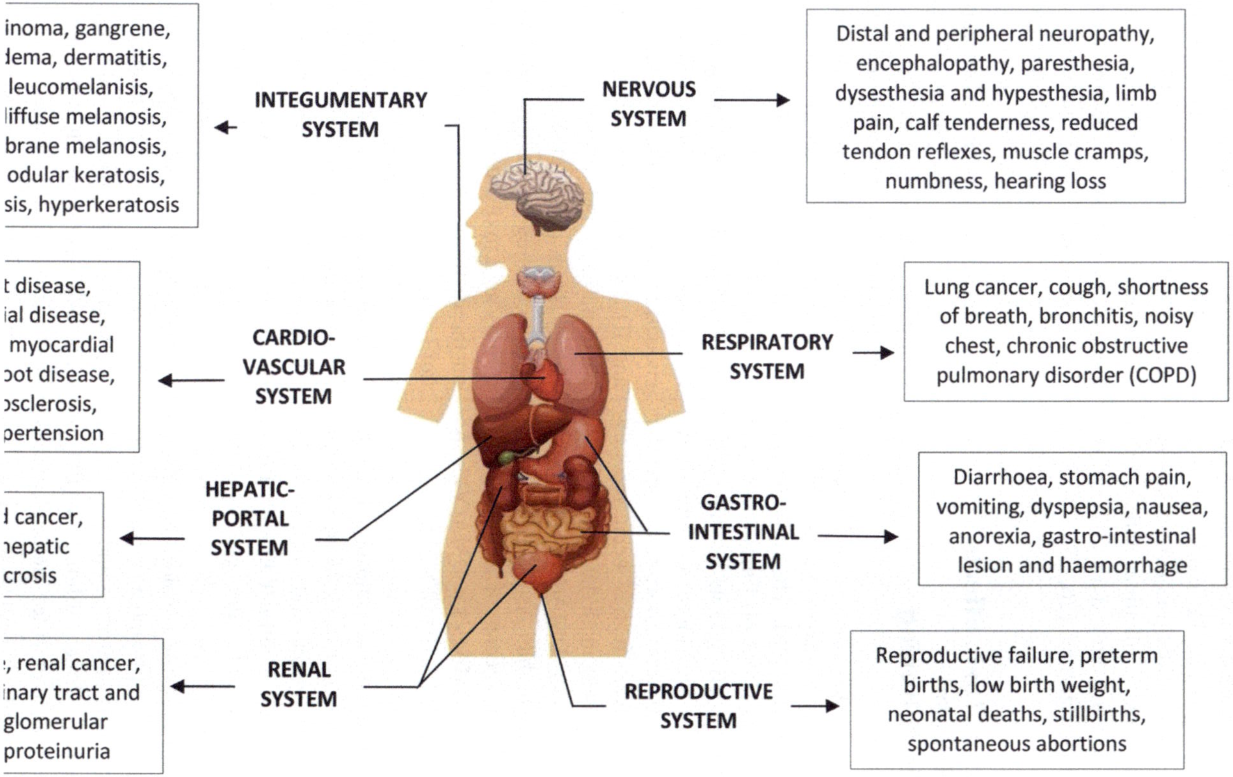

Fig. 1 Toxic effects of arsenic poisoning in different human organs [39, 40]; Image of the "Human body internal organ" is Designed by Freepik (www.freepik.com) [45]

skin scales of affected individuals in this stage. Patients with severe dermatological disorders have been predominantly diagnosed in the states of West Bengal, Bihar, Jharkhand and Uttar Pradesh of India [37]. Cutaneous manifestations are considered to be the most significant characteristic feature of patients suffering from As poisoning. Generally, diffuse melanosis, characterized by skin darkening on the palms and body, is the one of the earliest symptoms followed by spotted melanosis on the chest and limbs. The appearance of skin lesions is indicative of acute internal damage which occurs in the internal complications stage [38]. Different organs such as lungs, liver brain, eyes and muscles are affected and clinical symptoms including organ dysfunction, asthmatic bronchitis, enlargement of spleen and liver begin to manifest in the patients. The malignancy stage is characterized by cancers of different organs. Till date, no medicinal cure could be recommended for the treatment of As poisoning, only intake of vitamin enriched nutritious food and As-free water can help to improve the situation and can be considered as preventive measures [39, 40].

The mechanism of As induced toxic effects and health disorders have been comprehensively explored over the past few decades, once the expanse of the problem was realised along with the impending danger associated with it. Inside the body, As can trigger carcinogenesis by altering a number of cellular functions such as DNA repair and methylation, cell cycle, cell growth, proliferation and differentiation [36]. It can result in amplification of genes and chromosomal abnormalities and compromises with different signal transduction pathways e.g., MAPK, p53 and Nrf2-mediated redox signalling pathway [41, 42]. Increased level of As inside the cell promotes excessive ROS production by inducing the enzyme NADPH oxidase, which in turn damage cellular components like proteins and lipids and destroys the structure and function of mitochondria. Reactive oxygen species play a significant role in eliciting both cytotoxicity and genotoxicity. Formation of ROS by As give rise to oxidative stress, mutagenesis as well as other non-cancerous cytotoxic ill-effects including, apoptosis, altered cell signalling, increased cytokine production, inflammation etc. [43]. It has been found that trivalent forms of both inorganic and organic As distort glucose metabolism pathway by inhibiting α-ketoglutarate and pyruvate dehydrogenases, ultimately resulting in diabetes [44].

6 Contamination of Groundwater by Arsenic—Extent of the Problem in the Indian Subcontinent

Arsenic exists in excessive concentration in the groundwater aquifers of the fluvio-deltaic plains of the rivers Ganga and Brahmaputra or the Indo-Gangetic plains. In the northern states of the country, mostly the alluvial aquifers are affected, whereas in the case of Chhattisgarh and Karnataka, As is found in the hard rock aquifers [2, 40]. The affected states in the country include Gujarat, Rajasthan, Himachal Pradesh, Punjab, Haryana, Uttar Pradesh, Bihar, Jharkhand, West Bengal, Assam, Tripura, Manipur, Mizoram, Meghalaya, Nagaland, Sikkim, Madhya Pradesh, Chhattisgarh, Orissa, Andhra Pradesh, Telangana, Tamil Nadu, and Karnataka. Four Union territories—Puducherry, Jammu and Kashmir, Delhi, Daman and Diu are also not free from the clutches of this peril (Fig. 2) [39, 40]. West Bengal, lying on the banks of the river Bhagirathi-Hoogly, a distributary of the river Ganga in the lower part of the basin, Bihar (majorly in the Middle Ganga Plain) and Uttar Pradesh (Upper Ganga Plain) are the most affected states. The affected north-eastern states of Assam and Manipur lie in the plains of the rivers Brahmaputra, Imphal and Barak [46]. Around 50 million individuals are estimated to be vulnerable to groundwater As contamination in the Indian subcontinent [40].

The identification of As crisis in the country dates back to 1976, when its presence was confirmed in water samples from wells, hand pumps and springs from Chandigarh, Punjab and Haryana, in North India [47]. The problem was reported for the second time by dermatologist K. C. Saha of School of Tropical Medicine (STM),

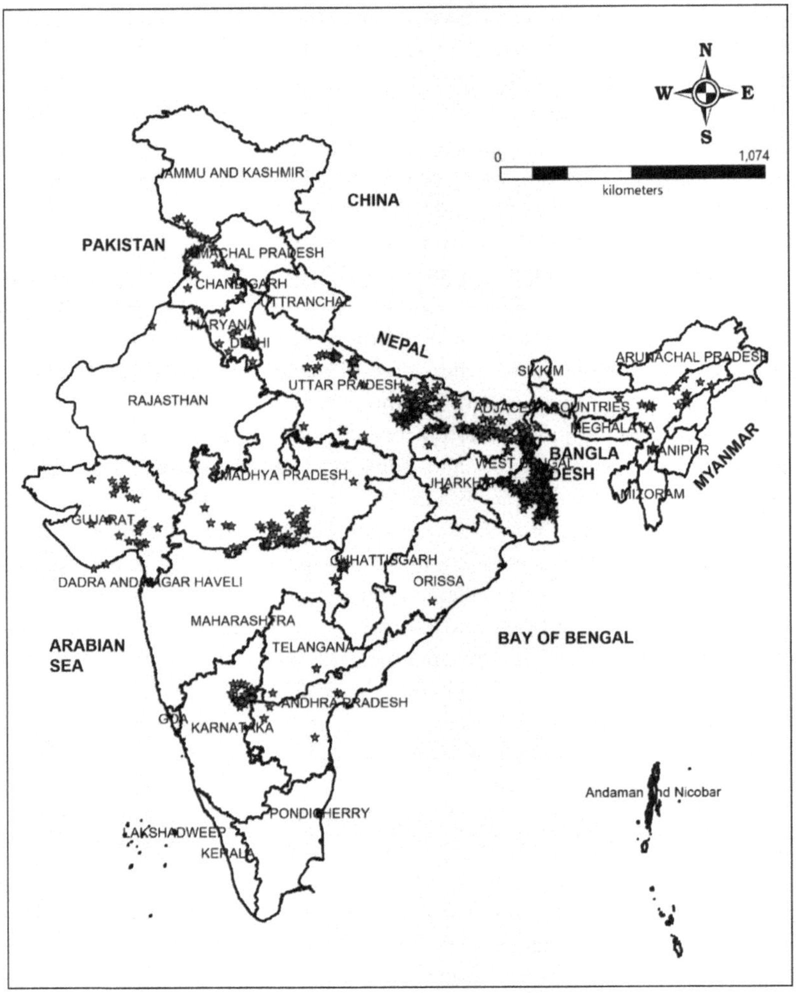

Fig. 2 Arsenic hotspots in India. Adapted from "Arsenic hot spot in ground water in India" [51] (http://jalshakti-dowr.gov.in/sites/default/files/ARSENIC.pdf)

Kolkata from South 24-Parganas, West Bengal. Other researchers also reported patients with As poisoning in the following years and warned about the imminent danger, if the issue is left unattended [48]. Research on groundwater As contamination gained momentum when School of Environmental Studies (SOES), Jadavpur University, India took up this issue in 1988. In collaboration with other research groups, SOES have carried out substantial research in neighbouring countries of Nepal, Tibet and Bangladesh. After 7 years of persistent work and with the publication of various local and national reports, the magnitude of the calamity was finally established.

Since its inception in the late 70's, the crisis has escalated exponentially. The results of the studies have been exceedingly devastating, as new areas kept on being added to the existing list of affected spots (Table 2) [49]. SOES has categorized the 19 districts of West Bengal into 3 groups according to the concentration of As found and severity of the poisoning—(a) highly affected 9 districts, where As concentrations in tube wells is above 300 μg/L (Malda, Murshidabad, Nadia, 24-Parganas (North and South), Bardhaman, Howrah, Hoogly and Kolkata), (b) mildly affected 5 districts where the concentration of the metalloid is mostly below 50 μg/L; however there are some instances when

Table 2 Major arsenic affected regions of India (concentration of the metalloid found to be above 50 μg/L)

Jha and Tripathi, 2021 [2]				Ministry of Jalshakti (according to SOES report, Jadavpur University) [51]		
Affected states	Affected districts	Max conc of As (μg/L)	Sources	Affected states	Affected districts	Max conc of As (mg/L)
Assam	Barpeta, Dhemaji, Karimganj, Darrang, Sonitpur	569	Reductive dissolution, high sedimentary deposition due to surface erosion, agricultural and industrial pollution	Assam	Nagaon, Cachar, Jorhat	0.1467
Bihar	Bhojpur, Patna, Bhagalpur, Siwan, Buxer, Begusarai, West Champaran, Samastipur	1805	Excessive groundwater extraction, weathering of minerals, river Sone alluvium, hydrolysis of FeAsS, reductive dissolution of calcite, dolomite, goethite and Fe oxyhydroxide, oxidation of organic matters, surface and groundwater interactions	Bihar	Godda, Dhanbad	0.06
Chhattisgarh	Rajnandgaon	985	Reductive dissolution	Chhattisgarh	Rajnandgaon	0.095
Jharkhand	Dehari, Sahebgunj	115	Soil erosion, industrial effluents, agricultural run-offs containing phosphate fertilizers	Jharkhand	Sahebgunj	0.068
Karnataka	Chitradurga, Raichur	235	Mining, sulphide mineralization	Karnataka	Raichur	0.063
Punjab	Malwa, Pathankot, Firozpur, Amritsar, Gurdaspur, Malva, Ropar	1000	Excessive groundwater withdrawal, reductive dissolution of As-minerals, oxidation of Fe hydroxide and sulphides, rampant use of fertilizers and pesticides, high levels of ammonium, iron and phosphate		Ropar, Amritsar, Tarn Taran	0.2
Tamil Nadu	Puducherry	620	Dissolution of rocks and ions	–	–	–
–	–	–	–	Haryana	Ambala, Jhajjar	0.36
Uttar Pradesh	Ballia, Unnao, Bahraich, Gautam Buddh Nagar, Ghazipur, Gonda, Lakhimpur Kheri, Moradabad	3192	Reductive dissolution, dissolution from Tibetan plateau and Himalayan rocks, weathering and sediment deposition by river Ghaghara and its	Uttar Pradesh	Azamgarh, Maunath Bhanjan, Deoria, Bahraich, Lakhimpur	0.811

(continued)

Table 2 (continued)

Jha and Tripathi, 2021 [2]				Ministry of Jalshakti (according to SOES report, Jadavpur University) [51]		
Affected states	Affected districts	Max conc of As (μg/L)	Sources	Affected states	Affected districts	Max conc of As (mg/L)
			tributaries, uncontrolled groundwater extraction, agricultural and industrial pollution, use of pesticides			
West Bengal	Burdwan, Malda, Murshidabad, Howrah, Hooghly, Nadia, Kolkata, 24 Parganas (north and south)	4622	Groundwater extraction for irrigation, As dissolution triggered by soil alkalinity, reductive dissolution of Fe-oxyhydroxides, sedimentation, Ion exchange	West Bengal	Malda, Nadia, Murshidabad, 24 Parganas (North and South), Hooghly	0.761

concentration above 50 μg/L has been found but not beyond 100 μg/L (Cooch Behar, Jalpaiguri, Darjeeling, Dinajpur (North and South) and (c) arsenic safe or unaffected 5 districts where the concentration is below 10 μg/L (Bankura, Birbhum, Purulia, East and West Medinipur) [50].

7 Technological Options for the Mitigation of Arsenicism and Supply of Clean Water

7.1 Identifying Alternative As-Safe Water Sources

Finding substitute water sources includes some long term measures such as tapping alternative aquifers, switching to surface waters, rainwater harvesting, well switching etc.

7.2 Deeper Aquifers

Arsenic contamination is restricted mainly to the shallow groundwater sources, whereas the deeper aquifers are still comparatively As-free. The depth of the deeper aquifers varies with regions. Literature suggest that in the Bengal Delta Plains, the groundwater reserves at a depth greater than100 m are devoid of As contamination, however in many cases the depth can go below 150–200 m [13, 40]. The contaminated and uncontaminated aquifers are separated by impermeable layers. Installation of hand pump fitted with deep tube wells that can be operated manually to draw water from these deeper groundwater reserves can be a reliable option for acquiring As-free water. Infiltration of contaminated water containing As can be avoided by sealing the annular spaces of the boreholes of the tube wells [52]. Still, this solution suffers from some limitations such as probability of over exploitation of the water reserves in absence of effective groundwater recharge, the possibility of intrusion of saline water enriched in dissolved Mn and Fe in coastal areas and high installation costs restricting its application only at the community level [53]. Water samples from these deep tube wells should be periodically monitored for contamination to ensure the safety of the local population.

7.3 Well Switching

The concentration of As in the shallow waters is not constant, thus it is not impossible to identify an uncontaminated tube well in the vicinity of a contaminated one. Switching of wells is a viable option and most preferred mitigation measure among other techniques [54]. One of the prominent drawbacks of well switching is the extent of spatial and temporal variation of the metalloid in the groundwater. The alteration in the concentration of As in the tube wells with change of seasons is another shortcoming which often make this option a bit unpredictable [55]. Moreover, similar to deep tube well, regular assessment of the tube wells is necessary to guarantee its safety.

7.4 Dug Wells

These are one of the primitive constructions for the extraction of groundwater, practiced since ancient times. Dug wells were largely used in the Bengal Delta as an alternative source of water before tube wells became the new preferred technique. The open wells with large diameters are able to supply As-free potable water even from shallow aquifers in regions where tube wells are affected. Groundwater recharge in the rainy season, aerobic environment and precipitation of iron are mainly responsible for the low concentration of As in these wells [56]. However, bacteriological contamination is one of the major disadvantages that reduce its potential. Drinking water from these well without disinfection, especially during the monsoon when microbial contamination increases, can result in water-borne diseases such as diarrhoea, typhoid, cholera, dysentery and hepatitis [57]. Moreover, turbidity, unwanted odour and taste and the tedious efforts behind regular disinfection make it all the more unfavourable.

7.5 Surface Water

The concentration of the metalloid in rivers, ponds and lakes is usually very low. Often these sources have been found to be devoid of the toxic element, making them suitable as an alternative water source. However, they are prone to microbial contamination which was the major cause, compelling the switching over to the groundwater reserves for supply of clean water [13]. The surface water sources need thorough disinfection or antimicrobial treatment prior its reintroduction into the affected areas.

7.6 Rainwater Harvesting

This method collects, stores and hygienically maintains rainwater from annual rainfall in large tanks or ponds for further domestic uses. This is a widely employed technique all over the world, which is mostly found at the household level. It is suitable in regions with an average annual rainfall of 1600 mm or above [58]. Installation of this technique at the community level in the affected areas is mainly constrained because of the huge cost of installation of special storage tanks with roofs for collecting and preserving the rainwater. Other limitations include uncertainty of average rainfall, availability of catchment area and contamination by microbes [59, 60]. The roof serves as the catchment area for the collection of rainwater in bigger households, which is connected to the tank with the help of a gutter system. In case of the under privileged class, this technique is unsuitable as they have houses without proper roofs or thatched roofs. The materials of the storage tanks and catchment area should be carefully chosen; as partially acidic rainwater can dissolve some metal or impurities from the materials [6].

7.7 Piped Water Supply

Collection, transportation, distribution and quality control is highly convenient in piped water systems. It is a long term measure and easy to keep free from any contamination. Water from any source such as deep tube wells, rivers, lakes, ponds, and groundwater can be collected and delivered to the consumers through a piped supply, after proper disinfection [61]. Even As contaminated water after proper treatment in an arsenic removal plant (ARP) can be supplied to the local population.

7.8 Arsenic Removal Technologies

Removal of As from the groundwater sources depends on its speciation and concentration and the physico-chemical characteristics of the contaminated water. Till date, there is no ultimate way to sequester As from polluted sources and provide clean water. Almost all the current treatment technologies for As mitigation follow some conventional principles (Table 3), with few advanced and innovative ones being developed, as researchers persistently explore new ideas.

Development of various in-situ and ex-situ technological solutions for the alleviation of high As concentration from groundwater sources and supply of As-free water to the affected population

Table 3 Comparison between different methods of arsenic removal [2, 8, 13, 62]

Process of As removal	Merits and demerits
Adsorption • Common adsorbents include different types of activated carbon, activated alumina, oxides and hydroxides of different elements, iron compounds (hydrous ferric oxide, granular ferric hydroxide, ferrihydrite, goethite, magnetite, hematite and siderite), red mud, low-cost agricultural and industrial wastes such as coconut shell, rice husk, biochar, bone char and fly ash and clay minerals (kaolinite, illite and montmorillionite) • Adsorption of As(V) requires acidic pH, whereas As (III) is maximally adsorbed around an alkaline pH of 8.5. Due to the difference in pH requirements, removal of total As from aqueous systems becomes difficult • Fe based adsorbents are majorly used for As remediation • Saturated adsorbent is regenerated using 4% NaOH • The residual NaOH is then neutralized by treating with 2% H_2SO_4 solution which prevents generation of caustic As rich sludge and loss in adsorbent capacity • It is regulated by several parameters including surface area, density, particle size, porosity, mineralogy, surface functional groups, zeta potential of the adsorbent, speciation of the metalloid and temperature, ionic strength, pH of the raw solution	Merits • Simple process • Cost effective adsorbents • Effective and efficient removal at point of entry level Demerits • Adsorbent need to be regenerated • Further treatment and disposal of saturated adsorbent is an additional problem • Involves use of additional chemicals
Chemical coagulation-flocculation • Coagulants destabilize the colloids by neutralizing their negative charges, which then combine to form larger flocs • The flocs formed as a result of colloidal interlinking, due to the different polymeric materials used in flocculation, aggregate to form larger masses • Soluble As is co-precipitated as insoluble precipitates of hydroxides, carbonates and sulfides, which can be filtered out or removed by sedimentation. The metal hydroxides, formed from the coagulants, act as adsorbents • Conventionally sand filtration is used, however for smaller flocs, microfiltration (MF) and ultrafiltration (UF) can be employed	Merits • Simple and economical • Effective for As(V) removal • Low operating and maintenance costs • Widely used for As removal Demerits • Requires chemical coagulants and pH adjustment • Generate huge amount of As rich sludge • Concentration of the residual coagulant in the media increases with its complexation with Natural Organic Matter (NOM)

(continued)

Table 3 (continued)

Process of As removal	Merits and demerits
• Fe and Al salts are mostly used coagulating agents, however Fe salts are more effective as Fe hydroxides are more stable over a wide pH range than Al ones	
Electro-coagulation • Complex physico-chemical process comprising of electrochemical reactions, flotation and coagulation • Include interfacial and surface phenomena • Metal ions are generated from the electrodes on application of direct current • These forms different oxides, hydroxides, sulphides and phosphates on reaction with other ions present in water • As(V) is co-precipitated with the insoluble metal salts or adsorbed onto their surface • Amorphous Fe oxyhydroxides and magnetite particles produced during the process can effectively remove both As(III) and As(V) • Electro-coagulation combined with air injection enhances the removal efficiency with minimum residual mass and reduced reactive retention time, thereby further reducing the operating costs	Merits • Simple process • Low operating and maintenance costs • No requirement of additional chemicals • Generation of less sludge Demerits • Batch mode of operation • High energy input • Electrodes gets coated with passive oxide films
Oxidation • Oxidation of soluble As(III) to sparingly soluble As(V) with subsequent precipitation, facilitates easy and effective removal • As(III) can undergo slow atmospheric oxidation, passive sedimentation or can be oxidized rapidly using O_3, $KMnO_4$, ClO^-, H_2O_2, Cl^-, ClO_2 etc • $KMnO_4$ and ClO^- are commonly used oxidants • Can serve as a pre-treatment in the adsorptive removal of the metalloid or separation by ion exchange • Other oxidative treatments include TiO_2 photo-catalyzed oxidation, photocatalytic oxidation, UV oxidation, iron based photooxidation, solar and photochemical oxidation	Merits • Very fast reaction • High removal efficiency • Common oxidants easily available Demerits • Expensive • Requires high energy input and additional chemicals • Formation of toxic byproducts • Hindered by the presence of Total Organic Carbon (TOC), Dissolved Organic Carbon (DOC), NOM and interfering ions
Ion exchange • Sophisticated technique, quite similar to adsorption • A synthetic resin, with specific ion exchange capacity is used in place of an adsorbent • Resins need to be regenerated once exhausted • Anion exchange resins are mostly used for As contaminated waters • The resins are rinsed with HCl solution when Cl^- ions are implanted on their surface • During flushing of As containing water oxy-anionic As (V) species are exchanged with the previously coated Cl^- ions	Merits • Efficient for As(V) removal • Exchange resins are locally available Demerits • Comparatively high cost input • Complex Operation and Maintenance • Generation of toxic waste

is a part of constant, meticulous research. These options are mostly based on the above mentioned principles and have been implemented in regions of severe As poisoning [8]. Decontamination of As containing aquifers with the help of in-situ techniques is very expensive. The process becomes complicated without proper knowledge about the size and geochemistry of the aquifer. Ex-situ technologies for the removal of As are short term solutions and best suited for both household and community levels. Tapped groundwater can be decontaminated with the help of these techniques but removal of the metalloid from the aquifer system is not possible. Moreover treatment of water by these methods for irrigation purposes can prove to be quite costly. The ex-situ techniques are mostly employed in the As affected India states for the supply of clean water [63]. The technologies for As removal can be categorized as centralized (community based), supplying water to the population of an entire city or community or point-of-use (POU) domestic systems that works at the household level. The POU systems can be used with tube wells also. The As removal schemes can be grouped into Arsenic Treatment Units (ATUs) and Arsenic Removal Units (ARUs) and Arsenic Removal/Treatment Plants (ARPs/ATPs), depending on their size and capacity [64, 65].

7.8.1 Arsenic Treatment Units (ATUs)

The ATUs are complete units that supply clean water to a small population. Their inlets are directly attached to a tube well or hand pump. In As prone distant rural regions, where surface water treatment plans cannot be implemented, pumped groundwater containing high concentration of the metalloid is treated in the ATUs using the conventional principles of As remediation mentioned above [66]. In some cases, advanced techniques such as membrane separation, phytoremediation and microbial bioremediation are also employed for the treatment of As containing water [67, 68]. The filtered water is supplied to the locals through piped water schemes or in large jars. The metalloid is adsorbed and co-precipitated on being treated with the adsorbent containing Al and Fe containing minerals [69, 70]. The treatment media is fabricated in such a way so as to provide a large number of adsorption sites on which As can be adsorbed even if other competing anions are present in the raw water [71]. The ATUs possess very high removal efficiency; it can be around 90% or more for feed water containing upto 50 ppm of the metalloid [72, 73].

7.8.2 Arsenic Removal Units (ARUs) and Arsenic Removal/Treatment Plants (ARPs/ATPs)

ARUs have a greater capacity to treat As contaminated water and supply potable water to around 600 households. ARPs can serve even a larger populace than ARUs. The As rich sludge generated from backwashing of the system in an ARU from time to time must be stabilized at a different site to avoid disposal issues and further contamination [74]. The adsorbent in the ARPs usually comprises of multiple media. Sludge separation beds are provided in ARPs for the in-situ stabilization of the generated sludge, which is one of the advantages of these bigger units. The backwash water, after getting treated in the separation beds, is recycled with the adsorbent media as a part of wastewater treatment. The base and sides of the beds are specifically sealed to prevent the leaching of arrested As sludge and its percolation into the water sources [75]. During the day, keeping the bed open allows sufficient aeration for quick stabilization of As. The oxidized As-Fe complexes get confined within the filter with the evaporation of the excess water present in the sludge. However, ARPs need large space for setting up elaborate infrastructure and efficient manpower for operation and maintenance, which become quite difficult in remote areas or villages [76, 77]. Different ARPs installed in As affected areas of West Bengal have been mentioned in Table 4.

7.9 Domestic and Community Filters

In India, several household and community filters have been developed and commercialized to supply As free water in the As prone rural areas (Table 5). Most of the projects have been supported by the national organization, Department of Science and Technology (DST) under Government of India, which encourages and supports novel research ideas for the development of economical solutions [79]. These filters mainly remove As via adsorption, coagulation and

Table 4 Comparison between different APRs installed in arsenic affected regions of India [63, 78]

Name of the technology	Mode of operation	Media	Capacity	Manufacturers' claim for efficiency of As removal	System performance
RPM Marketing Pvt. Ltd, New Delhi, Representing Alcan chemicals, USA	Adsorption	Activated Alumina + AAFS-50	Media capacity based on 2,00,000 L/yr with 0.5 ppm/L intake	Below detection level	Efficiency reduced after media replacement
Water systems International, USA represented by Harmonite Impex (Pvt) Ltd. Kolkata	Ion exchange	Bucket of resin (BOR)—mostly combination of metal oxides (patented)	3, 00,000 L based on 1 ppm	Less than 0.05 mg/L	Inconsistent and unsatisfactory field performance
Pal Trokner (P) Ltd., Kolkata	Adsorption	AdsorpAs ® (Granular ferric hydroxide—GFH)	9,00,000 L based on 1 ppm	Arsenic free water	Satisfactory with low O & M cost
Oxide India (Catalysts) Pvt. Ltd. Durgapur, representing B.E college model	Adsorption	Activated Alumina AS-37	Based on two years base performance with 600 L/hr capacity, 4000L/day for 0.5 ppm raw water	Arsenic free water	Satisfactory
Adhicon Kolkata	Catalytic precipitation/Ion exchange	AFDWS-2000	Requires periodical sludge removal or backwash	As prescribed by WHO	Unsatisfactory field performance
All India Institute of Hygiene & Public Health	Oxidation + Coagulation + Flocculation/ precipitation/filtration	Chlorinating agent (BP) + Ferric Alum	1000 L/hr (Requires periodical sludge	Remove 95%	System requires constant monitoring

(continued)

Table 4 (continued)

Name of the technology	Mode of operation	Media	Capacity	Manufacturers' claim for efficiency of As removal	System performance
			removal and/ or cleaning)		
Ionochem, Kolkata	Ion exchange	Ferric hydroxide	Based on 0.5 ppm, 14,40,000 L with 600 L/hr	Less than 0.01 ppm	Satisfactory
Apyron Technologies India (P) Ltd, Representing Apyron Technologies Inc. USA	Adsorption	Aqua bind (Activated Alumina mostly in combination with metal oxide)	Based on 1 ppm, 1000 L/day, periodic media replacement at 6–8 months	Less than 10 ppb	Satisfactory. Reduced high As concentration of 3500 ppb to below 50 ppb in treated water
School of Fundamental Research, Kolkata	Adsorption	Aluminum silicate + ferric hydroxide	Based on 0.5 ppm, 2000L/day, media recharging at 4–6 months	Less than 50 ppb	NA
Public Health Engineering Department, Govt. of West Bengal	Adsorption	Red Hematite (Fe_2O_3) lumps + quartz + sand + activated alumina	600–1000 L/hr (requires periodical sludge and/or backwash/cleaning)	Below permissible limit of 0.05 mg/L	Reduces very high As concentration but produces insufficient quantity of filtered water
Anir Engineering, Collaboration with ITP Gmbh, Germany	Adsorption	Slurry/granular ferric hydroxide (S/GFH)	600–900 L/h (requires periodic backwashing twice/thrice per week)	NA	NA

Table 5 Domestic filters for arsenic removal, used in India [79]

Product name	Type of filter	Cost of unit	Technology used	Cost of delivered water	Contaminants removed	Reject management	Locations of field trials
Low cost Laterite based Arsenic Filter (IIT Kharagpur)	Domestic and Community	Domestic: Rs.2500 Community: Rs. 15000 for 100 L/hr	Domestic: Single stage-Adsorption Community: Double stage-Sedimentation, Adsorption	Rs.30/m^3	As, Fe and Bacteria	Spent media meet TCLP protocol	25 household filters in West Bengal
IITB Arsenic Filter(IIT Bombay)	Community	Rs. 60000–75,000	Dual Stage—gravel filtration with Fe nails and Jali that supplements Fe for As removal. Removal process is based on the dissolution of Fe to Fe^{2+} from ZVI (Fe nails + Jali) and co-oxidation of Fe^{2+} and As^{3+} which is co-precipitated with Fe^{3+}. The gravel media has a long life but reactive media (Fe nail + Fe Jali) needs to be annually supplemented	Rs.3.3/m^3	As	A brick masonry tank is specially designed for sludge storage	53 units in Bihar, Uttar Pradesh, West Bengal and Assam
AMRIT-Arsenic and Metal Removal by Indian Technology (IIT Madras)	Domestic and community	Domestic: Rs. 1500 Community: Rs. 99,000	Multiple stages—1st stage: Surface Filtration, 2nd stage: Colloidal Iron adsorption, 3rd stage: Arsenic adsorption, 4th stage: Metal-based disinfection	Domestic: Rs. 50–70/m^3 Community: Rs. 50 /m^3	As, Fe and turbidity	Media can be easily disposed as it is prepared with eco-friendly materials. It can also be used for brick making as it is composed of iron oxides	200 household units in Karnataka, West Bengal and Bihar 160 community filters in West Bengal
Arsiron Nilogon	Domestic and community	Domestic: Rs. 600 Community: Rs. 9000 for 200 and Rs. 13,000 for	Multiple stages: 1st stage: oxidation-coagulation, adsorption-sedimentation,	Domestic: Rs. 31.67/m^3 Community: Rs.	As and Fe	The small solid sludge, which passes the TCLP test, can be disposed	25 household units in West Bengal and Assam

(continued)

Table 5 (continued)

Product name	Type of filter	Cost of unit	Technology used	Cost of delivered water	Contaminants removed	Reject management	Locations of field trials
		500 L/hr batch capacity	2nd stage: slow sand filtration	48/m^3(200 L) and Rs. 37/m^3(500 L)		safely. There is no reject water	
DRDO arsenic removal filter	Domestic filter	Rs. 2000	Multiple stages—1st stage: co-precipitation and adsorption, 2nd Stage: Filtration	Rs. 0.015/m^3	As	Used media is converted into non leachable cement bricks	Total ~2400 families
ARI groundwater arsenic treatment plant	Community filter	Rs. 74,000	Multiple stages (assembled single integrated unit)—1st stage: microbial oxidation of As^{3+} to As^{5+} using *Microbacterium lacticum* immobilized on brick pieces, 2nd stage: Adsorption of As^{5+} on alumina, 3rd stage: filtration using charcoal to remove odour, color, microbial cells. 4th stage: UV Radiation for disinfection	Rs. 10/m^3	As	The wash water from the columns is collected in a separate container and 3% $FeCl_3$ is added to form an As-Fe complex The dried sludge is disposed off in the concrete pit	5 systems in Chhattisgarh
Hand pump attached arsenic removal unit (Jadavpur University)	Community Filter	Rs. 175,000	Double stage technology—Co-precipitation and adsorption method	Rs. 3.60/m^3	As and Fe	As-rich sludge is stored in underground reservoir. As sludge (1%) needs to be mixed with concrete for reject disposal. TCLP test shows leaching of As within permissible limit	4 units in West Bengal

Fig. 3 Arsenic removal systems—**A** AMRIT domestic arsenic filter; **B** AMRIT community water purification unit; **C** IIT B Arsenic removal unit in West Bengal; **D** community model for Arsiron Nilogon arrangement; **E** Household Arsiron Nilogon system with sand gravel filter alone (HAN); **F** DRDO Water Filters (Stainless steel, Clay and Plastic. Adapted from "R&D Activities related to Arsenic Contamination in Drinking Water—Salient Efforts of Department of Science and Technology (DST), 2015" [79] (https://dst.gov.in/sites/default/files/Arsenic-Compendium.pdf)

filtration (Fig. 3). The adsorbents with greater surface area are manufactured in such a way so as to achieve enhanced adsorption capacity over a wide range of pH and temperature. The successful prototypes were implemented in As affected regions and gained community acceptance with promising results (Fig. 4).

A

B

Fig. 4 Arsenic removal systems—**A** ARI groundwater Arsenic removal unit; **B** Arsenic removal unit. Adapted from "R&D Activities related to Arsenic Contamination in Drinking Water—Salient Efforts of Department of Science and Technology (DST), 2015" [79] (https://dst.gov.in/sites/default/files/Arsenic-Compendium.pdf)

7.10 Emerging Technologies for Arsenic Removal from Contaminated Waters

7.10.1 Subterranean Arsenic Removal Technology (SAR)

Subterranean Arsenic Removal (SAR) is a recent technology, developed jointly by a group of European and Indian scientists, which aims at in-situ remediation of groundwater As contamination [80]. It is a simple technology and consists of several steps. Groundwater aquifer is treated as a biochemical reactor which aids in the adsorptive removal of the cations (As, Fe and Mn) with increasing concentration of dissolved oxygen (DO). Groundwater is extracted with the help of a submersible pump, a fraction of which (15–20%) is aerated and returned back to the aquifer under gravity in the sequence—delivery, intermission and infiltration. The DO level of the water is increased up to 6 mg/L.

The plant set up comprises of an oxidation unit, spray nozzle, storage tank and pipelines for water extraction and infiltration, contained in a well of depth 37 m and sealed well head to prevent water leaking during pressure infiltration. All the plant parts are locally available. Oxygenated water converts Fe^{2+} adsorbed onto soil grains to insoluble $Fe(OH)_3$ which furthers eliminates As as a co-precipitation product. The autotrophic microbes enhance the rate of oxidation [81]. This is a chemical free technology which does not generate sludge. Six SAR plants have been implemented in Kasimpore village in the district North 24 Parganas of West Bengal. The region lies in the lower Gangetic plains of India. The concentration of As in the treated water was found to be less than 10 ppm. The operating cost of the plant for supplying 2000 L of potable water is 1.0 US$/day.

7.10.2 Electro-Chemical Arsenic Remediation (ECAR)

The Electro-Chemical Arsenic Remediation (ECAR) technology applies electro-coagulation for the removal of As from groundwater sources. This is a business model of a community level As removal unit [82]. Ferric oxyhydroxides are produced as precipitates as a result of electrolytic oxidation of a Fe anode in As contaminated water. Fenton-type reactions generate strong oxidants which oxidizes As^{3+} to As^{5+} which are readily adsorbed on to the ferric oxyhydroxides. Binuclear As-Fe complexes are formed through inner sphere complexation, which further agglomerates as larger flocs. Alum is used as an easily available, low cost coagulant to separate the flocs from the water through gravitational settling [83]. In-situ generation of the adsorbent eliminates the necessity for separate manufacturing plant for the production of adsorbent media or media import from abroad. There is no requirement for additional hazardous chemicals and media regeneration which not only simplifies the system operation but also reduces maintenance costs.

It is a small, simple unit requiring relatively less input of electricity. The technology can be automated which improves reliability, reduces system downtime and manual errors and allows operation during night using intermittent power supply. The efficiency of ECAR technology has been tested using both synthetic groundwater and real contaminated samples from Bangladesh and Cambodia, before its demonstration in the As affected areas of West Bengal [82]. Two 100 L batch reactors for short field trials and a 600 L reactor for over 3.5 months have been implemented in West Bengal. The results are by far quite promising as ECAR has been found to constantly reduce As concentrations from 3000 μg/L to below 5 μg/L. It is expected that the operating costs of this technology can vary within the range of \$0.83–\$1.04/m^3 based on the results obtained from field trials [84].

7.10.3 Arsenic Removal by Solar Driven Inline-Electrolytic Induced Co-Precipitation and Filtration

This is an autonomous technology, driven by solar energy. An inline-electrolytic cell is used for the in-situ generation of chlorine utilizing the chloride content of water. Chlorine gas produced as a result of the electrolytic reactions, dissociates readily in water to produce hypochlorous acid (HClO), which substitutes for any external oxidizing agents. Hypochlorous acid oxidizes Fe^{2+} to Fe^{3+} and As^{3+} to As^{5+}. Oxidation of As (III) to As(V) is prerequisite for its adsorption on the surface of freshly formed Fe hydroxides. The metalloid is also entrapped within the growing particles and is removed by co-precipitation and filtration with a Manganese Greensand Plus® automatic filter media.

This is a user friendly, independent As removal technology which does not require any heavy infrastructure and is devoid of several operational challenges such as exchange of adsorbents, external supply of oxidants, dosing etc. The unit was installed in Chandamari village in Nadia District, one of the dangerously As affected sites of West Bengal, India. The field trial lasted for 10 months during which the unit treated approximately 393 m^3 groundwater in continuous flow. The maximum percentage of removal obtained for arsenic, iron, manganese, phosphate and ammonium were found to be 94%, >99%, 96%, 81% and 84% respectively. The average loading density of As on Fe hydroxides was 31 μgAs/mgFe [85].

7.10.4 Arsenic and Iron Removal from Groundwater by Oxidation-Coagulation at Optimized pH (OCOP)

It is a low-cost technique, which can be employed both at domestic and community level. The technology is easy to operate and does not require extensive infrastructure and external energy input. Groundwater is treated in simple steps for the removal of As and Fe. The contaminated water first undergoes an alkaline treatment with $NaHCO_3$, which acts as a pH conditioner [86]. As^{3+} and Fe^{3+} ions are oxidized by $KMnO_4$, which precipitates as insoluble MnO_2. $FeCl_3$ is used as the coagulating agent. The pilot experimental set up comprised of two PET containers for coagulation and sedimentation respectively. A sand-gravel filter was used to filter the treated water after settling for 1–2 h. Field trials of the method were conducted in different villages of Jorhat district, in Assam, India. The concentration of As and Fe in the raw water samples from these sites were found to be within the range of 100–240 μg/L and 1–20 mg/L respectively. The pilot systems in batch mode were set up for both household and small community scale in different capacities (10 L systems for 10 households, 25 L for 5 schools and a 200 L system for 1 school). The sludge in concentrated form was filtered using earthen pot sand filters of 1–3 L capacity. In the treated water, the concentration of As and Fe were found to be 5 μg/L and <0.1 mg/L respectively [87].

7.10.5 Management of Arsenic Containing Wastes Generated from Treatment Units/Plants

Safe disposal of As-rich sludge after groundwater treatment is an indispensable aspect of As remediation plants. Both liquid and solid As wastes are generated by every small or large scale As removal units of which solid ones are considered to be less harmful. Solid wastes are actually the spent adsorbents or spent ion-exchange resins used in the water treatment. Liquid waste is backwash, produced due to media washing and contains higher concentration of the metalloid than the solid ones [88]. Toxicity Characteristic Leaching Procedure (TCLP) test is mostly adopted to understand the leaching potential of As from these wastes. The economic output of recycled raw As from As bearing wastes for further industrial applications is very low, so these wastes are mainly mixed with easily available materials such as polymer resins, mineral oxides, activated carbon, industrial and agricultural by-products etc. for stabilization, prior to disposal [89, 90]. The extent of stabilization is dependent on the speciation and concentration of the metalloid in a specific waste. This technique prevents the percolation of As in the environmental matrices from the wastes by arresting it within a cementitious system.

In the rural areas of the Bengal Basin, As sludge was mixed with cow dung, poultry wastes or rice straw in a biogas plant or closed, bottom sealed pit in order to confine it. This method was frequently employed for the disposal of As sludge few years back [91, 92]. Recent sustainable management of As wastes focuses on using them with construction materials such as cement, concrete and mortars. Extensive laboratory experiments have been conducted to analyse the binding efficiency of these materials towards As and the long term leaching characteristics of the metalloid after stabilization [93, 94]. However definite results supporting the leaching of As from these materials, during production, weathering or chemical treatment, is not yet available. As-rich sludge has also been used in the manufacturing of clay bricks which is considered to be economic and environmentally sustainable [95]. The As-laden bricks are safe in terms of leaching and can be used for even heavy-duty constructions beside ornamental purposes [96].

8 Conclusions

Arsenic contamination in the groundwater aquifers of the Ganga–Brahmaputra basin in India can be regarded as the worst groundwater crisis in history. The identification of the problem may not be too old; however, it has escalated to a level of endemic in India, inflicting deleterious effects upon the mankind. People exposed to elevated levels of As for long duration are potentially at risk for developing arsenicosis [1, 6]. Surfacing of arsenic polluted regions with every new survey is a major concern. Although the origin of the problem is considered to be mostly geogenic, still definitive information regarding the sources, mobilization and behaviour of the toxic metalloid, once it is introduced into the sub-surface layers is lacking, in spite of extensive research. The problem has been gaining due attention in recent times, with different governing bodies acknowledging the seriousness and magnitude of the issue. Different State Pollution Control Boards, State Public Health Engineering Departments and Central Government regulatory bodies (Central Pollution Control Board, Central Ground Water Board (CGWB), Ministry of Water Resources etc.,) are working jointly, guided by international forums such as WHO, UNICEF etc., to come up with sustainable, effective and socially acceptable mitigation plans [78, 79].

In order of severity and prevalence of the problem, several counteractive steps have been taken by the Government of West Bengal (GoWB), Bihar and Uttar Pradesh, whereas well-structured plans to combat the crisis in other states are insignificant. The GoWB has roped in experts from different disciplines for constituting an 'Arsenic Task Force' to formulate a master plan and deal with the issue. Although a master plan of piped water supply scheme for the affected villages is ready, it is taking inappropriately long time for its implementation [63]. Other remedial measures in West Bengal include identification of infected tube wells and restricting their use, tapping alternative aquifers, construction of As free deep wells, hand pump fitted tube wells by CGWB and setting up of laboratories in districts for the detection of As in groundwater samples by West Bengal Public Health Engineering Department (WBPHED). Precautionary steps by the Government of Bihar consist of construction of deep tube wells and dug wells for villagers, sanitary wells in schools and structures for rain water harvesting. Supply of surface waters after proper treatment through piped water supply scheme has been operationalized in As contaminated areas, where feasible and these schemes have been successfully providing safe potable water to the affected masses [63].

In addition to the actions put forward by the Government, various Non-Governmental Organizations (NGOs), academic institutions and research organizations have lent their valuable support to fight against this disaster. Numerous research interventions have been extended to elucidate about the cause, source and magnitude of the problem, the geo-chemical processes and mechanisms of As dissolution. Significant investigations regarding its effect on the food chain and human health have been carried out. Water Technology Initiative (WTI) programme, launched by the Department of Science and Technology (DST) in 2007–08, aims to address different R&D issues related to clean water supply as well as encourage and support novel ideas for the development of sustainable treatment plans along with proper management of generated sludge [79]. It also promotes the development of cost effective adsorbents from locally available resources, superior in quality than the commercial ones. Attempts are also made at manufacturing field test kits for the efficient detection of As in water sources. The R&D activities prioritize on devising improved techniques for rapid oxidation of As(III) to As (V), since removal of the latter is easier.

However, the presence of the metalloid in the Indian groundwater reserves in an excessive concentration makes the task daunting. Moreover, no universal plan can be implemented throughout the affected areas because As is present in varied concentration in these regions. Arsenic removal technologies based on the conventional principles have been developed for both domestic and community use, which can meet As-free water demand even in the rural outskirts. The ex-situ treatment techniques based on adsorption and coagulation-precipitation are widely practiced in India and have shown promising results; however different ARUs and ATUs or filters have proven to be unsatisfactory on extensive field trials and face problems with sludge disposal, due to improper co-ordination and poor operation and maintenance [64]. The successful implementation of a treatment plan is governed to a large extent, by the geo-political issues and socio-economic conditions of the affected region. Simplicity of the device and its user-friendliness determines the reliability and sustainable running of the unit while efficiency of As removal depends on the concentration and speciation of As. The local awareness, user-friendly operation and community acceptance play a predominant role in the successful operation of an As remediation unit.

Till now, the technologies developed to alleviate the crisis of excessive As contamination of groundwater, mainly focus on providing As-free water to the affected populace, to meet their drinking water needs. However, majority of the rural population depend on agriculture for which they exploit substantial amount of groundwater. Thus care should be taken to address the water demand for both the purposes in order to come up with a comprehensive and successful mitigation plan. Development of a strategy or implementation of a technology for As removal alone cannot ensure the efficacy of a treatment plant. It should be monitored periodically in order to derive long term benefits and prevent the system from crashing down due to improper management. Large scale awareness programs should be carried out in order to educate the ignorant mass about the problem as well as its severity. They should be involved and made knowledgeable about their responsibility. Collective efforts from every segment of the society are indispensable in resolving this issue.

References

1. Jadhav SV, Bringas E, Yadav GD, Rathod VK, Ortiz I, Marathe KV. Arsenic and fluoride contaminated groundwaters: a review of current technologies for contaminants removal. J Environ Manage. 2015;162:306–25.
2. Jha PK, Tripathi P. Arsenic and fluoride contamination in groundwater: a review of global scenarios with special reference to India. Groundw Sustain Dev. 2021;13: 100576.
3. Mukherjee A, Bhattacharya P, Fryar AE. Arsenic and other toxic elements in surface and groundwater systems. Appl Geochem. 2011;26:415–20.
4. George CM, Sima L, Arias MHJ, Mihalic J, Cabrera LZ, Danz D, et al. Arsenic exposure in drinking water: an unrecognized health threat in Peru. Bull World Health Organ. 2014;92:565–72.
5. International Agency for Research on Cancer (IARC). A review of human carcinogens. Part C: Arsenic, metals, fibres, and dusts. IARC monographs on the evaluation of carcinogenic risks to humans, 100c: 1–54; 2012. PMID: 23189751.
6. Guidelines for Drinking-Water Quality, vol 4, World Health Organization; 2011.
7. Arsenic in Drinking Water. United States Environmental Protection Agency; 2013. http://water.epa.gov/lawsregs/rulesregs/sdwa/arsenic/index.cfm.
8. Bhardwaj A, Rajput R, Misra K. Status of arsenic remediation in India. In: Ahuja S, editors. Advances in water purification techniques: meeting the needs of developed and developing countries. Elsevier; 2019. p. 219–58.
9. Guidelines for drinking-water quality: fourth edition incorporating the first Addendum. ISBN 978-92-4-154995-0. World Health Organization; 2017.
10. Edition of the Drinking Water Standards and Health Advisories Tables, Office of Water, United States Environmental Protection Agency; 2018.
11. Indian Standard (IS 10500). Drinking water—specification (2nd Revision). New Delhi: Bureau of Indian Standards (BIS); 2012
12. Singh R, Singh S, Parihar P, Singh VP, Prasad SM. Arsenic contamination, consequences and remediation techniques: a review. Ecotoxicol Environ Saf. 2015;112:247–70.
13. Shankar S, Shanker U, Shikha. Arsenic contamination of groundwater: a review of sources, prevalence, health risks, and strategies for mitigation. Sci World J. 2014;304524.
14. Smedley P, Kinniburgh D. A review of the source, behaviour and distribution of arsenic in natural waters. Appl Geochem. 2002;17:517–68.
15. Fendorf S, Michael HA, van Geen A. Spatial and temporal variations of groundwater arsenic in South and Southeast Asia. Science. 2010;328:1123–7.
16. Sikdar PK, Dey S, Ghosal U, Chakraborty S. Development and management of base flow of a sand-dominated alluvial aquifer of a large ephemeral river for drinking water supply in semi-arid and fluoride affected Areas: example of the River Mayurakshi, Birbhum District, West Bengal, India. J Geol Soc India. 2019;94:249–59.
17. Ravenscroft P, Burgess WG, Ahmed KM, et al. Arsenic in groundwater of the Bengal Basin, Bangladesh: distribution, field relations, and hydrogeological setting. Hydrogeol J. 2005;13:727–51.

18. Shakoor MB, Nawaz R, Hussain F, Raza M, Ali S. Human health implications, risk assessment andremediation of As-contaminated water: a critical review. Sci Total Environ. 2017;601–602:756–69.
19. Welch AH, Westjohn DB, Helsel DR, Wanty RB. Arsenic in ground water of the United States: occurrence and geochemistry. Groundwater. 2000;38:589–604.
20. Madhavan N, Subramanian V. Environmental impact assessment, remediation and evolution of fluoride and arsenic contamination process in groundwater. In: Thangarajan M, editor. Groundwater: resource evaluation, augmentation, contamination, restoration, modeling and management. Dordrecht: Springer; 2007. p. 128–55.
21. Kumar M, Das A, Das N, Goswami R, Singh UK. Co-occurrence perspective of arsenic and fluoride in the groundwater of Diphu, Assam, North-eastern India. Chemosphere. 2016;150:227–38.
22. McArthur JM, Ravenscroft P, Safiulla S, Thirlwall MF. Arsenic in groundwater: testing pollution mechanisms for sedimentary aquifers in Bangladesh. Water Resour Res. 2001;37:109–17.
23. Bhattacharya P, Jacks G, Ahmed KM, Khan AA, Routh J. Arsenic in groundwater of the Bengal Delta plain aquifers in Bangladesh. Bull Environ Contam Toxicol. 2002;69:538–45.
24. Das S, Bora SS, Lahon JP, Chetia M, Barooah M. Groundwater arsenic contamination in north eastern states of India. J Environ Res Develop. 2015;9:621.
25. Santra BK. Arsenic contamination of groundwater in West Bengal: awareness for health and social problems. Int J Appl Sci Eng. 2017;5:43–6.
26. Guo H, Zhang B, Li Y, Berner Z, Tang X, Norra S, et al. Hydrogeological and biogeochemical constrains of arsenic mobilization in shallow aquifers from the Hetao basin, Inner Mongolia. Environ Pollut. 2011;159:876–83.
27. Bhowmick S, Nath B, Halder D, Biswas A, Majumder S, Mondal P, et al. Arsenic mobilization in the aquifers of three physiographic settings of West Bengal, India: understanding geogenic and anthropogenic influences. J Hazard Mater. 2013;262:915–23.
28. Pinheiro BS, Gimenes LLS, Moreira AJ, Freschi CD, Freschi GPG. Arsenic speciation in environmental samples using different acid concentrations and ultrasonic extraction for the determination by HG-FAAS. At Spectrosc. 2016;37:83–9.
29. Petrick JS, Ayala-Fierro F, Cullen WR, Carter DE, Aposhian HV. Monomethylarsonous acid (MMA III) is more toxic than Arsenite in Chang human hepatocytes. Toxicol Appl Pharmacol. 2000;163:203–7.
30. Rathor R, Verma M, Misra K, Brar SK. Arsenic poisoning and remedial measures: a global scenario. In: Brar SK, editor. Hazardous materials: types, risks and control. New York: Nova Science Publishers; 2011. p. 225–57.
31. Taylor V, Goodale B, Raab A, Schwerdtle T, Reimer K, Conklin S, et al. Human exposure to organic arsenic species from seafood. Sci Total Environ. 2017;580:266–82.
32. Huang RN, Lee TC. Cellular uptake of trivalent Arsenite and pentavalent arsenate in KB cells cultured in phosphate-free medium. Toxicol Appl Pharmacol. 1996;136:243–9.
33. Ratnaike RN. Acute and chronic arsenic toxicity. Postgrad Med J. 2003;79:391–6.
34. Jomova K, Jenisova Z, Feszterova M, Baros S, Liska J, Hudecova D, et al. Arsenic: toxicity, oxidative stress and human disease. J Appl Toxicol. 2011;31:95–107.
35. Khairul I, Wang QQ, Jiang YH, Wang C, Naranmandura H. Metabolism, toxicity and anticancer activities of arsenic compounds. Oncotarget. 2017;8:23905–26.
36. Naujokas MF, Anderson B, Ahsan H, Aposhian HV, Graziano JH, Thompson C, et al. The broad scope of health effects from chronic arsenic exposure: update on a worldwide public health problem. Environ Health Perspect. 2013;121:295–302.
37. Kumar R, Kumar R, Mittal S, Arora M, Babu JN. Role of soil physicochemical characteristics on the present state of arsenic and its adsorption in alluvial soils of two agri-intensive region of Bathinda, Punja, India. J Soils Sediments. 2016;16:605–20.
38. Chakraborti D, Rahman MM, Das B, Nayak B, Pal A, Sengupta MK, et al. Groundwater arsenic contamination in Ganga-Meghna-Brahmaputra plain, its health effects and an approach for mitigation. Environ Earth Sci. 2013;70:1993–2008.
39. Chakraborti D, Singh SK, Rahman MM, Dutta RN, Mukherjee SC, Pati S, et al. Groundwater arsenic contamination in the ganga river basin: a future health danger. Int J Environ Res Public Health. 2018;15:180.
40. Shaji E, Santosh M, Sarath KV, Prakash P, Deepchand V, Divya BV. Arsenic contamination of groundwater: a global synopsis with focus on the Indian Peninsula. Geosci Front. 2021;12: 101079.
41. Wang L, Kou MC, Weng CY, Hu LW, Wang YJ, Wu MJ. Arsenic modulates heme oxygenase-1, interleukin-6, and vascular endothelial growth factor expression in endothelial cells: roles of ROS, NF-κB, and MAPK pathways. Arch Toxicol. 2012;86:879–96.
42. Sinha D, Biswas J, Bishayee A. Nrf2-mediated redox signaling in arsenic carcinogenesis: a review. Arch Toxicol. 2013;87:383–96.
43. Shi H, Hudson GL, Ding W, Wang S, Cooper KL, Liu S, et al. Arsenite causes DNA damage in keratinocytes via generation of hydroxyl radicals. Chem Res Toxicol. 2004;17:871–8.
44. Navas-Acien A, Silbergeld EK, Streeter RA, Clark JM, Burke TA, Guallar E. Arsenic exposure and type 2 diabetes: a systematic review of the

experimental and epidemiologic evidence. Environ Health Perspect. 2006;114:641–8.
45. Human body internal organs. Freepik Company. Available at https://www.freepik.com/premium-vector/human-body-internal-organs_6958731.htm
46. Shewale M, Bhandari D, Garg RK. A review of arsenic in drinking water: Indian scenario. Int J Sci Res Sci Eng Technol. 2017;3:300–4.
47. Datta DV, Kaul MK. Arsenic content of drinking water in villages in Northern India. A concept of arsenicosis. J Assoc Physicians India. 1976;24:599–604.
48. Garai R, Chakraborty AK, Dey SB, Saha KC. Chronic arsenic poisoning from tube-well water. J Indian Med Assoc. 1984;82:34–5.
49. Singh MK, Kumar A. A global problem of arsenic in drinking water & its mitigation-a review. Int J Adv Eng Technol. 2012;3:196–203.
50. Dey TK, Banerjee P, Bakshi M, Kar A, Ghosh S. Groundwater arsenic contamination in West Bengal: current scenario, effects and probable ways of mitigation. Int Lett Nat Sci. 2014;8:45–58.
51. Arsenic hot spot in ground water in India. Department of water resources. River development & ganga rejuvenation. Ministry of Water Resources. Government of India. Available at http://jalshakti-dowr.gov.in/sites/default/files/ARSENIC.pdf.
52. Hoque MA, Burgess WG. 14C dating of deep groundwater in the Bengal Aquifer System, Bangladesh: implications for aquifer anisotropy, recharge sources and sustainability. J Hydrol. 2012;444–445:209–20.
53. Hug SJ, Gaertner D, Roberts LC, Schirmer M, Ruettimann T, Rosenberg TM et al. Avoiding high concentrations of arsenic, manganese and salinity in deep tubewells in Munshiganj District, Bangladesh. Appl Geochem 26:1077–85.
54. Ahmed MF, Ahuja S, Alauddin M, Hug SJ, Lloyd JR, Pfaff A, et al. Ensuring safe drinking water in Bangladesh. Science. 2006;314:1687–8.
55. Rahman MM, Mandal BK, Roy CT, Sengupta MK, Chowdhury UK, Lodh D, et al. Arsenic groundwater contamination and sufferings of people in North 24-Parganas, one of the nine arsenic affected districts of West Bengal, India. J Environ Sci Health A: Toxic/Hazard Substances Environ Eng. 2003;38:25–59.
56. Bennett HB, Shantz A, Shin G, Sampson ML, Meschke JS. Characterisation of the water quality from open and rope-pump shallow wells in rural Cambodia. Water Sci Technol. 2010;61:473–9.
57. Hira-Smith MM, Yuan Y, Savarimuthu X, Liaw J, Hira A, Green C, et al. Arsenic concentrations and bacterial contamination in a pilot shallow dug well program in West Bengal, India. J Environ Sci Health A: Toxic/Hazard Substances Environ Eng. 2007;42:89–95.
58. Union Wise Water Technology Mapping [Dhaka Circle]. DPHE Publication # 1, GOB-UNICEF Project. Department of Public Health Engineering (DPHE), Dhaka, Bangladesh; 2008.
59. Islam MA, Sakakibara H, Karim MR, Sekine M, Mahmud ZH. Bacteriological assessment of drinking water supply options in coastal areas of Bangladesh. J Water Health. 2011;9:415–28.
60. Karim MR. Microbial contamination and associated health burden of rainwater harvesting in Bangladesh. Water Sci Technol. 2010;61:2129–35.
61. Inauen J, Hossain MM, Johnston RB, Mosler HJ. Acceptance and use of eight arsenic-safe drinking water options in Bangladesh. PLoS ONE. 2013;8: e53640.
62. Mondal P, Bhowmick S, Chatterjee D, Figoli A, der Bruggen BV. Remediation of inorganic arsenic in groundwater for safe water supply: a critical assessment of technological solutions. Chemosphere. 2013;92:157–70.
63. Ghosh NC, Singh RD. Groundwater arsenic contamination in India: vulnerability and scope for remedy. In: Technical papers included in the special session on ground water in the 5th Asian regional conference of INCID, December 9–11, 2009 held at Vigyan Bhawan. New Delhi; 2009.
64. Hossain MA, Sengupta MK, Ahamed S, Rahman MM, Mondal D, Lodh D. Ineffectiveness and poor reliability of arsenic removal plants in West Bengal, India. Environ Sci Technol. 2005;39:4300–4306.
65. Panda BK. A brief review on the scenario of ground water pollution by arsenic in West Bengal. Int J Ecosyst. 2015;5:75–85.
66. Dominguez-Ramos A, Chavan K, Garcia V, Jimeno G, Albo J, Marathe K, et al. Arsenic removal from natural waters by adsorption or ion exchange: an environmental sustainability assessment. Ind Eng Chem Res. 2014;53:18920–7.
67. Kumari N, Rana A, Jagadevan S. Arsenite biotransformation by *Rhodococcus* sp.: characterization, optimization using response surface methodology and mechanistic studies. Sci Total Environ. 2019;687:577–89.
68. Maity JP, Chen CY, Bhattacharya P, Sharma RK, Ahmad A, Patnaik S, et al. Advanced application of nano-technological and biological processes as well as mitigation options for arsenic removal. J Hazard Mater. 2021;405: 123885.
69. Kobya M, Soltani RDC, Omwene PI, Khataee A. A review on decontamination of arsenic-contained water by electro-coagulation: reactor configurations and operating cost along with removal mechanisms. Environ Technol Innov. 2020;17: 100519.
70. Ordonez D, Valencia A, Chang NB, Wanielista MP. Synergistic effects of aluminum/iron oxides and clayminerals on nutrient removal and recovery in water filtration media. J Clean Prod. 2020;275: 122728.
71. Carneiro MA, Pintor AMA, Boaventura RAR, Botelho CMS. Current trends of arsenic adsorption in continuous mode: literature review and future perspectives. Sustainability. 2021;13:1186.

72. Luong VT, Cana's Kurz EE, Hellriegel U, Luu TL, Hoinkis J, Bundschuh J. Iron-based subsurface arsenic removal technologies by aeration: a review of the current state and future prospects. Water Res. 2018;133:110–122.
73. Iriel A, Marco-Brown JL, Diljkan M, Trinelli MA, dos Santos Afonso M, Fernandez CA. Arsenic adsorption on iron-modified montmorillonite: kinetic equilibrium and surface complexes. Environ Eng Sci. 2020;37:22–32.
74. Koley S. Future perspectives and mitigation strategies towards groundwater arsenic contamination in West Bengal, India. Environ Qual Manage. 2021;1–23.
75. Ghosh D, Sarkar S, Sengupta AK, Gupta A. Investigation on the long-term storage and fate of arsenic obtained as a treatment residual: a case study. J Hazard Mater. 2014;271:302–10.
76. Singh SK. An analysis of the cost-effectiveness of arsenic mitigation technologies: implications for public policy. Int J Sustain Built Environ. 2017;6:522–35.
77. Odling G, Chatzisymeon E, Karve P, Ogale S, Ivaturi A, Robertson N. Naturally derived carbon for *E. coli* and arsenic removal from water in rural India. Environ Technol Innov. 2020;18:100661.
78. Arsenic Contamination in Groundwater and its Control. Information manual on pollution abatement and cleaner technologies series: IMPACTS14/2000–2001. Central Pollution Control Board, Ministry of Environment and Forests. New Delhi; 2002.
79. R & D Activities related to Arsenic Contamination in Drinking Water—Salient Efforts of Department of Science and Technology (DST). Ministry of science and technology. Government of India; 2015. Available at https://dst.gov.in/sites/default/files/Arsenic-Compendium.pdf.
80. Subterranean Arsenic Removal Technology for in-situ groundwater Arsenic and Iron treatment. Available at www.insituarsenic.org.
81. Sen Gupta B, Chatterjee S, Rottb U, Kauffman H, Bandopadhyay A, DeGroot W. A simple chemical free arsenic removal method for community water supply—a case study from West Bengal, India. Environ Pollut. 2009;157:3351–3.
82. Amrose SE, Gadgil A, Srinivasan V, Kowolik K, Muller M, Huang J, et al. Arsenic removal from groundwater using iron electrocoagulation: effect of charge dosage rate. J Environ Sci Health—Part A Toxic/Hazard Substances Environ Eng. 2013;48:1019–30.
83. van Genuchten CM, Addy SEA, Peña J, Gadgil AJ. Removing arsenic from synthetic groundwater with iron electrocoagulation: an Fe and As K-edge EXAFS study. Environ Sci Technol. 2012;46:986–94.
84. Amrose SE, Bandarua SRS, Delaire C, van Genuchten CM, Dutta A, Deb Sarkar A, et al. Electro-chemical arsenic remediation: field trials in West Bengal. Sci Total Environ. 2014;488–489:539–46.
85. Otter P, Malakar P, Jana BB, Grischek T, Benz F, Goldmaier A, et al. Arsenic removal from groundwater by solar driven inline-electrolytic induced co-precipitation and filtration-a long term field test conducted in West Bengal. Int J Environ Res Public Health. 2017;14:1167.
86. Bordoloi S, Nath M, Dutta RK. pH-conditioning for simultaneous removal of arsenic and iron ions from groundwater. Process Saf Environ Prot. 2012;91:405–14.
87. Bordoloi S, Nath SK, Gogoi S, Dutta RK. Arsenic and iron removal from groundwater by oxidation–coagulation at optimized pH: laboratory and field studies. J Hazard Mater. 2013;260:618–26.
88. Zhu Y, Yang K, Jia R, Shao Y, Zhang A, Yang X. Effects of different additives and ambient conditions on acceleration of sludge stabilization: laboratory-scale simulated landfill study. J Environ Eng. 2020;146:04020012.
89. Kumar A, Joshi H, Kumar A. Remediation of arsenic by metal/metal oxide based nanocomposites/nanohybrids: contamination scenario in groundwater, practical challenges, and future perspectives. Sep Purif Rev. 2020;50:235–6.
90. Lata S, Prabhakar R, Adak A, Samadder SR. As(V) removal using biochar produced from an agricultural waste and prediction of removal efficiency using multiple regression analysis. Environ Sci Pollut Res. 2019;26:32175–88.
91. Maˇnakova B, Kuta J, Svobodova M, Hofman J. Effects of combined composting and vermicomposting of waste sludge on arsenic fate and bioavailability. J Hazard Mater. 2014;280:544–51.
92. Vašiˇckova J, Maˇnakova B, Šudoma M, Hofman J. Ecotoxicity of arsenic contaminated sludge after mixing with soils and addition into composting and vermicomposting processes. J Hazard Mater. 2016;317:585–92.
93. Lei J, Peng B, Min X, Liang Y, You Y, Chai L. Modeling and optimization of lime-based stabilization in high alkaline arsenic-bearing sludges with a central composite design. J Environ Sci Health, Part A. 2017;52:449–58.
94. Roy A, van Genuchten CM, Mookherjee I, Debsarkar A, Dutta A. Concrete stabilization of arsenic-bearing iron sludge generated from an electrochemical arsenic remediation plant. J Environ Manage. 2019;233:141–50.
95. Thakur SL, Goyal H, Mondal P. Simultaneous removal of arsenic and fluoride from synthetic solution through continuous electrocoagulation: operating cost and sludge utilization. J Environ Chem Eng. 2018;7: 102829.
96. Clancy TM, Hayes KF, Raskin L. Arsenic waste management: a critical review of testing and disposal of arsenic-bearing solid wastes generated during arsenic removal from drinking water. Environ Sci Technol. 2013;47:10799–812.

Assessment of the Changing Landscape and Overall Aquifer Condition Due to Saline Water Intrusion: A Review

Bhabani Prasad Mukhopadhyay, Ananya Chakraborty, Rima Roy, and Amitava Bandyopadhyay

Abstract

Seawater intrusion in the low-lying coastal areas is increasing drastically around the globe. Most people of the coastal areas live in extreme poverty and practice agriculture for their livelihood. Anthropological activities have altered the overall landscape of the coastal areas worldwide thereby affecting other characteristics of the associated aquatic sub-systems. These coastal areas are being affected frequently by severe cyclonic storms along with acute flooding. In addition to this, most of the coastal aquifers are under the threat of saline water intrusion and excessive groundwater extraction from the coastal fresh-water wells is one of the major causes. The coastal region faces constant pressure from several water-related threats such as water and soil-salinization, floods etc. Salinity intrusion along with flooding leads to lesser land fertility and greater health hazards in the coastal areas. If this continues, saline water will eventually intrude the shallow coastal aquifers entirely reducing the availability of fresh water in the coastal areas. Hence both the soil and water might get completely salinized over time. This adversely affects the production of vegetables, fruits, crops etc. The widespread growth of mangroves is also stunted as a result of rising humidity and salinity. Proper scientific studies of environmentally significant physicochemical parameters of groundwater are required in the coastal aquifers. This will help to identify the aquifer condition along with the seasonal salinity changes and to develop future planning to minimize the region's vulnerability to natural hazards as well as gradual landscape changes. The proposed study aims to analyze data related to the aquifer property of different coastal aquifer systems focusing on the shallow aquifers. The analyses have been done using secondary data and interpretation of the secondary data shows that salinity of the aquifers increases near coasts. This study is significant in terms of assessment of the risks associated with aquifer contamination and landscape changes. The present study can be used for the management, training and development of the socio-economic condition and the ecosystem of the coastal area.

B. P. Mukhopadhyay (✉) · A. Chakraborty · R. Roy
Hydrogeology Laboratory, Department of Earth Sciences, Indian Institute of Engineering Science and Technology, Howrah, West Bengal, India
e-mail: bpm.earth@faculty.iiests.ac.in

A. Bandyopadhyay
Department of Chemical Engineering, University of Calcutta, Kolkata, West Bengal 700009, India

A. M. Bahadir et al. (eds.), *Hydrology and Urban Water Supply*, Water and Wastewater Management,
https://doi.org/10.1007/978-3-031-72589-0_15

Keywords

Coastal areas · Freshwater quality · Landscape change · Salinity · Saline water intrusion

1 Introduction

Saline water intrusion in the freshwater aquifers of the coastal areas is common problem around the globe. Simultaneously the landscapes are also gradually changing in those areas due to the saline water intrusion. In recent years, landscape approach has become multi-dimensional and different components are being considered. Earlier landscape approach was partial at micro-level in practical aspects. The landscape approach now includes human–environment relationship to protect the social interest and environment as well. It is not only practical but realistic approach at macro-level [29]. Considering the modern landscape approach, inter-relationship among three components, namely saline water, subsurface fresh water and anthropological activities have been discussed in this study to address the problems of the coastal areas.

Groundwater is a valuable natural resource occurring beneath the surface in the land areas around the globe. But this is finite renewable resource and due to lack of understanding on its finiteness, a major portion of the globe is facing serious threats. Various problems of groundwater have become serious topic of research. Coastal aquifers along the stretch of South-East and South Asia act as significant sources of water supply [4]. India is also having an extremely long coastline which also acts as the framework for its national economy. Owing to the ever-increasing urbanization along the Indian coastline, there exists an immediate need for rational usage of groundwater in these regions [22]. Most of these coastal areas are densely inhabited, particularly the areas along a river delta due to the easy and widespread availability of good soil and abundant water resources which are required in order to support a huge number of population [1]. Coastal aquifers are affected by different groundwater problems. Saltwater or saline water intrusion is the migration of seawater into the freshwater coastal aquifers. According to Basack et al. [3], seawater intrusion may also be defined as that coastal hazard and it occurs in two different processes. The seawater may be intruded into the river channels through storm tide and on the other hand, the saline seawater contaminates freshwater coastal aquifers through infiltration. Whatever may be the process, the mixing of saline and fresh waters creates serious problems for human health and socioeconomic conditions as well. These are very common and recurring problems in almost every coastal aquifer. Some factors, in particular, have been recognized which might be the possible reason behind this intrusion. Few of these problems are entirely natural and not so convenient to control, while others are mostly anthropogenic and can be minimized. Of these, the primary cause is that of excessive groundwater extraction from the coastal aquifers as compared to their comparatively low rate of recharge with fresh water. This results in lateral encroachment of seawater from the oceans or seas into the coastal aquifers. Freshwater usually floats on top of seawater due to its lower density [14]. However, the boundary between freshwater and saltwater is not sharp or distinct and a mixing zone occurs, and its thickness depends on the aquifer's hydrodynamics. It is commonly seen that saltwater intrusion takes place owing to the differences in the density between the denser saline water and comparatively less dense freshwater. Seawater intrusion also occurs where groundwater is being extracted from the aquifers which are hydraulically linked to the sea [5]. In such cases, the induced gradient leads to the movement of seawater from the sea into the well, thereby rendering the freshwater almost unusable. The extent of seawater intrusion varies depending on the geographic location as well as the hydro-geologic settings of the area. Contamination of saltwater may be restricted to only a small portion of the aquifer and thereby having little or almost no effect upon the wells being pumped for groundwater supply. On the contrary, contamination might also occur in regional extent while having substantial effects on the groundwater supplies. Water having higher salinity levels adversely affects plant growth, germination, and crop yields [28]. Deviation in the groundwater quality is a consequence of all the physical and chemical parameters which, in turn, are entirely influenced by the groundwater recharge-discharge techniques, geological formations as well as anthropogenic activities. Often overexploitation of groundwater resources from coastal aquifers accounts for seawater intrusion into those aquifers [30]. It is important to mention that there exists no direct connection between excessive groundwater extraction and the extent of groundwater pollution. Interference among the different extraction sources may create large cone of depression. Lateral transport of pollutants in the flow path approaching the cone of the depression in the pumping wells is a commonly observed phenomenon [26]. At present, intrusion of saltwater occurs in most of the coastal aquifers around the globe. With the growing usage of groundwater, especially for drinking and agricultural purposes, it is important to analyze the migration of seawater into the coastal aquifers in order to predict their movement and find out appropriate remedial measures [24] depending on the geologic and geographic aspects of the area. Industrial usages of groundwater have not been considered in the study as industries are not normally found in coastal areas, especially in the present study area. However, the pattern as well as the degree of saltwater intrusion into a coastal aquifer greatly depends on numerous factors as already stated resulting in highly deteriorated subsurface water quality. Salinity of groundwater can be categorized into two distinct types such as coastal salinity which occurs from seawater intrusion in the coastal aquifers and inland salinity which occurs from processes other than seawater ingression. The present study aims emphasizes mainly on the former type of groundwater salinity occurring due to seawater intrusion.

2 Study Area

The study area lies between 21°32′ N to 22°40′ N latitudes and 88°05′ E to 89°51′ E longitudes, covering an area of about 10,000 km^2, of which nearly 62% of the areas fall in Bangladesh and the rest lies in West Bengal, India. The Sundarbans form the biggest contiguous mangrove reserve found on the earth and is considered to be a deltaic mangrove forest which was formed years ago by sediment deposition from the Himalayas which were carried downhill by the Ganges River [19]. This mangrove ecosystem occupies a major portion in the southern region of West Bengal, India as well as the entire southwest Bangladesh (Fig. 1).

The Sundarbans experiences humid tropical monsoon climate due to its proximity to the sea. Temperature ranges from about 34 °C in summer to nearly 20 °C in winter with an annual temperature range of 8 °C [8]. It further experiences a dry season from November to April while a wet monsoonal period exists during the other months. The total annual precipitation ranges between 1500 and 2000 mm [8]. Small tides and tropical cyclones are very common during the monsoon season which regularly hit the area thereby causing severe floods and other damages. This mangrove wetland is spread over a vast portion of the Ganges–Brahmaputra-Meghna (GBM) delta having a mean elevation of nearly 0.9–2 m above the mean sea level [12, 19]. In Bangladesh, Sundarbans occupies the districts of Bagerhat, Khulna and Satkhira while in India it mainly lies in the districts of North and South 24 Parganas in West Bengal. Widespread erosion along with rapid accretion of these weathered materials has resulted in the formation of various landscapes which are mostly fluvial and tidal [8]. There exist numerous variables for evaluating the salinity in the waters of the Sundarbans. However, the present study focuses on assessing the changing landscape and overall aquifer condition due to saline water intrusion in the Sundarbans. The re-appraisal of the cause of salinity intrusion and its adverse impacts upon the mangrove wetland ecosystems using secondary data will

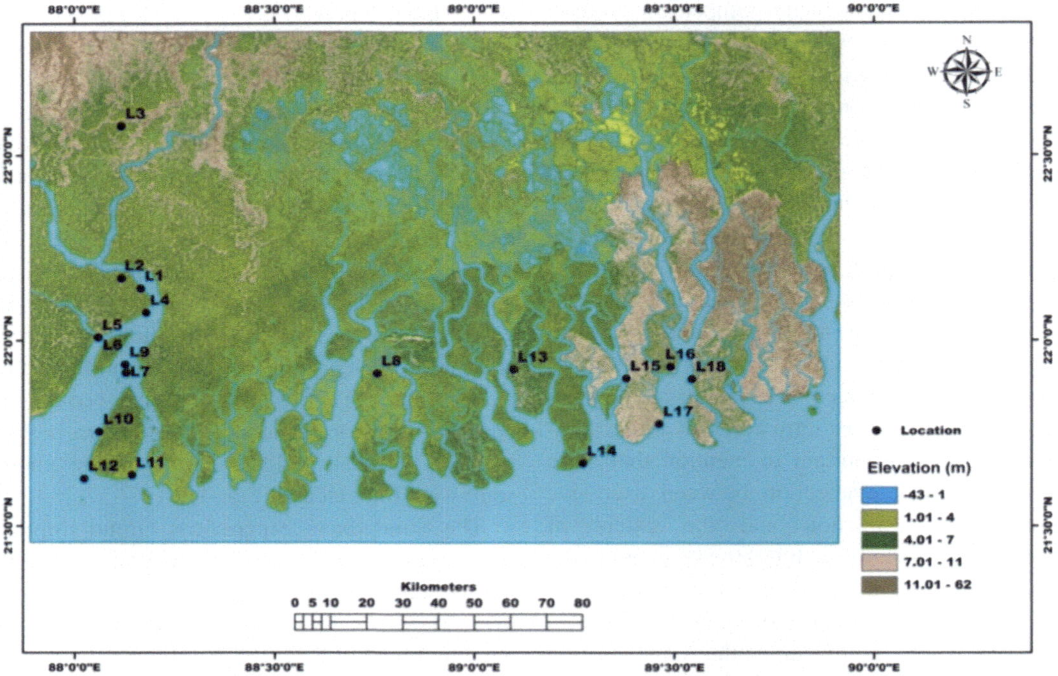

Fig. 1 Location map of the study area

add inputs on the overall understanding about the aquifer systems and in planning for conservation of the mangrove ecosystems of Sundarbans.

3 Results and Discussion

3.1 Lithology and Aquifers of the Study Area

Seawater intrusion into the coastal aquifers is controlled by different geological factors. The inland flow of saltwater inside the coastal aquifer depends primarily on the lithological associations and their properties. Geological formations with varying porosity and permeability have distinctive water bearing capacities [13]. The seawater flooding into the inland areas may additionally get trapped inside porous litho-units while the impervious clay layers which are found in small patches in coastal regions may retain the paleo-seawater. These patches of clay layers may create perched aquifers in non-seawater flooding regions. Hence, the interplay of freshwater and seawater relies upon the presence of pervious and impervious layers in the lithological associations in coastal regions.

The formation of the Sundarbans-delta began in the Tertiary period. A major portion of the delta was formed by the deposition of various sediments of recent origin. These sediments were derived from widespread soil erosion from Himalayas [13]. The basement of the delta consists of sand, silt and other sediments from the Quaternary Era together with various marine deposits and clay [6]. Deltaic formations dominate almost the entire Sundarbans together with various splays, levees as well as tidal flats.

The porosity of the aquifers in the Sundarbans varies depending on the sediment composition and the degree of compaction. Generally, the porosity of the lithological units in the Sundarbans varies from 20 to 35% [6]. The permeability is relatively low due to the fine-grained sediment and high degree of compaction. The permeability of the Sundarbans litho-associations ranges from

10–4 to 10–6 cm/s which indicates low movement of groundwater through these aquifers [16]. However, the porosity and permeability of the lithological formations hosting aquifers of Sundarbans can vary significantly depending on the location, depth, and geological history of the aquifer. Additionally, human activities such as extraction of groundwater and changes in land-usages can also effects the porosity and permeability of these aquifers.

3.2 Physical Parameters

3.2.1 pH

In general, the pH in the study area exhibits a slight variation. The spatial variation of pH in the study area is presented in Fig. 2. The values range from 7.6 to 8.20 with a mean value of 7.9 [20]. The comparatively greater pH values are mainly noticed during high tide which indicates the possibility of saline water intrusion from the adjacent Bay of Bengal in selected locations. The development of funnel shaped delta towards the Bay of Bengal allows maximum infiltration of seawater into the upstream area producing an overall alkaline effect even in the extremely uppermost locations. However, the range of pH towards the downstream zone is nearly 8–8.15, which is absolutely normal at the zone of confluence with the Bay of Bengal [25].

3.2.2 Alkalinity

The alkalinity of saline water maybe considered equal to the stoichiometric sum of all the bases present in solution. However, in the present study area, the major constituents responsible for alkalinity within the current geographical locale mostly include few carbonate rocks and other chemical components like phosphates, nitrates, ammonia etc. which are mainly contaminated from the nearby sewage and municipal wastes. The spatial variation of alkalinity in the study area is presented in Fig. 3.The values of

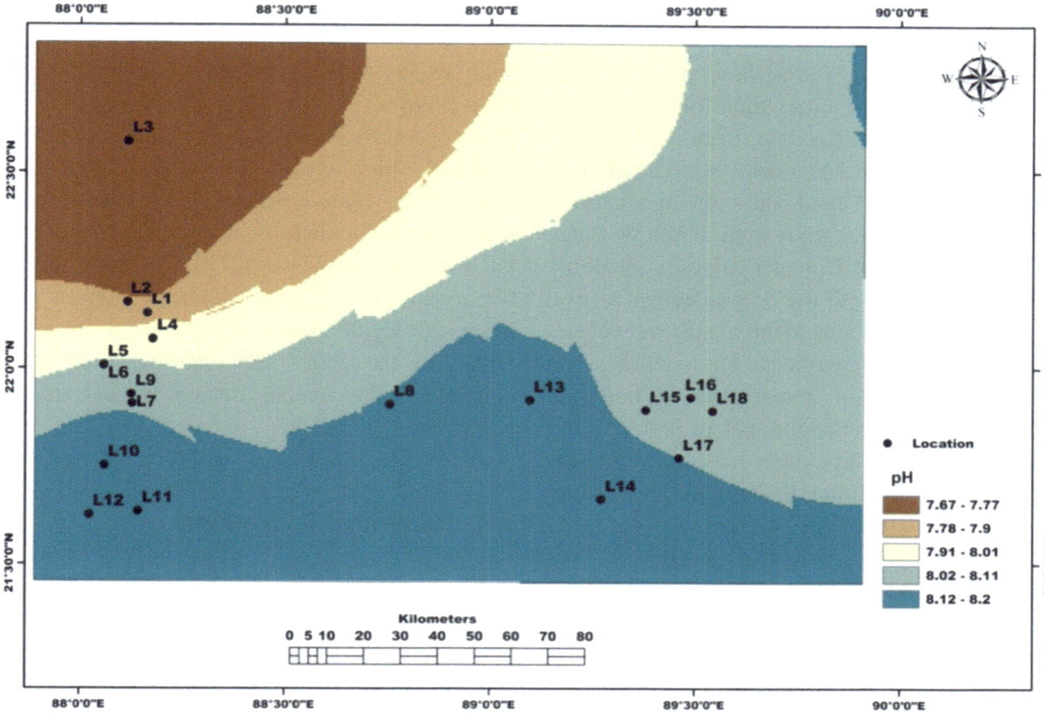

Fig. 2 Spatial variation of pH in the study area

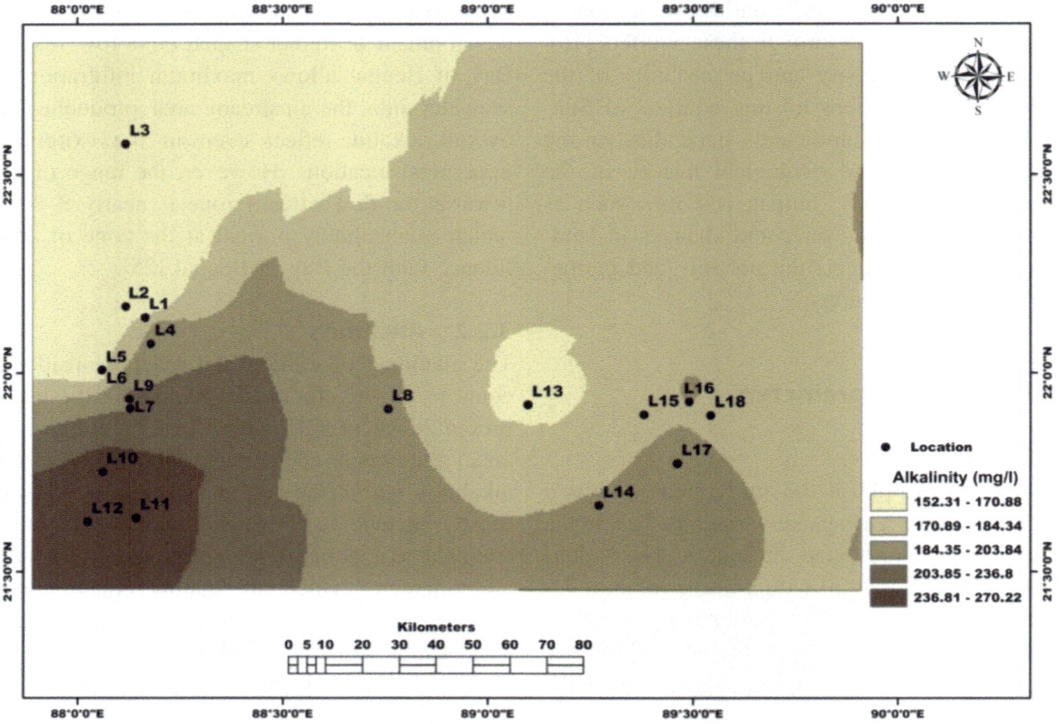

Fig. 3 Spatial variation of Alkalinity in the study area

alkalinity in the study area varies from a minimum of 155 mg/l to a maximum of 270 mg/l as measured in high tides and from 137 mg/l to around 258 mg/l in low tides [20]. Therefore, a notable spatial and tidal variation in alkalinity was confirmed in our study area. It can be stated that the comparatively greater values of alkalinity indicate a proximity to the Bay of Bengal as well as the presence of abundant mangrove vegetation. Alkalinity is significant for the fishes and other aquatic life as it protects them by buffering against the rapidly changing pH values [23]. For living organisms, particularly for aquatic life, the best survival condition is with the pH values ranging from 6.0 to 9.0.

3.2.3 Temperature

In the study area, the variation of temperature in different locations and the spatial variation of temperature are presented in Figs. 4 and 5 respectively. The temperature of water in the study region varies between 35 and 36.5 °C with an average of 35.75 °C [20]. The homogeneity in the temperature of water is specifically due to the extremely high specific heat capacity of water that, in turn, allows water to withstand extreme temperature fluctuations as compared to that of the adjacent landmasses [21]. In addition to this, the aquatic sub-systems within the current geographical locale thus function as a balancing factor on the entire temperature profile while conserving the overall deltaic biodiversity of severe thermal shock.

3.3 Chemical Parameters

3.3.1 Sodium

The value of sodium in the aquifer system of the study area varies from 715 to 6619 mg/l at high tides and from 436 to 6100 mg/l at low tides [21]. The spatial variation of sodium in the study area is

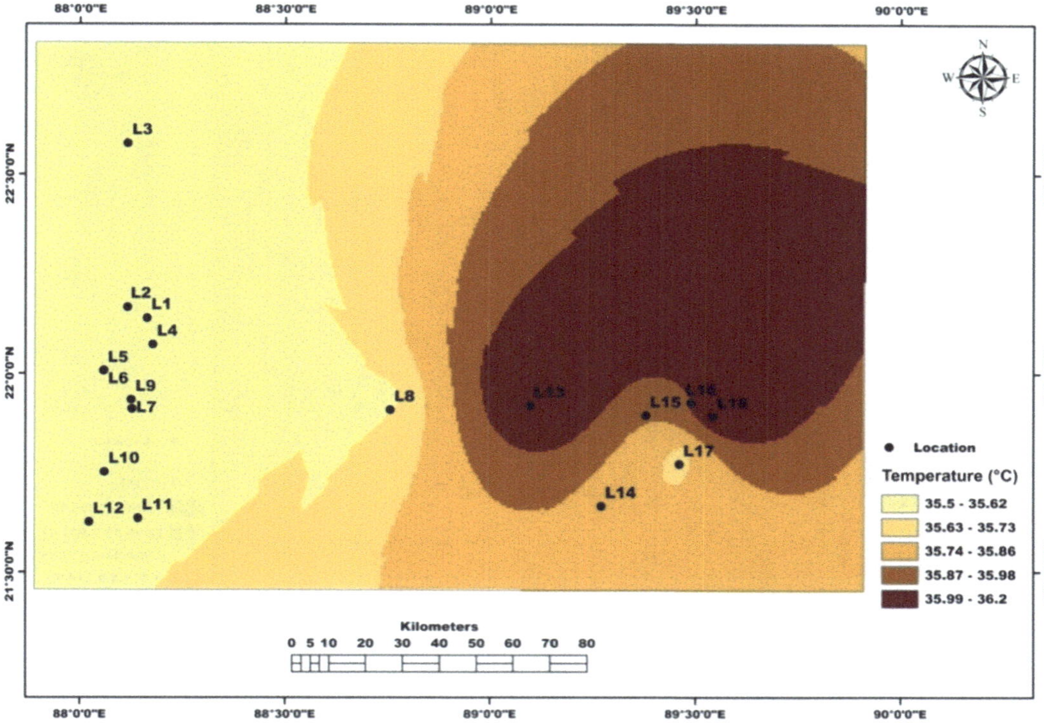

Fig. 4 Spatial variation of Temperature in the study area

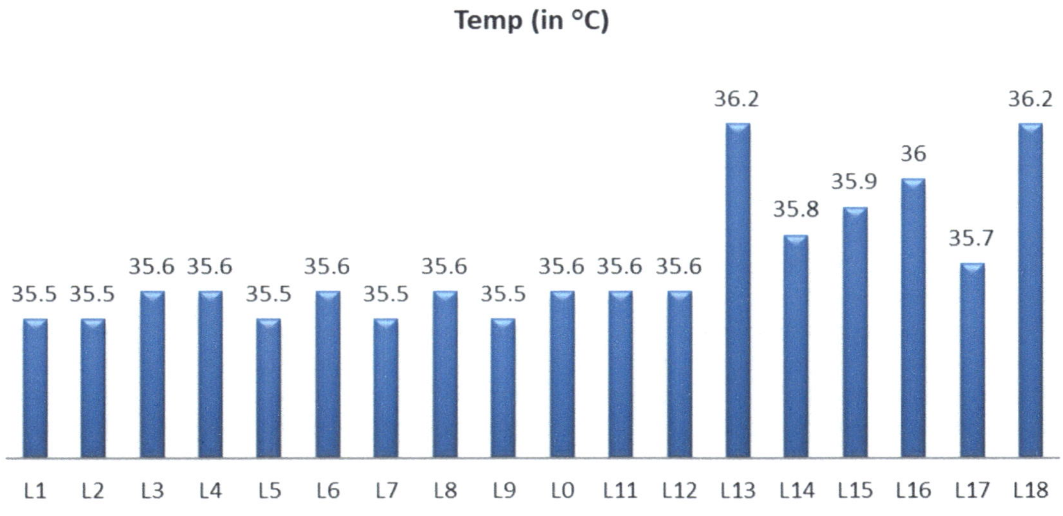

Fig. 5 Temperature variation in different locations in study area

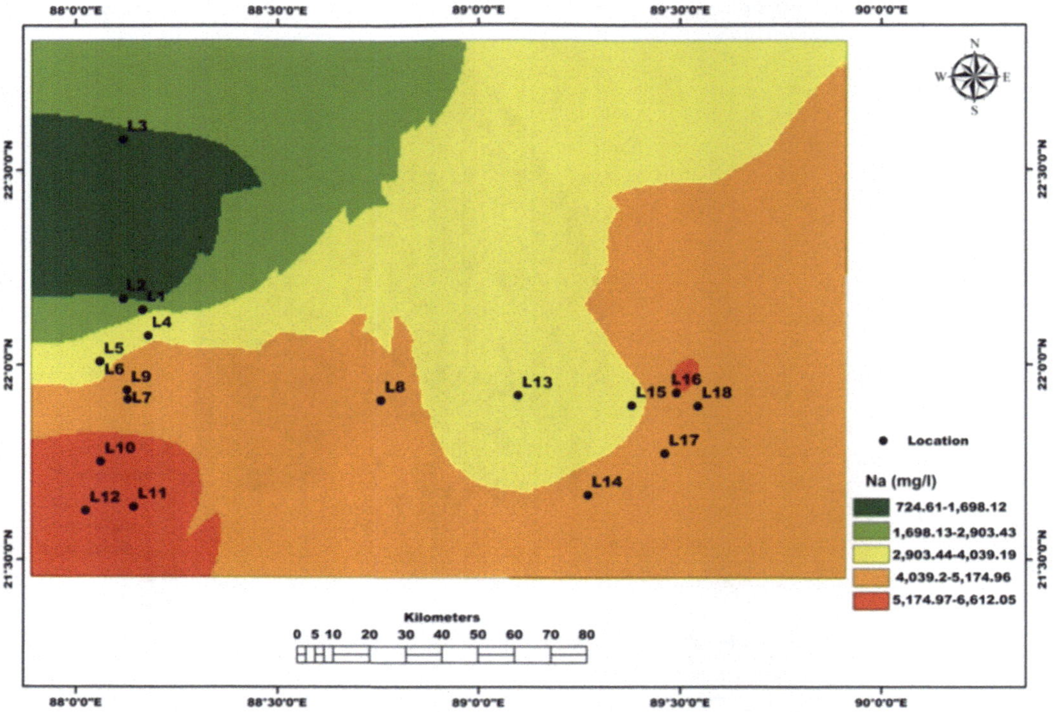

Fig. 6 Spatial variation of sodium in the study area

presented in Fig. 6. It demonstrates how the major cation present within the aquifer system contributes to nearly 28% of the total surface water salinity. According to Gopinath et al. (2019), these values of sodium can be easily compared with the sodium level as found in the groundwater of the Pichavaram mangrove or the Vellar estuary as well as the Coleroon estuary of southeastern coast in India. However, the progressively growing value of sodium in different locations can be attributed to the increasing proximity of those locations to the Bay of Bengal thus resulting in widespread intrusion of the seawater into the coastal aquifers from the nearby bay.

3.3.2 Potassium

Generally seawater contains nearly 400 ppm of potassium which mostly settles and finally accumulates in the sediments [18]. On the contrary, river water usually contains around 2–3 ppm of potassium which is extremely low in comparison to seawater. This huge difference can be caused mainly due to the huge concentration of potassium within the oceanic basalts that ultimately gets incorporated within seawater through various weathering activities [17]. In addition to this, various other clay minerals containing high quantities of potassium are mixed up with seawater by different natural processes and ultimately settle in the sediments. The spatial variation of potassium in the study area is presented in Fig. 7. The value of potassium is seen to range from 27 to 280 mg/l [18]. Unlike most of the other variables, it is found that there is a considerable spatial variation of potassium as a cation but the overall difference in values of potassium content between high tides and low tides is not considerable. Higher values of potassium within locations in lower estuary occur owing to additional clay deposition from Bay of Bengal. The widespread erosion within the lower stretch maybe regarded as a significant source of several clay particles which are enriched in potassium.

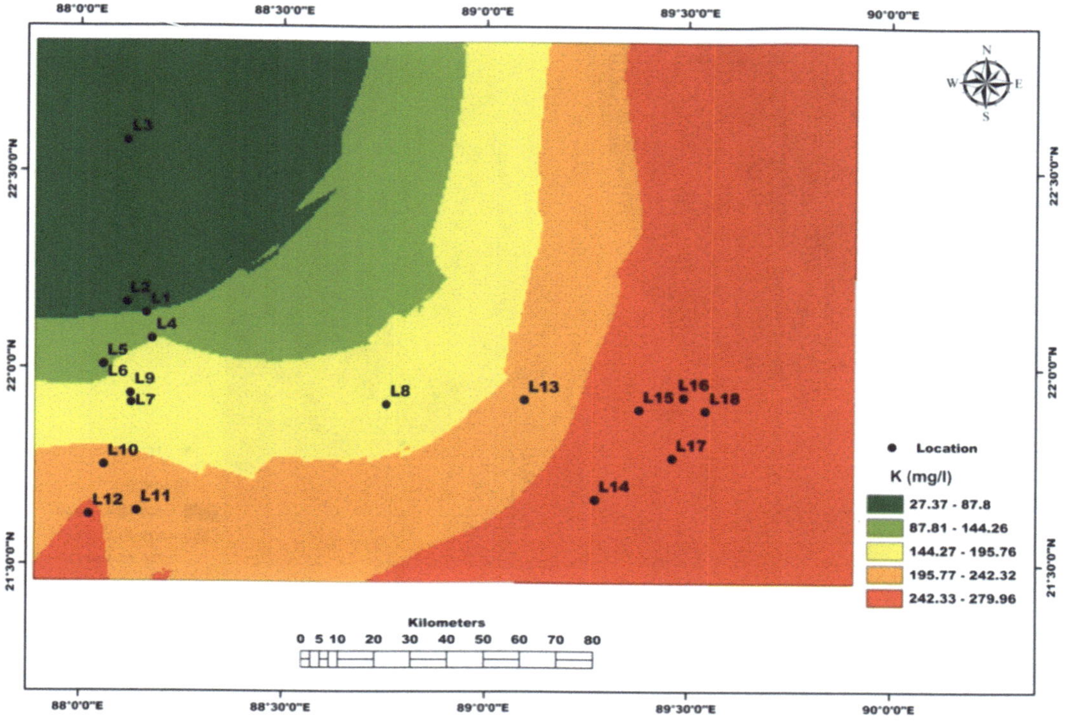

Fig. 7 Spatial variation of potassium in the study area

3.3.3 Chloride

Chloride can be regarded as one of the major ions as present in seawater. It regulates the overall salinity of the estuarine system. It can be said that the overall level of chloride present within the seawater strongly exerts a significant influence upon the total biotic community of both the marine as well as estuarine system [15]. In case of saline water, the fishes and other aquatic organisms regulate their body fluids to prevent unnecessary water loss by osmo-regulation. According to Islam and Gnauck (2008), chloride ions cause damage to the aquatic lives at a concentration of 0.1 mg/l. The spatial variation of chloride in the study area is presented in Fig. 8. The value of chloride ranges from 1305 to 15,967 mg/l [15]. There exists a significant tidal and spatial variation in the chloride ions within the study area which confirm the strong influence of saltwater intrusion into coastal aquifers of the Gangetic delta region. In addition to that, the high salinity of soil and groundwater near the Sundarbans is a significant reason for high levels of chloride as found in the study area.

3.3.4 Salinity

The overall value of surface water salinity varies from 2.48‰ at L1 to 27.30‰ at L14 as found in figure no 8 [11]. The spatial variation of salinity and electrical conductivity in the study area are presented in Figs. 9 and 10. It is found that the value of salinity decreases from downstream to upstream zone with a considerable spatial variation. High freshwater discharge by rivers may be responsible for very low salinity values at L1. According to Islam and Gnauck (2011), the declining salinity values in the aquifers as we move away from the coastal areas may be regarded due to riverine contribution. In the current study area, high freshwater discharge from nearby rivers such as Rupnarayan, Hugli and the Damodar is the major reason for most locations experiencing lower salinity levels even during the pre-monsoon. In addition to this, high

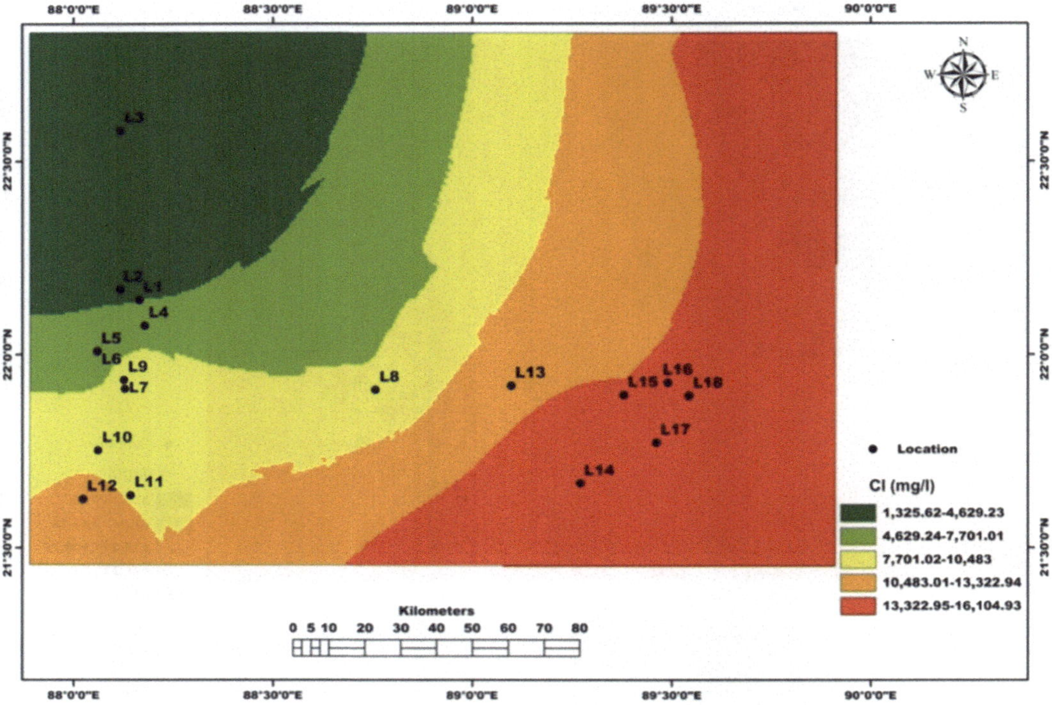

Fig. 8 Spatial variation of chloride in the study area

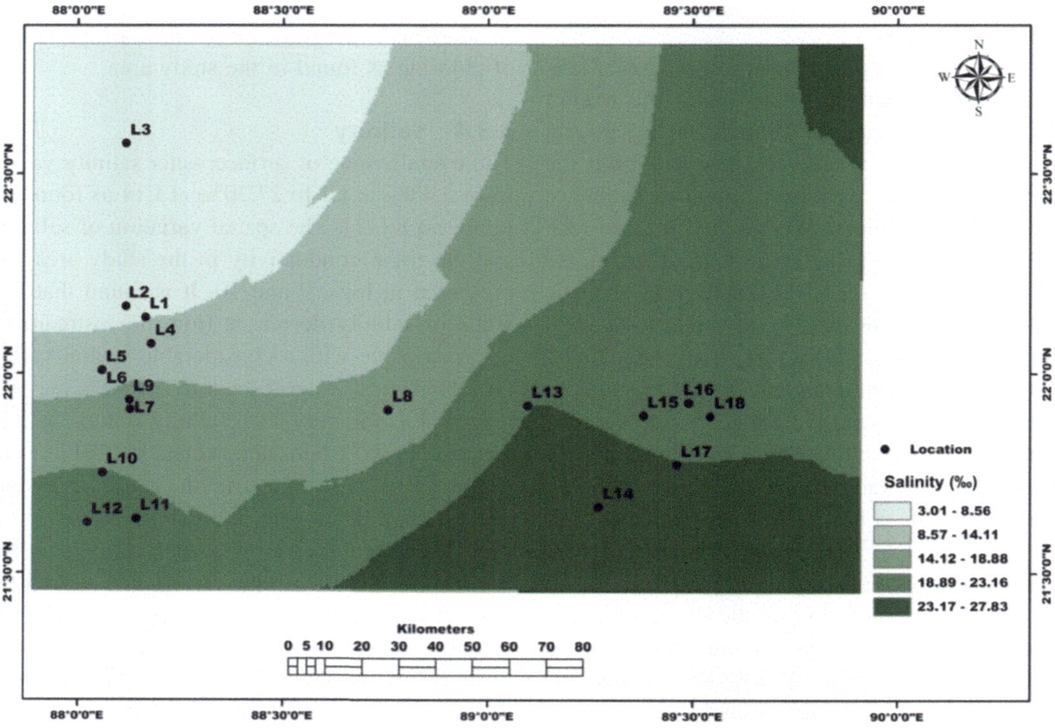

Fig. 9 Spatial variation of salinity in the study area

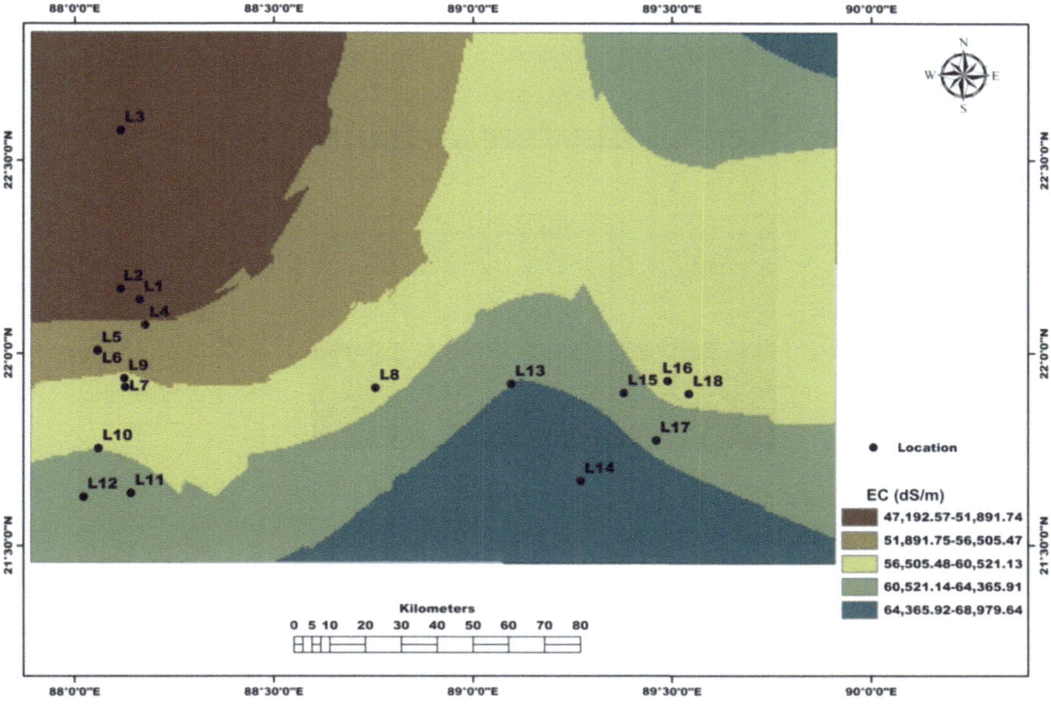

Fig. 10 Spatial variation of electrical conductivity in the study area

level of freshwater discharge near Farakka Barrage significantly impacts the overall salinity profile within the study region. During conditions of high tide, huge quantity of saline water enters the coastal aquifers from the Bay of Bengal thus rising the overall salinity of the coastal aquifers due to saline water intrusion [7]. Nevertheless, under conditions of low tide, high level of freshwater discharge near Farakka causes significant lowering of the overall salinity in the coastal aquifers of the study area.

4 Evaluation of the Effect of Lithology on Hydrogeochemistry of Groundwater in Sundarbans

The effect of lithology on the hydro geochemistry of groundwater is a complex process that depends on the physico-chemical properties of the rocks and soils through which the groundwater flows [21]. The geology and the lithology of the rocks and soils present in the study area should be considered as important aspects to understand the effect of lithology on the hydro-geochemistry of the groundwater. This can be done by field observation, geological mapping or studying well logs.

The lithology of Sundarbans is predominantly alluvial consisting of various sedimentary rocks, namely sand, silt and clay as shown in Fig. 11. These sediments have deposited within the period of Tertiary to Recent. The groundwater samples which have been collected from the wells were analyzed to assess a range of chemical parameters as shown below in Fig. 12. In the study area, sand is dominant [12]. Sand acts as a natural filter through which groundwater can easily flow thereby removing contaminants or impurities present. Sand is predominantly composed of silica, calcium and magnesium particles which easily dissolve into groundwater [23]. In Sunderbans, the sand is mostly silica rich which

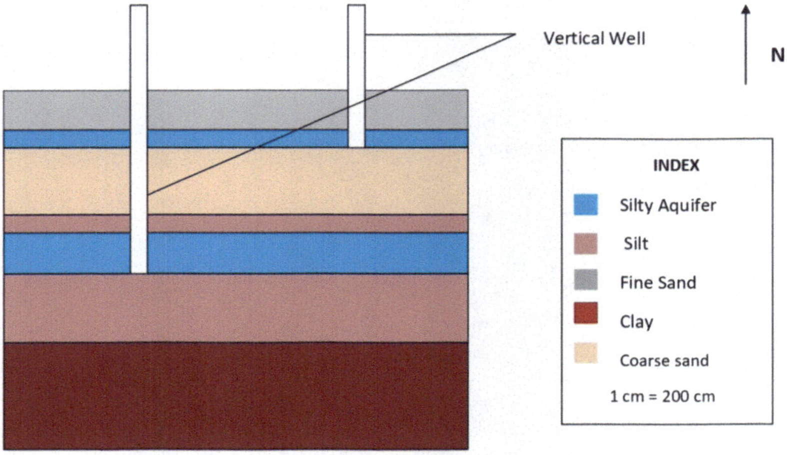

Fig. 11 Geological cross section showing the aquifers in the study area

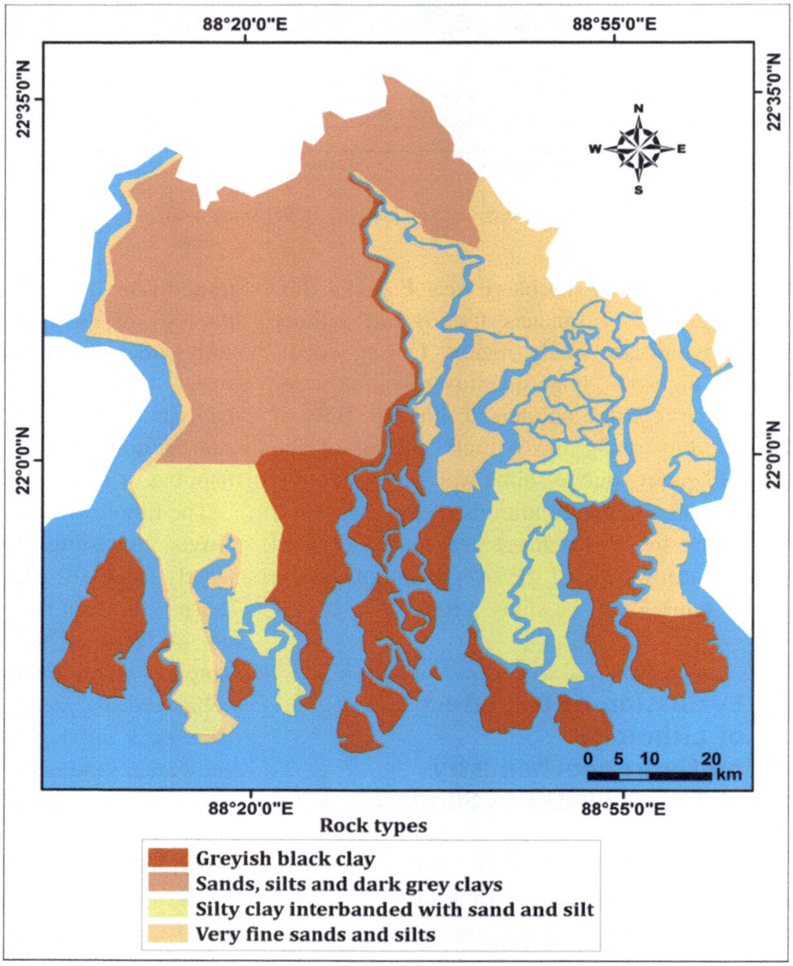

Fig. 12 Geological map of the study area

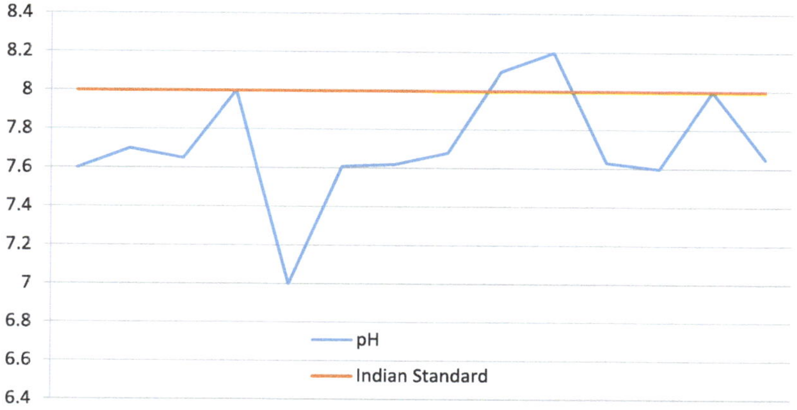

Fig. 13 pH variation in different locations in study area

results in low pH values of groundwater discussed earlier (Fig. 13).

The silt and sand particles further act as adsorbents for the dissolved ions and other organic compounds in groundwater. As a result, the overall concentration and distribution of various chemical compounds in groundwater is changed. Similar to sand, both silt and clay particles also affect the pH of groundwater by either absorbing or releasing hydrogen ions. Most importantly, the presence of silt and clay particles can increase the concentration of certain ions and minerals in groundwater. It is already mentioned that the weathering of clay mineral releases aluminium, iron and other metals into the groundwater which are responsible for acidic nature of water. They also act as a major source of energy and nutrients for the microorganisms present in groundwater. The chemical properties of groundwater in Sundarbans are influenced not only by its lithology but also by the complex interplay of riverine processes, tidal dynamics and organic accumulations. Other factors which affect groundwater chemistry in Sundarbans are the existing land use practices, climatic and the hydrologic conditions of the study area.

5 Conclusion

The Sundarbans are regarded as an invaluable component of world's biodiversity. Conservation of this biodiversity hotspot is required with effective measures [10]. Considering the present scenario, it is important to plan for a proper long-term management to utilize sustainably the mangrove ecosystems along with its various services. The increasing salinity owing to lack of abundant flow of freshwater inside the trans-boundary river basin have resulted in widespread damage to entire Sundarbans as well as the coastal wetlands ecosystems [3]. Groundwater constitutes a very significant component of the environment and it is extremely essential to identify the availability of groundwater resources within the study area along with the assessment of quality of groundwater. In these coastal areas specifically, extraction of groundwater for several purposes such as domestic and agricultural usage degrades the groundwater quality by giving way to the ingression of saline water in response to the withdrawal of freshwater [2]. At times, this might lead to serious adverse consequences. The growing saline water intrusion in the coastal aquifers from the Bay of Bengal is a matter of concern as it degrades the overall water quality making it unfit for drinking purposes [6]. Groundwater levels in the coastal aquifers are seen to be the highest during the months of August to October while it is lowest between the months of April to May [27]. A sharp rise in the groundwater levels occurs usually in May which continues until the month of July. The consistent reduction as well as deterioration of the water quality in the riverine freshwater catchment has been identified as the root cause for increased salinity intrusion in the coastal aquifers thus causing huge damage to the entire Sundarbans ecosystem. According to Gobster and Xiang [9] landscape is an integrative concept which focuses on land use, ecological system, human–environment interaction, mitigation measures through design, planning and management. Following the views of Gobster and Xiang [9], Sundarban-delta is now vulnerable with respect to the landscape and it is obvious from the discussions that saline water intrusion is responsible for the vulnerability. Habitable and agricultural lands are decreasing rapidly due to quantitative and qualitative degradation of fresh water as a consequence of saline water contamination. The characteristic coastal wetland ecosystem of Sunderban-delta is also under threat. As a whole the landscape of the study area is changing keeping pace with the depletion of freshwater aquifer. Studying the current situation of salinity intrusion within different coastal aquifers in and around the Sundarbans, it may be proposed that a proper management, integrated research as well as awareness program are needed to protect and conserve the entire mangrove wetland ecosystem from further degradation.

Acknowledgements The authors are grateful to the Central Groundwater Board, Govt. of India for providing information related to the saline water intrusion in the coastal areas.

References

1. Aziz A, Paul AR. Bangladesh Sundarbans: present status of the environment and biota. Diversity. 2015;7(3):242–69. https://doi.org/10.3390/d7030242 .
2. Barua S, Mukhopadhyay BP, Bera A. Hydrochemical assessment of groundwater for irrigation suitability in the alluvial aquifers of Dakshin Dinajpur district, West Bengal, India. Environ Earth Sci. 2021;80:1–14. https://doi.org/10.1007/s12665-021-09832-y
3. Basack S, Loganathan MK, Goswami G, Baruah P, Alam R. Review of risk assessment and mitigation measures of coastal aquifers vulnerable to saline water intrusion. Pol J Environ Stud. 2022;31(2). https://doi.org/10.15244/pjoes/142382
4. Basu JP, Basu A. Climate change vulnerability and resource dependent communities: an empirical study in coastal Sunderban, West Bengal, India. Ecol Environ Conserv. 2020;26(2):776–772.
5. Bhadra T, Hazra S, Ray SS, Barman BC. Assessing the groundwater quality of the coastal aquifers of a vulnerable delta: a case study of the Sundarban Biosphere Reserve, India. Groundw Sustain Develop 11:100438. https://doi.org/10.1016/j.gsd.2020.100438
6. Das GK. Changing environmental scenario of Sunderbans, India. Int Res J Environ Sci. 2021;10(2):70–7.
7. Dasgupta S, Akhter Kamal F, Huque Khan Z, Choudhury S, Nishat A. River salinity and climate change: evidence from coastal Bangladesh. In: World scientific reference on Asia and the world economy; 2015. p. 205–42. https://doi.org/10.1142/9789814578622_0031
8. Ghosh S, Mistri B. Coastal agriculture and its challenges: a case study in Gosaba Island, Sundarban, India. Space Cult. 2020;8(2):140–154. https://doi.org/10.20896/saci.vi0.747
9. Gobster PH, Xiang WN. What do we mean by "landscape"? Landsc Urban Plan. 2012;3(106):219–20.
10. Islam SN. Sundarbans a dynamic ecosystem: an overview of opportunities, threats and tasks. The Sundarbans: a disaster-prone eco-region; 2019. p. 29–58. https://doi.org/10.1007/978-3-030-00680-8_2
11. Islam SN, Gnauck A. Water salinity investigation in the Sundarbans rivers in Bangladesh. Int J Water. 2011;6(1–2):74–91.
12. Islam SN, Gnauck A. Mangrove wetland ecosystems in Ganges-Brahmaputra delta in Bangladesh. Front Earth Sci China. 2008;2(4):439–48. https://doi.org/10.1007/s11707-008-0049-2.
13. Jeyapaul VK, Antony Jebamalai AR, Selvam RA, SudarsananKrishnaveni A, Antony Johnson AAP, A case study of freshwater discharge in porous calcarenite formation in coastal terrace at Manapad, South India. Carbonates Evaporites. 2020;35:1–14. https://doi.org/10.1007/s13146-020-00641-1
14. Kanagaraj G, Elango L, Sridhar SGD, Gowrisankar G. Hydrogeochemical processes and influence of seawater intrusion in coastal aquifers south of Chennai, Tamil Nadu, India. Environ. Sci Pollut Res. 2018;25(9):8989–9011. https://doi.org/10.1007/s11356-017-0910-5
15. Khan AF, Srinivasamoorthy K, Prakash R, Rabina C. Hydrochemical and statistical techniques to decode groundwater geochemical interactions and saline water intrusion along the coastal regions of Tamil Nadu and Puducherry, India. Environ Geochem Health. 2021;43(2):1051–1067, 2021.https://doi.org/10.1007/s10653-020-00713-0
16. Krishan G, Prakash J, Dasgupta P, Mackenzie A, Devi TT. Groundwater salinity—impacts and possible remedial measures and management solutions. In: Transforming coastal zone for sustainable food and income security: proceedings of the international

symposium of ISCAR on coastal agriculture, March 16–19, 2021. Cham: Springer International Publishing; 2022. p. 697–703.
17. Kumar P, Tiwari P, Biswas A, Acharya T. Geophysical and hydrogeological investigation for the saline water invasion in the coastal aquifers of West Bengal, India: a critical insight in the coastal saline clay–sand sediment system. Environ Monitor Assess. 2020;192(9):1–22. https://doi.org/10.1007/s10661-020-08520-x
18. Loc HH, Lixian ML, Park E, Dung TD, Shrestha S, Yoon YJ. How the saline water intrusion has reshaped the agricultural landscape of the Vietnamese Mekong Delta, a review. Sci Total Environ. 2021;794: 148651. https://doi.org/10.1016/j.scitotenv.2020.143919.
19. Maity PK, Das S, Das R. A geochemical investigation and control management of saline water intrusion in the coastal aquifer of Purba Midnapur district in West Bengal, India. J Indian Chem Soc. 2018;95 (3):205–10.
20. Mitra A, Mondal K, Banerjee K. Spatial and tidal variations of physico-chemical parameters in the lower Gangetic delta region, West Bengal, India. J Spatial Hydrol. 2011;11(1).
21. Mondal M, Biswas A, Haldar S, Mandal S, Mandal P, Bhattacharya S, Paul S. Climate change, multi-hazards and society: an empirical study on the coastal community of Indian Sundarban. Nat Hazards Res. 2022. https://doi.org/10.1016/j.nhres.2022.04.002
22. Mukhopadhyay BP, Chakraborty A, Bera A, Saha R. Suitability assessment of groundwater quality for irrigational use in Sagardighi block, Murshidabad district, West Bengal. Appl Water Sci. 2022;12 (3):38. https://doi.org/10.1007/s13201-021-01565-4.
23. Nath A, Samanta S, Banerjee S, Danda AA, Hazra S. Threat of arsenic contamination, salinity and water pollution in agricultural practices of Sundarban Delta, India, and mitigation strategies. SN Appl Sci. 2021;3(5):1–15. https://doi.org/10.1007/s42452-021-04544-1
24. Park E, Loc HH, Van Binh D, Kantoush S. The worst 2020 saline water intrusion disaster of the past century in the Mekong Delta: impacts, causes, and management implications. Ambio. 2022;51(3):691–9. https://doi.org/10.1007/s13280-021-01577-z.
25. Rashid B, Ahsan K, Khan Majlis AB, Ahsan K, Mahmud A. Sedimentation and coastal area management in the human-modified Ganges-Brahmaputra tidal delta plain of Bangladesh. Int J River Basin Manage. 2022. p. 1–37. https://doi.org/10.1007/s12524-022-01558-z
26. Sabareshwari V, Ramya A. Coastal saline soils of India: a review. Agric Rev. 2018;39(1):86–8.
27. Saha S, Saha S. Toward sustainable conservation plan for fragile ecosystem in Indian Sunderban. In: Urban and transit planning. Cham: Springer; 2020. p. 193–202. https://doi.org/10.1007/978-3-030-17308-1_18
28. Sahana M, Rehman S, Ahmed R, Sajjad H. Analyzing climate variability and its effects in Sundarban Biosphere Reserve, India: reaffirmation from local communities. Environ Develop Sustain. 2021;23 (2):2465–2492. https://doi.org/10.1007/s10668-020-00682-5
29. Sayer J, Sunderland T, Ghazoul J, Pfund JL, Sheil D, Meijaard E, Buck LE. Ten principles for a landscape approach to reconciling agriculture, conservation, and other competing land uses. Proc Natl Acad Sci. 2013;110(21):8349–8356. https://doi.org/10.1073/pnas.121059511
30. Tran DA, Tsujimura M, Pham HV, Nguyen TV, Ho LH, Le Vo P, Ha KQ, Dang TD, Van Binh D, Doan QV. Intensified salinity intrusion in coastal aquifers due to groundwater overextraction: a case study in the Mekong Delta, Vietnam. Environ Sci Pollut Res. 2022;29(6):8996–9010. https://doi.org/10.1007/s11356-021-16282-3.

Drinking Water

A Case Study on Controlling Real Losses in Large-Scale Water Distribution Systems

Tugba Akdeniz, Cem Cakmak, and I. Ethem Karadirek

Abstract

Real losses in water distribution systems stand for the water that is physically lost. In this study, results of implementing intervention tools to reduce real losses in a large-scale water distribution system are presented. Findings proved that pressure management and active leakage control studies are effective tools to reduce real losses. On the other hand, pipeline and asset management are crucial for sustainable management of water losses, as around 76% of pipe failures in 2021 occurred in service connection pipes. Water loss management strategies should be based on a continuous process as water distribution systems are dynamic systems. The rate of water losses in 2021 is around 43% of the system input volume. After comprehensive work to reduce real losses, a significant amount of real losses was reduced.

Keywords

Leakage · Real losses · Water losses · Water distribution systems

1 Introduction

The water loss problem in water distribution systems is a crucial issue that should be tackled for sustainable management of water resources, which are currently under pressure due to climate change, population growth, and urbanization. Real and apparent losses are the main components of water losses in water supply systems [1–3]. Apparent losses in water distribution systems stand for the water used but not paid for, whereas real losses are the amount of water that is physically lost [3]. Reducing water losses in any water distribution system allows for reducing the energy that is required for the abstraction and distribution of water, reducing consumption of chemicals used for treatment/disinfection, and reducing the pressure on water demand [3, 4]. There are four basic methods to control the real losses in water distribution systems: (i) pressure management, (ii) active leakage control, (iii) speed and quality of repairs, (iv) pipeline and asset management [2].

Antalya is one of the largest tourism destinations of Türkiye, which is located along the Mediterranean coast. The amount of water losses

T. Akdeniz · C. Cakmak
Antalya Water and Wastewater Administration of Antalya Metropolitan Municipality (ASAT), Antalya, Turkey

I. E. Karadirek (✉)
Department of Environmental Engineering, Akdeniz University, Antalya, Turkey
e-mail: ethemkaradirek@akdeniz.edu.tr

A. M. Bahadir et al. (eds.), *Hydrology and Urban Water Supply*, Water and Wastewater Management,
https://doi.org/10.1007/978-3-031-72589-0_16

is relatively high and real losses due to pipe failures in Antalya cannot be easily detected, as the province is located in a karstic region. Many studies have been carried out in Antalya city to control water losses [5–8]. This chapter aims to provide studies carried out to control real losses in Antalya city.

2 Pilot Study Area

Antalya province consists of 19 districts and the Antalya Water and Wastewater Administration of Antalya Metropolitan Municipality (ASAT) provides water services to the whole province. In this study, five central districts (Aksu, Dosemealti, Kepez, Konyaalti, and Muratpasa) were selected as pilot study area (PSA) for the evaluation of water losses management strategies. The PSA has highly varying elevations from sea level up to 320 m. Due to highly varying elevations, the PSA has been divided into eight pressure zones for better management of drinking water services (Fig. 1).

Water supplied to the PSA is abstracted from 148 groundwater wells and a spring namely Gurkavak. The abstracted water is pumped to the PSA after disinfection and distributed to around 1.5 million people (TUİK 2021). The total length of the water distribution system is around 4800 km consisting of different types of pipes such as PVC, HDPE, steel, and ductile. Each

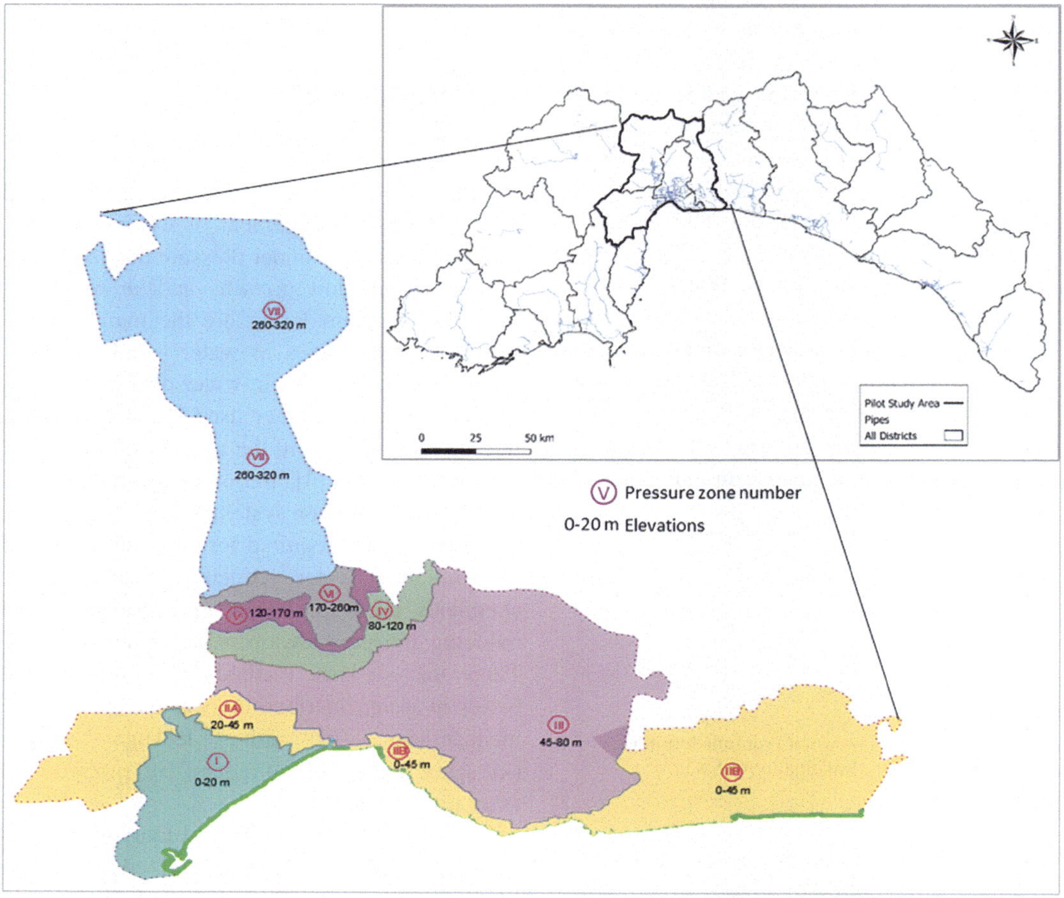

Fig. 1 Map showing pilot study area (PSA) and pressure zones of the PSA

Table 1 Water sources, capacity, and number of wells of each source

Water source	Capacity (l/s)	Number of wells
Duraliler	3125	52
Bogacay	750	13
Candir	398	8
Termessos	2150	39
Kovanlik	1200	15
Kutukcu	350	7
YSE	158	2
Ermenek	45	3
148 Depo	100	2
Gocerler	120	3
Cakirlar	10	1
Hurdacilar	100	3

pressure zone has been started to be divided into DMAs for better management practices. Physical and/or other features such as bridges and canals are used to determine the borders of DMAs considering the number of connections. Furthermore, hydraulic modeling studies have been carried out in the PSA. Currently, there are 110 DMAs in the PSA. Water sources, capacity, and the number of wells of each source providing water to the PSA are given in Table 1.

A total of 20 main water storage tanks with a total capacity of 140 050 m3 serve the system not only to balance differences in supply and demand but also to help in supplying water up to higher elevations. For instance, abstracted water from Bogacay and Candir wells is firstly supplied to pressure zone I, then pumped to another water storage tank to supply water to pressure zone II. A schematic flow chart of the water supply system of PSA is given in Fig. 2.

The supervisory control and data acquisition system (SCADA) of ASAT was established in 2006 for real-time monitoring and controlling of physical components of water distribution system such as pumps, valves, storage tanks and pipes, and flow rates, water pressures, and water levels at storage tanks. All wells, storage tanks, pumping stations, and water mains were integrated into the SCADA system. Moreover, all currently available DMAs were also equipped with flow meters and water pressure sensors and connected to.

Abstracted and pumped water from all sources of the PSA is continuously monitored and recorded which helps in analyzing the efficiency of the whole system. In 2021, around 151 million m3 of water was supplied to the PSA, and the annual level of water losses was determined as 43% of the system input volume. Monthly records of water supplied to the PSA are given in Fig. 3.

3 Control of Real Losses

3.1 Active Leakage Control

3.1.1 On-Line Monitoring

The SCADA system is very efficient for water loss control in the PSA, as the volume of real losses is the function of awareness time, location time, and repair time [9]. In 2021, a total of 1734 interventions were carried out and a total of 135 pipe breaks were determined and repaired with the help of the SCADA System. As a result of interventions carried out in 2021, an average of 11,713 m3/hour real losses was reduced. Flow rates and water pressure levels in the event of pipe burst and after repairment are given as an example in Fig. 4. Flow rates and water pressure levels of each DMA equipped with flow meters and water pressure sensors are online monitored. A sudden increase in flow rate and a sudden decrease in water pressure were identified in a DMA with an average minimum night flow of 85 m^3/h and a maximum daily flow rate of around 125 m^3/h, as depicted in Fig. 4. A relatively little pressure drop from around 4.5 to 4.2 bar was also determined. The awareness time of pipe burst was less than one hour with the help of the SCADA system. Location and repair times were also relatively short in that case.

The existing SCADA system is also used for long-term evaluation of the efficiency of pumps. In Fig. 5, a linear decay in flow rates pumped to the system was observed. The flow rate started to decrease from around 200 m^3/h down to 150 m^3/h

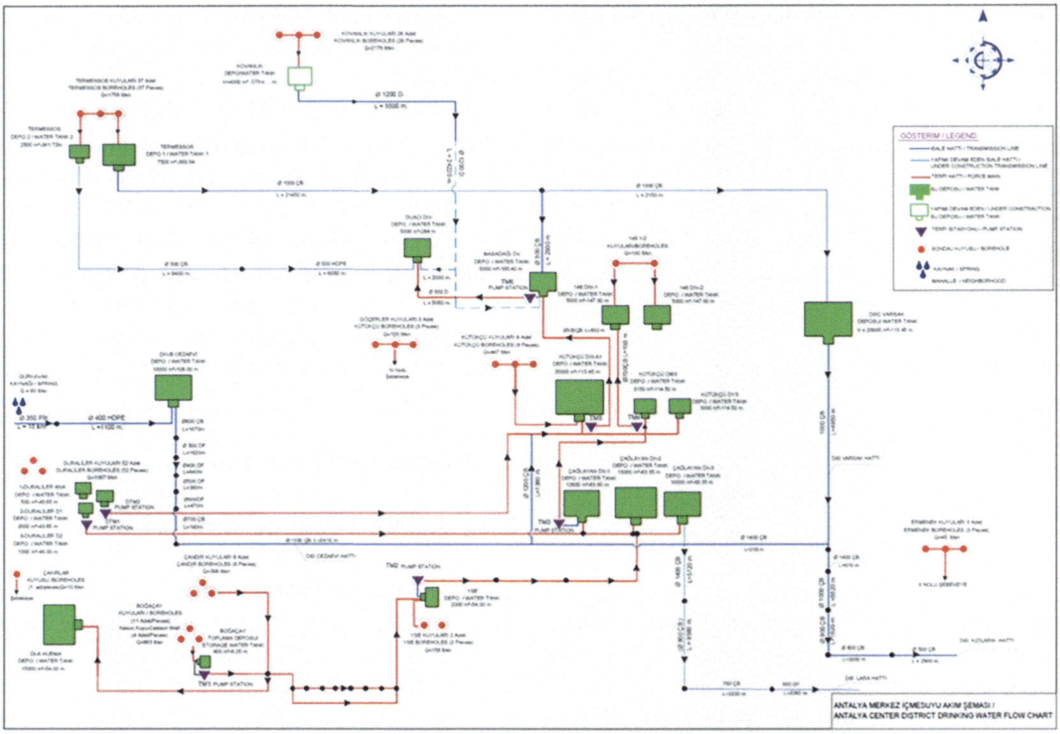

Fig. 2 Schematic flow chart of the water supply system of the PSA

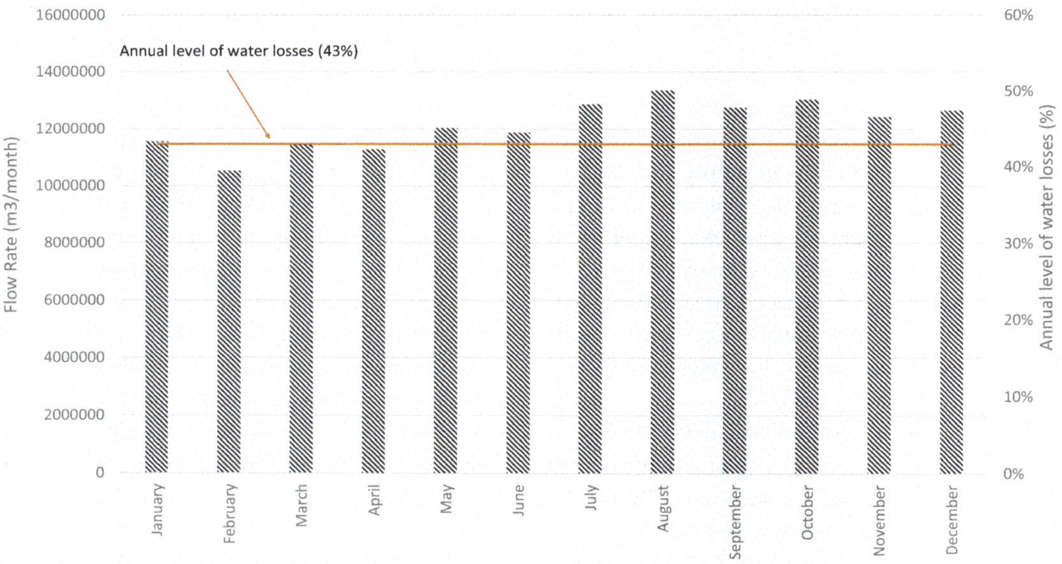

Fig. 3 Monthly records of water supplied to the PSA in 2021

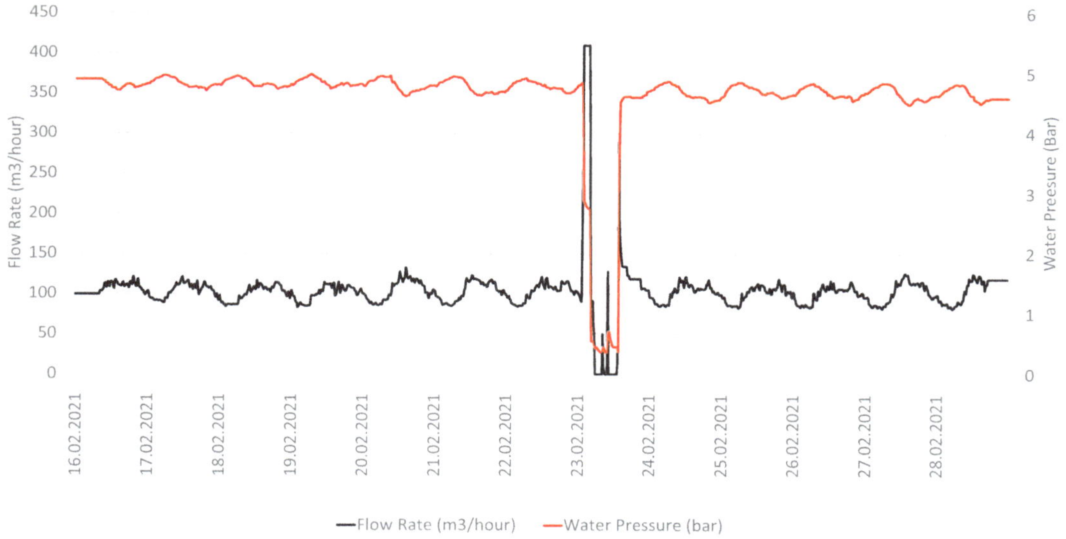

Fig. 4 Flow rates and water pressure levels in the event of pipe burst and after repairment

within two months. Afterward, intensive work to determine the efficiency loss of the pump was carried out and a pipe crack at the outlet of the pumping station was determined as the cause of pump efficiency loss (Fig. 5). Sudden drops in flow rates in Fig. 5 occur when the pump is off. Reducing the real losses is not only the benefit of the online monitoring system, but it also helps in reducing the energy consumption for the production and distribution of water.

3.1.2 Acoustic Leakage Detection

Acoustic leakage detection to control real losses is generally based on the detection of unreported and background leakages [10, 11]. In 2021, acoustic leakage detection studies were carried out in the PSA, and a total of 1,612,042 m distribution pipes and service connection pipes were inspected. Ultrasonic portable flow meters and acoustic microphones to detect and amplify the noise created by leaks are extensively used in the PSA (Figs. 6 and 7).

As a result of acoustic leakage detection studies, 1182 pipe failures on distribution mains and 3000 failures on service connection pipes were detected and repaired. The minimum night flow analysis of each DMA in the PSA is frequently monitored and evaluated. As an example, given in Fig. 8, a study was carried out in 2021 in Molla Yusuf DMA and the minimum night flow of the DMA was reduced from 125 to 45 m^3/h. According to the annual evaluation of flow rates, real losses in the PSA were reduced by 1392 m^3/h, which leads to an annual saving of 12,193,920 m^3 water.

3.2 Passive Leakage Control

Passive leakage control studies, in addition to active leakage studies, are carried out in daily routine to provide water services to the PSA. There is a call center to provide reporting services to the community within the PSA. When a notice is reported by consumers, staff evaluates the notice and informs the related department for action. Afterward, the technical staff takes action for the repair of failures (Fig. 9). According to the reports in 2021, a total of 73,855 notice was received. Around 50% of received notices (around 37,540 notices) were related to pipe failures whereas the rest of them were complaints

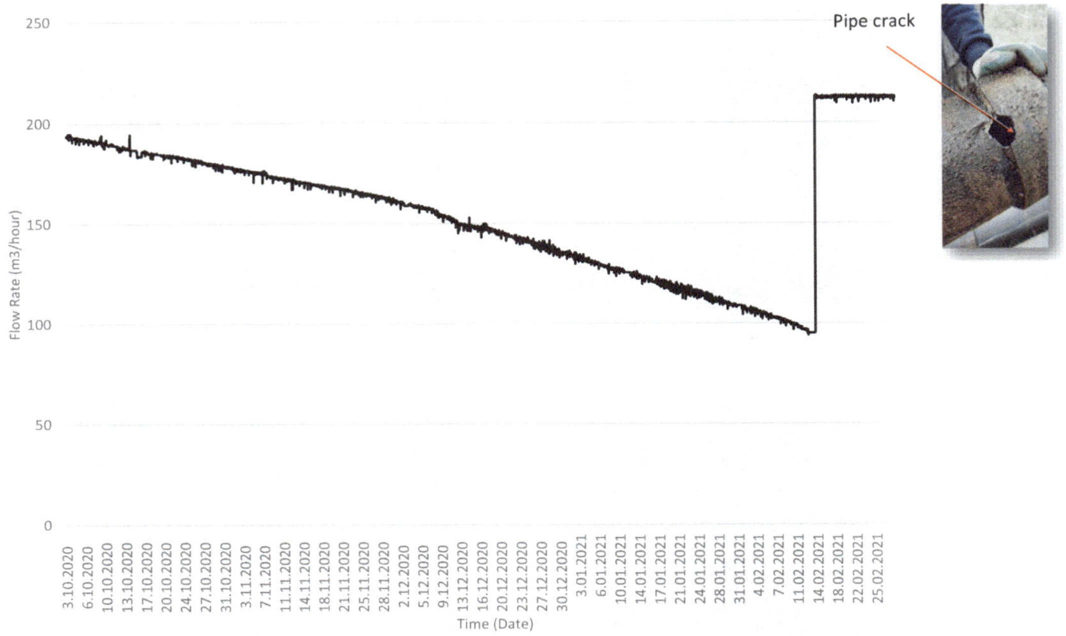

Fig. 5 Decreasing flow rates of a pump in the PSA due to a pipe crack

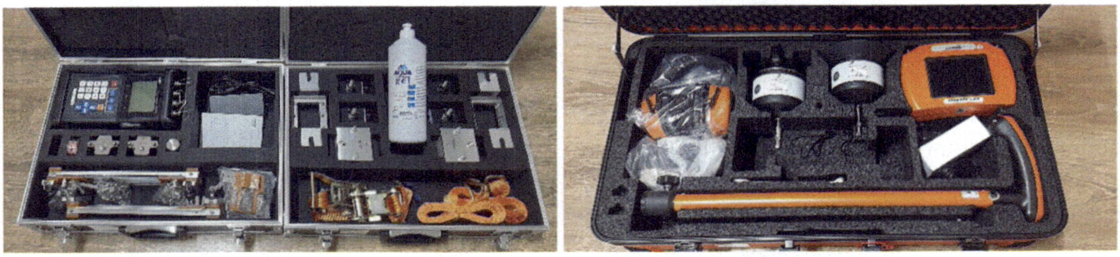

Fig. 6 Ultrasonic portable flow meters and acoustic microphone

about insufficient pressure and/or no water at taps. The most of pipe failures (28,783 failures) reported in 2021 occurred in service connection pipes while 8,757 failures were inspected on distribution pipes. The number and location of pipe failures that occurred in 2021 are given in Table 2.

3.3 Pressure Management

Pressure management is one of the intervention tools to control real losses. Management of pressure is a key to control real losses as water pressure flexes the volume of real losses [3, 4, 12]. The focus on pressure management is to reduce real losses, pipe failures, and consumption of water that is associated with excess pressure, and to increase the life span of components of distribution systems [3]. The relationship between water pressure and leakage [13], and the benefits of pressure management in water distribution systems are discussed in the literature [3–5].

Pressure management was started to be implemented successfully in the PSA [5]. The optimum pressure at the entrance of each DMA of the PSA is determined by implementing a

Fig. 7 Acoustic leakage detection studies carried out in the PSA

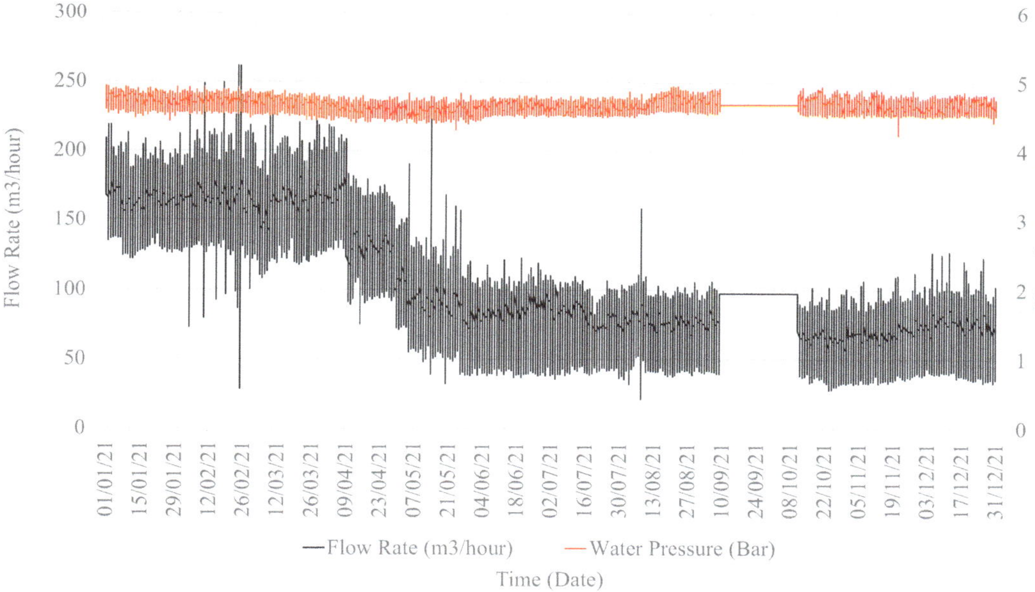

Fig. 8 Impacts of acoustic leakage detection on minimum night flow

Fig. 9 Pipe failure repairment in the PSA

Table 2 Number and location of pipe failures that occurred in 2021

District	Failures on service connections	Failures on distribution pipes
Aksu	6416	1522
Dosemealti	2900	1264
Kepez	12,108	3574
Konyaalti	2427	949
Muratpasa	4932	1488
Total	28,783	8757

hydraulic model. The main objective to determine the optimum pressure is to provide a minimum water pressure of around 2 bars at all points of distribution pipes of the DMA. A pressure management study carried out in DMA-Kultur of the PSA (Fig. 10) showed that reducing water pressure from 5.4 bars (average) to 4.1 bars (average) revealed a reduction in MNF from 145 to 89 m^3/h [9].

4 Conclusion

With many factors such as climate change, population growth and urbanisation threatening the sustainability of water resources, it has also become all the more important to increase the control of water losses in water distribution systems. Many countries such as Türkiye suffer from excessive water losses in water distribution systems. In this study, comprehensive work was carried out to control real losses. Pressure management, and reactive and proactive leakage

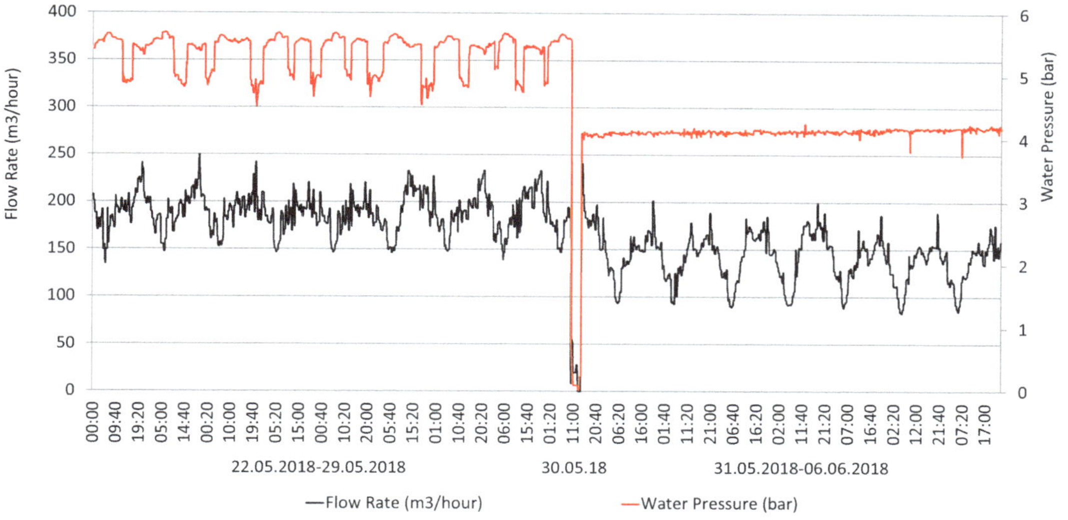

Fig. 10 Impacts of water pressure on flow rates [8]

control studies showed that a significant amount of real losses can be reduced by implementing on a regular basis. On the other hand, pipeline and asset management is another issue that should be tackled in further studies, as pipe failures occurred on service connection pipes play an important role. Water loss management strategies should be implemented broadly, comprehensively, and regularly including management strategies for apparent losses.

Acknowledgements The authors would like to thank the Antalya Water and Wastewater Administration (ASAT) of Antalya Metropolitan Municipality, Akdeniz University, Antalya, Türkiye, and EXCEED Swindon project funded by DAAD (German Academic Exchange Service).

References

1. Alegre H, Hirner W, Baptista JM, Parena R. Performance indicators for water supply services. 1st ed. London: IWA Publishing; 2000.
2. Lambert Aom F, Tveit OA, Abdin NAZ, Lazzari L, Lorenze H, Lee H, Farley M, Masakat E, Suphani R, Esko H, Johnson K, Rapinat M, Dohnal P, McKenzie R, Manesc A, Weimer D, Lai SKS, Somos E, Monteir A, Davis S, Martinez F, Lo SL. Onep, International report: water losses management and techniques. In: Wilderer PAG, Arvin E, Blackwell L, Hamoda MF, Mikkelsen PS, Mino T, Morgenroth E, Otterpohl R, Pons MN, Rauch W, Stephenson T, Ujang Z, Jianrong Z, editor. 2nd world water congress: water distribution and water services management. London: IWA Publishing; 2002. p. 1–20.
3. Karadirek IE, Aydin ME. Water losses management in urban water distribution systems. In: Bahadir M, Haarstrick A, editors. Water and wastewater management: global problems and measures. Cham: Springer International Publishing; 2022. p. 53–65.
4. Karadirek IE. Non revenue water management: current trends and future prospects. Fresenius Environ Bull. 2019;28(7):5226–33.
5. Karadirek IE, Kara S, Yilmaz G, Muhammetoglu A, Muhammetoglu H. Implementation of hydraulic modelling for water-loss reduction through pressure management. Water Resour Manag. 2012;26(9):2555–68.
6. Kanakoudis VMH. Urban water pipe networks management towards non-revenue water reduction: two case studies from Greece and Turkey. Clean-Soil Air Water. 2014;42(7):880–92.
7. Kara S, Karadirek IE, Muhammetoglu A, Muhammetoglu H. Real time monitoring and control in water distribution systems for improving operational efficiency. Desalin Water Treat. 2016;57(25):11506–19.
8. Akdeniz T. Physical water loss management in drinking water networks of karst regions using leak noise correlators and pressure optimization. Afyon Kocatepe Univ J Sci Eng. 2021;21(1):113–21.
9. Fanner P, Thornton J. The importance of real loss component analysis for determining the correct intervention strategy. In: Conference the importance of real loss component analysis for determining the

correct intervention strategy, Halifax, Nova Scotia, Canada.
10. Thornton J, Sturm R, Kunkel GA. Water loss control. New York: McGraw-Hill; 2008.
11. Aboelnga H, Saidan M, Al-Weshah R, Sturm M, Ribbe L, Frechen FB. Component analysis for optimal leakage management in Madaba, Jordon. J Water Supply Res Technol-Aqua. 2018;67(4):384–96.
12. Vicente DJ, Garrote L, Sanchez R, Santillan D. Pressure management in water distribution systems: current status, proposals, and future trends. J Water Resour Plan Manage-ASCE. 2016;142(2):13.
13. van Zyl JE, Malde R. Evaluating the pressure-leakage behaviour of leaks in water pipes. J Water Supply Res Technol-Aqua. 2017;66(5):287–99.

Occurrence, Fate, and Treatment of Micro/Nano Plastics in Drinking Water Sources

Sehnaz Sule Kaplan-Bekaroglu, Gokhan Civelekoglu, and Nuray Ates

Abstract

Plastics have become ubiquitous and essential raw material in various industries since their invention in the 1930s. Plastic waste in the environment can be classified into five categories: nanoplastics (NP), microplastics (MP), mesoplastics, macroplastics, and megaplastics. Micro- and nano-plastics (MNPs) have become a threat to the environment in recent years and interest in the environmental occurrence, fate and elimination of MNPs has shifted towards surface waters. MNPs have been detected in water resources around the world, including tap water, streams, rivers, and lakes. Scientific literature indicates that MNPs destroy water quality, affect source water quality and pose potential adverse impacts on human health directly and indirectly. MNPs are subjected to various mechanical, chemical, and biological weathering processes (such as UV radiation, oxidants, wave actions, and biofilm formation) in aquatic environment. And weathered MNPs may show different physicochemical properties from the original ones. Several techniques and processes such as coagulation, sedimentation, magnetic extraction; sand filtration, membrane filtration and adsorption, have been attempted to remove MNPs from drinking water with total removal efficiency ranging from <10 to 100%. This chapter is aimed to provide review about current knowledge of occurrence, fate and transport, and removal of MNPs in drinking water sources.

Keywords

Drinking water sources · Fate · Fragmentation · Human health · Microplastics · Nanoplastics · Water treatment

1 Introduction

Plastics are petrochemicals that possess synthetic compounds of high molecular weight as well as chemical additives to improve their properties to suit their intended use and performance [1]. Plastics are widely used, and plastic production has increased exponentially since the 1960's [2]. Plastics are made up of a large number, with a carbon backbone, of similar repeating units called monomers that bind together to form long

S. S. Kaplan-Bekaroglu (✉)
Environmental Engineering Department, Suleyman Demirel University, Isparta, Turkey
e-mail: sulebekaroglu@sdu.edu.tr

G. Civelekoglu
Environmental Engineering Department, Akdeniz University, Antalya, Turkey

N. Ates
Environmental Engineering Department, Erciyes University, Kayseri, Turkey

A. M. Bahadir et al. (eds.), *Hydrology and Urban Water Supply*, Water and Wastewater Management,
https://doi.org/10.1007/978-3-031-72589-0_17

chains (polymers) of very high molecular mass. The unique molecular structure of these plastic polymers directly causes distinctive properties in plastics. In fact, the diversity in characteristic properties of different plastics is related to the variance in their structure [3]. Antioxidants, fillers, reinforcements, anti-degradants, stabilizers, flame retardants, and plasticizers are all additives used in plastics, but type of polymer determines the fundamental properties of a plastic material [3, 4]. The various functional groups on the polymer are the cause of the fundamental differences in the mechanical, thermal, and chemical resistance qualities of plastic.

In aquatic environments, plastic pollution was detected for the first time in the 1970s [4]. It is estimated that about 250 million tons of plastics will have been accumulated in water bodies globally by 2025 with the ability to persist for up to hundreds of years [5]. Plastic waste in the environment can be classified into five categories: nanoplastics (NP) (<1 μm), microplastics (MP) (≥ 1 μm to <5 mm), mesoplastics (≥ 5 mm to 5 cm), macroplastics (>5 to 50 cm), and megaplastics (>50 cm) [6, 7] Micro- and nano-plastics (MNPs) can be divided as primary or secondary based on whether they are manufactured to be of micron/nano size or derived from fragmentation of macroplastics [8]. MNPs in the aquatic environment may be originated from aquatic and land-based sources [9, 10]. Type, composition, shape, and size of MNPs are incredibly diverse in surface waters [11, 12].

MNPs pollution has become a severe environmental threat because of their extensive distribution, small particle size, persistence, and harmful effects on ecosystems and human health [13, 14]. Ingestion, inhalation, or direct dermal contact are the pathways for MNPs, and ingestion is the primary exposure pathway [15, 16]. Aquatic ecosystems are well recognized as being at risk from MNPs, which poses risks in the form of sorbed contaminants, attached microbes, and MNPs particles themselves [3]. When in water, MNPs undergoes natural abiotic and biotic processes including mechanical, biological and chemical reactions. MNPs become fragmented as a result of these degradation processes, also known as weathering [17]. Due to their small particle size, they have high surface areas, enhancing their interactions and reactivity in water [18].

MNPs have been detected in water resources throughout the world, including our tap water, streams, rivers, lakes and oceans [3, 12–14]. In these waterways, the MNP end up in the drinking water. Although analysis and detection of MNPs in water are challenges and standardized methods and parameters for detection/characterization of MNPs in aqueous water are still critical gab for field and lab-scale experiments, Scanning Electron Microscopy, Fourier Transform Infrared Spectroscopy, Dynamic Light Scattering, Raman Spectroscopy, Mass Spectrometry and Thermal Analysis are commonly employed for identification of MNPs [19, 20]. Hundreds of review studies have been published underlying MNP characterization, behavior, and reactivity in the environment and water systems to reduce the burden and harmful effects of MNPs and to explore their reactivity and fate in the environment [3, 13]. MNPs are deteriorating water quality, changing the quality of source waters, and potentially having negative direct and indirect effects on human health [13].

2 Occurrence of Nano/Microplastics in Drinking Water Sources

2.1 MNPs Occurrence in Freshwater Bodies

The presence of MNPs is commonly found in surface waters such as lakes, reservoirs, and rivers. These pollutants have often been reported in urban and/or populated areas as well as in water bodies near remote areas [21]. About 80% of the MNPs found in the sea come from terrestrial sources [5, 22, 23]. The relative numbers of the detected MNPs depends on the location, sampling technique and analysis method [2]. Waste management, the economy, urbanization, particle size, population and hydrological conditions all affect the amount of MNPs in freshwater sources [24–27].

Many studies have reported that MNPs have been found worldwide in major [22], urban [24], and estuarine rivers [28, 29]. MNP numbers are ranged from 0.01 to 12.9 MP/L in Los Angeles River, San Gabriel River and the tributary Coyote Creek [24]. In a city river in Chicago, the number of MNPs in surface water has been reported to reach 730,341–6,698,264 MP/km^2 [30]. In China's Yangtze River, MNPs are 3,407,700–13,617,500 MP/km^2 in the mainstream and 192,500–11,889,700 MP/km^2 in the estuaries of the four tributaries [31]. In surface water samples collected from the four estuary tributaries of the Chesapeake Bay in the US, MNPs were found in all samples [28]. Gündoğdu and Çevik (2017) determined the level of MNPs pollution as 906 MP/m^2 at the mouth of the estuary of the Seyhan River in Mersin Bay (Türkiye) [29].

Baldwin et al. (2016) reported that MNP amounts ranged from 0.05 to 32 MP/m^3 in Great Lakes tributaries in the US [32]. Erdogan (2020) stated that the MNP pollution level in Cevdet Pond (Yozgat/Türkiye) as about 233 MP/m^3 [33]. In Switzerland, all surface samples collected contained MNPs and were seen in all the matrices investigated [34]. MNPs detected in other lakes ranged between 2,779 and 105,503 MP/km^2 [26, 27]. Mani et al. (2015) reported a maximum of 3.9×10^6 MP/km^2 in the Germany [22]. MNPs numbers have reached 20,264 MP/km^2 in Lake Hovsgol, Mongolia [27]. According to Dusaucy et al. (2021) plastic was detected in 98 lakes around the world, with plastic concentrations ranging from 0.27 to 34,000 MP/m^3 [12].

As well known, due to porous structure, groundwater is well protected from particle contamination. The amounts of MNPs in groundwater were determined in the average of 4.7 MP/L [13]. In various studies, MNPs in drinking water ranged from 0 to 104 MP/L and mean values were between 3–10 and 103 MP/L [2].

In the literature, a significant number of MNPs have been reported in tap and bottled water ((plastic and glass bottles) [35–38]. A study by Oßmann et al. (2018) reported the highest MNP counts ranging from 2,649 ± 2,857 MP/L in single-use plastic bottles to 6,292 ± 10,521 MP/L in glass bottles [37]. Table 1 provides a summary of the chosen studies on identified MNPs for the listed water types.

2.2 MNPs Shape, Size and Polymer Type in Freshwaters

A wide variety of MNP shapes and diameters were found in different types of water. Fibers and fragments and fibers were found to be the dominant MNPs [12, 13]. The studies in freshwater sources have reported MNPs in different shapes such as spherical MNP, parts, pellets, granules, films and styrofoam species [39–45]. Many studies generally analyzed the shapes of isolated MNPs. The dimensional analysis and its results include spatial and temporal changes due to differences in freshwater resource type or system. However, the wide variety of shapes limits the quantitative analysis and comparability of obtained data [13].

The most common types of MNP detected in environmental samples worldwide are polyethylene (PE) > polypropylene (PP) > polystyrene (PS) > Polyvinyl chloride PVC > polyethylene terephthalate (PET) (Table 1). It reflects their global use and production [11]. In the literature, different MNP material was identified in freshwater sources, PE and PP were observed in 80–90% of the works [33, 46]. In some literature sources, it has been reported that other materials such as polyamide (PA), polyacrylamide (PAM), polyacrylonitrile (PAN), polybutylacrylate (PBA), polyester (PES), polymethyl methacrylate (PMMA), p-phenylene terephthalamide (PPTA), polytrimethylene terephthalate (PTT) are found in the structure of MNPs. The most abundant MNPs type and color are fiber (91%) and blue (36%), respectively. PE and PP (90%) were the most prevalent type of polymers [11, 36, 37, 42–50]. In some studies, it's clearly seen that the size distribution of microplastics is mostly in the low range. Oßmann et al. (2018) indicated that most of the MNPs (>90%) were detected in the 1–5 μm size range

Table 1 Summary of selected studies on occurrence of MNPs in freshwater bodies

Water body	MNP counts	Particle size (μm)	Polymer type	References
River	Max: 297,927/km^2	300–2,000	PE	[28]
	Min: 3,407,700/km^2	500–1,600	PP, PE, PS	[31]
	Mean: 892,777/km^2	300–5,000	PS, PP, PVC	[22]
	1.6–9.0/L	50–2,000	PE, PP, PS, PVC	[53]
	Max: 17,383/m^3	50–5,000	PE, PP	[54]
Lake	Mean: 91,000/km^2	<1,000	PE, PP, PS	[34]
	0.05–32/m^3	355–4,750	PE, PP	[32]
	Max: 6,800,000/km^2	333–1,000	PET, PS, PP	[40]
	1–4.7/L	50–330	PE, PP	[42]
	5–34/L	50–500	PE, PP, PS	[45]
	0.27–1,600/m^3	20–300	PE, PP, PET, PAN	[50]
	Min: 23,000/km^2	355–1,000	PE, PP	[36]
Groundwater	<1/L	50–150	PVC, PES	[48]

[37]. However, the highest number of MNP observed in 100–250 μm (56%) size class. Moreover, it was found that 81–96% of MNPs were in the range of 1–20 μm [41, 51], 61–73% and >80% of MNPs were in the range of 10–500 μm [45, 52], and 50–2,000 μm [53], respectively.

3 Transport and Fate of Nano/Microplastics in Drinking Water Sources

Primary MNPs are specifically produced for a range of uses. Secondary MNP are produced indirectly from break down, abrasion or degradation of the primary MNPs [20]. Point, non-point sources, and tributaries are possible sources of MNPs in freshwater [22]. MNPs in the aquatic environment originates from aquatic and land-based sources [9, 10]. Discarding of fishing lines, spillage during marine transport, and direct littering are generally considered as aquatic-based sources [10]. Land-based sources are formed because of surface run-off and the effluent from the wastewater treatment plants to the receiving aquatic environment [21]. Effluent discharge of MNPs from wastewater treatment plants are predominant source of land-based sources and around 80% of the overall MNPs in aquatic environment [55]. Phuong et al. [56] also showed that domestic wastewater/runoff followed with industry emissions, fisheries wastewater, and aquaculture runoff were the primary sources of MNPs in Asian freshwater ecosystems. Atmospheric deposition should also take in account for as MNPs sources [14]. It is very hard to determine the origin of the MNPs due to its small size, fragment nature, and unknown range of possible sources in aquatic environment.

MNPs are released continuously into water and can travel through freshwater. Transport behaviors is crucial to evaluate the environmental fate of MNPs [57]. Since density of most MNPs are lower than those of freshwater or ocean water, MNPs frequently float at the water's surface and are carried by water along rivers and into oceans, a process called as surface transport (advective transport) [58]. Plastic emissions can be redistributed throughout media compartments through subsequent transport processes, which typically results in a flow from land to rivers and the sea. Also, wind and water movement are effective in transport [59]. Density, particle size, shape of MNPs as well as processes such as adsorption, aggregation have impacts on transport in water [59, 60]. The variety and complexity of MNPs characteristics (density, shape, size, polymer type) is reflected in the diversity and complexity of their transport and fate [57]. This complexity is further increased by the adsorption of organic and inorganic pollutants onto MNPs, and the presence of residual additive chemicals derived from the original polymer [3]. With respect to diversity and complexity of MNPs, it is difficult to evaluate their whole life cycle fate 'from cradle to grave', and all of fate and transport steps of MNPs [3, 13].

MNPs are fragmented as a result of chemical, biological and mechanical reactions during biotic and abiotic degradation processes known as weathering [17]. Photooxidation, hydrolytic degradation, and biodegradation processes are main weathering processes to determine fate of MNPs. Due to their small particle size, they have high surface areas, enhancing their interactions and reactivity in water bodies [18]. It was reported that the fate of MNPs in the aquatic environment can be affected by different environmental factors, such as surface circulation leading to vertical distribution and mixing of water as a reason of wind [61]. Nevertheless, the fate of the plastic material mainly depends on the density of the plastic polymer [62, 63]. It was stated that the density of microplastics in the aquatic environment can be affected by different environmental factors that can change the fate of microplastics, such as aggregation and interaction with organic materials [62]. The formation of biofilms and hetero-aggregation have a significant effect on fate of MNPs because of increasing density [63–65].

When MNPs reach an aquatic environment, natural weathering processes lead to a variety of short- and long-term changes on their surfaces and structural elements [66, 67]. In water, MNPs are subjected to aggregation and deposition, which are key factors in determining fate and transport and interactions with other pollutants [68]. Interactions between MNPs–MNPs, MNPs– pollutants, and MNPs–water play crucial roles to control aggregation and deposition of MNPs in aquatic environment [3, 69]. The aggregation mechanism of MNPs depends on hydrodynamic conditions (such as river geometry, water velocity, flood events, environmental factors (such as weathering, wave actions, heat, UV radiation, oxidants, microorganisms) and physiochemical properties of MNPs, water type [3, 56].

The fragmentation of MNPs structures occurs through a sequence of degradation reactions that can be classified as photo-oxidative, thermal, chemical, or microbiological [70, 71]. As a result of degradation reactions, physicochemical characteristics of MNPs (i.e., shape, color, size, functional groups, crystallinity, and densities) changes in short and long periods with depending on the UV radiation, oxygen, and MNPs/water type [3, 72–74]. Plastics manufactured for different uses and purposes are combined with additives such as antioxidants and UV stabilizers, which are not chemically bound to the parent polymer [49, 58, 66, 68, 75]. Thus, exposure to harsh environmental conditions (e.g., temperature, oxidants, UV light, and pH) has detrimental effects, mainly reduction and depletion of the antioxidant levels and many other structural and morphological properties [76, 77]. Consequently, aged plastics in water deform and degrade with time (breaking of the carbon-hydrogen or carbon–carbon bonds) while becoming less oxidative resistant as antioxidants are depleted. Previous studies concluded that some halogenated, volatile, and odorous compounds form as a result of oxidant reacting with plastic pipes and degradation of some of the plastics [68]. The most common degradation mechanism in water is biodegradation, but there is limited information about effects of environmental factors (such as temperature and water chemistry), characteristics of MNPs, and bacterial species in biodegradation process [78–80]. One of the major concerns with the fate of MNPs is about the interactions with other contaminants such as organic contaminants and inorganic contaminants onto MNPs under various environmental conditions [3, 81]. The adsorption of contaminants on MNPs may change fate and transport at different background water chemistries. Physicochemical properties of MNPs, characteristics of contaminants and aqueous media consequently influence the adsorption behavior on MNPs surface [82–85]. Adsorption of contaminants on MNPs affect fate, transport, and the associated risks to the aquatic environment.

4 The Removal of Nano/Microplastics in Drinking Water Treatment Plants

Although the removal of solids and colloids is a specified requirement to get potable water, drinking water treatment plants (DWTPs) are not designed primarily to remove MNPs. Several studies have documented removal of MNPs throughout the whole scale the entire drinking water treatment process [41, 48, 86, 87]. The concentration of MNPs in the raw water differed significantly because of population, and geographical location of cities and DWTPs [68]. Cherniak et al. [89] reported that a full-scale conventional DWTP removed around 70% of microplastics (>10 μm) from surface water [89]. Pivokonsky et al. [41] found that removal efficiency of MNP reached 70–83% on average, with concentration 338–628 MP/L for treated water from three DWTPs in Czech Republic. However, Mintenig et al. [48] and Johnson et al. [90] reported similar removal performances with lower concentration 0.7 MP/m^3 and 0.00011 MP/L, respectively. Namely Mintenig et al. [48] and Johnson et al. [90] were analyzed the size range of MNPs to 20 μm−5 mm and 25 μm–5 mm, respectively but significant part of MNP are <20 μm which are missing. Removal efficiencies in China's DWTPs, which were basically the same, were over 85% [86, 87, 91].

Wu et al. [92] found that only 80.96% of MNPs $\leq$ 20 μm was removed in the DWTP, while MNPs >20 μm reached above 99% removal. An evaluation of eight DWTPs in the UK using various treatment modalities (including filtration and adsorption) at various water sources revealed far better removal, often about 99.9 percent (rivers, groundwater, and reservoir sources) [93]. To date, most studies showed that the size of MNPs had a significant impact on removal efficiency in various treatment processes. Larger MNPs (>20 μm) can be entirely removed after treatment processes, but the removal efficiency of smaller sizes, especially <1 μm, was highly uncertain [92]. Pivokonsky et al. [94] also found that advanced drinking water treatment could remove MNPs smaller than 1 μm (88%) more effectively than conventional drinking water treatment (40%). In order to accurately compare findings, greater attention should be paid on differences of MNP size in array of similar treatment technologies.

Physical, chemical, and biological approaches are the three primary categories of MNPs removal technologies [94]. In conventional DWTPs, coagulation and granular media filtration are the primary processes for particle removal. Numerous laboratory studies have examined removal of MNPs at different treatment processes however, there are no systematic data to evaluate effect of full scale conventional operating conditions on MNPs removal for each treatment processes [13, 14, 21, 92]. Pivokonsky et al. [94] reported that coagulation removed 62% of microplastics from full scale DWTP, however Zhang et al. [95] found that coagulation removed microplastics from 0.1 to 1.2% in laboratory conditions. Ma et al. [96, 97] reported less than 15% removal for PE particles with using coagulation with ferric chloride and aluminum chloride at lab-scale tests. Zhang et al. [95] found that coagulation with $Al_2(SO_4)_3$ was insufficient on the removal of 180 nm polystyrene (PS) nanoplastics from surface water. Arenas et al. [98] suggested that polyaluminum chloride was more effective than iron chloride (FeCl3) than PS removal by coagulation. Despite different results, it concluded from several studies that positively charged MNPs are more challenging to remove by coagulation only from surface waters [92, 98].

According to studies, the removal of microplastics was also significantly removed by the filtration processes for MNPs with physical retention and adsorption mechanisms. Besides, the average MNPs removal in the sand filtration step ranged between 30 and 80% with coagulation [99]. Concentrations in the source, sample collection variables and analytical methods causes different removal performance in similar processes. Wang et al. [86] reported that up to 44% removal of residual microplastics by subsequent sand filtration in coagulated water. Zhang et al. [95] reported that filtration process removed nanoplastics in ranging between 87% to nearly complete. The size and form of the particles had a significant impact on removal efficiency in filtration especially in removing fibers and fragments >50 μm. Arenas et al. [98] showed that granular activated carbon (GAC) filters without coagulation achieved 88.1% removal for overall nanoplastic. The previous studies demonstrated the effectiveness of GAC in the removal of PS from untreated natural surface waters with removal efficiencies from 78 to 90%. While biological activated carbon (BAC) filtration effectively removed MNPs 20 μm with 63.8%, it provided a higher removal performance for 5–20 μm MNPs [92]. Other effective methods for MNP treatment from water include electrocoagulation, magnetic extraction, and membrane filtration according to Shen et al. [87]. The complete rejection of <0.5 mm PE particles with or without coagulation was observed by ultrafiltration in lab-scale tests [96, 97]. Even though reverse osmosis and nanofiltration are already employed to purify drinking water, but they are prone to fouling by the MNPs [100, 101].

According to some studies, MNPs were subjected to breakdowns into smaller fragments throughout the drinking water treatment [92]. According to Wang et al. [86], ozonation increased the amount of 1–5 m MNPs by 2.8–16.0%. This increase was most likely caused by the shearing flow, which broke up the MNPs into smaller particles and fibers. The difficulty of MNPs removal increases with fragmentation and fragmentation impairs process performance [8]. The biological pretreatment and ozonation caused a significant increase of 5–20 μm MNPs due to the fragmentation, which is difficult to remove in the conventional treatment plants. The main MNPs fragmentation products are in size of ≤ 20 μm [92]. Level and mechanism of fragmentation in drinking water treatment should be investigated in detail to optimize treatment processes. The removal of MNPs in individual drinking water treatment processes remain less understood and the mechanisms and side effect of MNPs urgently needs to be explained to design and/or modify DWTPs [100, 101]. Also, a method for producing trustworthy and reliable data must be developed in order to address the significant discrepancy between the reported removal rates of MNPs in drinking water sources.

5 Conclusion

Contamination by MNPs is ubiquitous in drinking water resources. The combination of ubiquity and special properties (such as density, hydrophobicity, surface functionalisation, particle shape and size) has recently attracted increased attention in scientific research. However, the occurrence, fate and removal of MNPs in complex (i.e. real) systems are still very difficult to study. Additionally, the collection, pretreatment, and detection of MNPs in surface water must be optimized and unified in order to more effectively comparison of the findings of various research. Although analysis and detection of MNPs in water samples are challenging and the available techniques are still in their early stages, studies have reported the presence of MNP in raw and tap waters. The types of MNPs that are found in freshwater are commensurate with plastic concentrations and production rates. There is a lack of knowledge about fate pathways of MNPs through photochemistry, oxidation, or microbial interactions. Additionally, the fate of MNPs may also be affected by the adsorption of organic and inorganic hazardous pollutants and/or the acquisition of biofilm coatings.

Regarding fate, aggregation, and sedimentation, nanoplastic particles may exhibit significantly distinct characteristics from the microplastic particles, consequently, characterization of the each MNPs group may be required. Several techniques and processes such as coagulation, sedimentation, magnetic extraction; sand filtration, membrane filtration, adsorption, have been attempted to remove MNPs from drinking water with total removal efficiency ranging from <10 to 100%. Indeed, conventional and advanced drinking water treatment is still inefficient in removing particles in the low micron-to-nanometer range such as nanoplastics. Although drinking water treatment can be effective in removing some of MNP particles, more research is needed to understand the fate of MNP across different treatment processes under different operational conditions. MNPs pollution should be taken into consideration, when designing and implementing water treatment technologies.

References

1. Ahmed R, Hamid AK, Krebsbach SA, He J, Wang D. Critical review of microplastics removal from the environment. Chemosphere. 2020;293: 133557.
2. WHO (World Health Organization). Microplastics in drinking-water. World Health Organization. https://apps.who.int/iris/handle/10665/326499. License: CC BY-NC-SA 3.0 IGO; 2019.
3. Ateia M, Ersan G, Alalm MG, Boffito DC, Karanfil T. Emerging investigator series: microplastic sources, fate, toxicity, detection, and interactions with micropollutants in aquatic ecosystems: a review of reviews. Environ Sci Process Impacts. 2022;24(2):172–95.
4. Carpenter EJ, Smith KL Jr. Plastics on the Sargasso Sea surface. Science. 1972;175(4027):1240–2124.
5. Jambeck JR, Geyer R, Wilcox C, Siegler TR, Perryman M, Andrady A, Narayan R, Law KL. Plastic waste inputs from land into the ocean. Science. 2015;347(6223):768–71.
6. Lebreton L, Slat B, Ferrari F, Sainte-Rose B, Aitken J, Marthouse R, Reisser J. Evidence that the great pacific garbage patch is rapidly accumulating plastic. Sci Rep. 2018;8(1):1–15.
7. Anik AH, Hossain S, Alam M, Sultan MB, Hasnine MT, Rahman MM. Microplastics pollution: a comprehensive review on the sources, fates, effects, and potential remediation. Environ Nanotechnol Monitor Manage. 2021;16: 100530.
8. Enfrin M, Dumée LF, Lee J. Nano/microplastics in water and wastewater treatment processes–origin, impact and potential solutions. Water Res. 2019;161:621–38.
9. Hammer J, Kraak MH, Parsons JR. Plastics in the marine environment: the dark side of a modern gift. Rev Environ Contam Toxicol. 2012;220:1–44.
10. Ziajahromi S. Identification and quantification of microplastics in wastewater treatment plant effluent: investigation of the fate and biological effects. Griffith School of Environment Griffith University. PhD Thesis. Gold Coast, Queensland, Australia; 2018.
11. Strungaru SA, Jijie R, Nicoara M, Plavan G, Faggio C. Micro-(nano) plastics in freshwater ecosystems: abundance, toxicological impact and quantification methodology. TrAC, Trends Anal Chem. 2019;110:116–28.
12. Dusaucy J, Gateuille D, Perrette Y, Naffrechoux E. Microplastic pollution of worldwide lakes. Environ Pollut. 2021;284: 117075.
13. Koelmans AA, Nor NHM, Hermsen E, Kooi M, Mintenig SM, De France J. Microplastics in freshwaters and drinking water: critical review and assessment of data quality. Water Res. 2019;155:410–22.
14. Eerkes-Medrano D, Thompson RC, Aldridge DC. Microplastics in freshwater systems: a review of the emerging threats, identification of knowledge gaps and prioritisation of research needs. Water Res. 2015;75:63–82.
15. Rahman I, Mujahid A, Palombo EA, Müller M. A functional gene-array analysis of microbial communities settling on microplastics in a peat-draining environment. Mar Pollut Bull. 2021;166: 112226.
16. Galloway TS. Micro-and nano-plastics and human health. In Marine anthropogenic litter. Cham: Springer; 2015. p. 343–66.
17. Song YK, Hong SH, Jang M, Han GM, Jung SW, Shim WJ. Combined effects of UV exposure duration and mechanical abrasion on microplastic fragmentation by polymer type. Environ Sci Technol. 2017;51(8):4368–76.
18. da Costa JP, Santos PS, Duarte AC, Rocha-Santos T. (Nano) plastics in the environment–sources, fates and effects. Sci Total Environ. 2016;566:15–26.
19. Shim WJ, Hong SH, Eo SE. Identification methods in microplastic analysis: a review. Anal Methods. 2017;9(9):1384–91.
20. Yusuf A, Sodiq A, Giwa A, Eke J, Pikuda O, Eniola JO, Bilkis A, Sambudi NS, Bilad MR. Updated review on microplastics in water, their occurrence, detection, measurement, environmental pollution, and the need for regulatory standards. Environ Pollut. 2022;292: 118421.
21. Eerkes-Medrano D, Leslie HA, Quinn B. Microplastics in drinking water: a review and assessment. Curr Opin Environ Sci Health. 2019;7:69–75.
22. Mani T, Hauk A, Walter U, Burkhardt-Holm P. Microplastics profile along the Rhine River. Sci Rep. 2015;5(1):1–7.
23. Wagner M, Scherer C, Alvarez-Muñoz D, Brennholt N, Bourrain X, Buchinger S, Fries E, Grosbois C, Klasmeier J, Marti T, Rodriguez-Mozaz S, Urbatzka R, Vethaak AD, Nielsen MW, Reifferscheid G. Microplastics in freshwater

ecosystems: what we know and what we need to know. Environ Sci Eur. 2014;26(1):1–9.
24. Moore C, Lattin G, Zellers A. Quantity and type of plastic debris flowing from two urban rivers to coastal waters and beaches of Southern California. J Integr Coastal Zone Manage. 2011;11:65–73.
25. Zbyszewski M, Corcoran PL. Distribution and degradation of fresh water plastic particles along the beaches of Lake Huron, Canada. Water Air Soil Pollut. 2011;220(1):365–72.
26. Eriksen M, Mason S, Wison S, Box C, Zellers A, Edwards W, Farley H, Amato S. Microplastic pollution in the surface waters of the Laurentian Great Lakes. Mar Pollut Bull. 2013;77(1–2):177–82.
27. Free CM, Jensen OP, Mason SA, Eriksen M, Williamson NJ, Boldgiv B. High-levels of microplastic pollution in a large, remote, mountain lake. Mar Pollut Bull. 2014;85(1):156–63.
28. Yonkos LT, Friedel EA, Perez-Reyes AC, Ghosal S, Arthur CD. Microplastics in four estuarine rivers in the Chesapeake Bay, USA. Environ Sci Technol. 2014;48(24):14195–202.
29. Gündoğdu S, Çevik C. Micro-and mesoplastics in Northeast Levantine coast of Turkey: The preliminary results from surface samples. Mar Pollut Bull. 2017;118(1–2):341–7.
30. McCormick A, Hoellein TJ, Mason SA, Schluep J, Kelly JJ. Microplastic is an abundant and distinct microbial habitat in an urban river. Environ Sci Technol. 2014;48(20):11863–71.
31. Zhang K, Gong W, Lv J, Xiong X, Wu C. Accumulation of floating microplastics behind the Three Gorges Dam. Environ Pollut. 2015;204:117–23.
32. Baldwin AK, Corsi SR, Mason SA. Plastic debris in 29 Great Lakes tributaries: relations to watershed attributes and hydrology. Environ Sci Technol. 2016;50(19):10377–85.
33. Erdogan S. Microplastic pollution in freshwater ecosystems: a case study from Turkey. Ege J Fish Aquat Sci. 2020;37(3):213–21.
34. Faure F, Demars C, Wieser O, Kunz M, De Alencastro LF. Plastic pollution in Swiss surface waters: nature and concentrations, interaction with pollutants. Environ Chem. 2015;12(5):582–91.
35. Kosuth M, Mason SA, Wattenberg EV. Anthropogenic contamination of tap water, beer, and sea salt. PLoS ONE. 2018;13(4): e0194970.
36. Mason SA, Welch VG, Neratko J. Synthetic polymer contamination in bottled water. Front Chem. 2018;407.
37. Oßmann BE, Sarau G, Holtmannspötter H, Pischetsrieder M, Christiansen SH, Dicke W. Small-sized microplastics and pigmented particles in bottled mineral water. Water Res. 2018;141:307–16.
38. Schymanski D, Goldbeck C, Humpf HU, Fürst P. Analysis of microplastics in water by micro-Raman spectroscopy: release of plastic particles from different packaging into mineral water. Water Res. 2018;129:154–62.
39. Zhao S, Zhu L, Wang T, Li D. Suspended microplastics in the surface water of the Yangtze Estuary System, China: first observations on occurrence, distribution. Mar Pollut Bull. 2014;86(1–2):562–8.
40. Su L, Xue Y, Li L, Yang D, Kolandhasamy P, Li D, Shi H. Microplastics in taihu lake, China. Environ Pollut. 2016;216:711–9.
41. Pivokonsky M, Cermakova L, Novotna K, Peer P, Cajthaml T, Janda V. Occurrence of microplastics in raw and treated drinking water. Sci Total Environ. 2018;643:1644–51.
42. Wang W, Yuan W, Chen Y, Wang J. Microplastics in surface waters of dongting lake and hong lake, China. Sci Total Environ. 2018;633:539–45.
43. Di M, Liu X, Wang W, Wang J. Manuscript prepared for submission to environmental toxicology and pharmacology pollution in drinking water source areas: microplastics in the Danjiangkou Reservoir, China. Environ Toxicol Pharmacol. 2019;65:82–9.
44. Luo W, Su L, Craig NJ, Du F, Wu C, Shi H. Comparison of microplastic pollution in different water bodies from urban creeks to coastal waters. Environ Pollut. 2019;246:174–82.
45. Yuan W, Liu X, Wang W, Di M, Wang J. Microplastic abundance, distribution and composition in water, sediments, and wild fish from Poyang Lake, China. Ecotoxicol Environ Saf. 2019;170:180–7.
46. Bordós G, Urbányi B, Micsinai A, Kriszt B, Palotai Z, Szábo I, Hantosi Z, Szoboszlay S. Identification of microplastics in fish pond and natural freshwater environments of the Carpathian basin, Europe. Chemosphere. 2019;216:110–6.
47. Di M, Wang J. Microplastics in surface waters and sediments of the three Gorges Reservoir, China. Sci Total Environ. 2018;616:1620–7.
48. Mintenig SM, Löder MGJ, Primpke S, Gerdts G. Low numbers of microplastics detected in drinking water from ground water sources. Sci Total Environ. 2019;648:631–5.
49. Novotna K, Cermakova L, Pivokonska L, Cajthaml T, Pivokonsky M. Microplastics in drinking water treatment—current knowledge and research needs. Sci Total Environ. 2019;667:730–40.
50. Uurasjärvi E, Hartikainen S, Setälä O, Lehtiniemi M, Koistinen A. Microplastic concentrations, size distribution, and polymer types in the surface waters of a northern European lake. Water Environ Res. 2020;92(1):149–56.
51. Triebskorn R, Braunbeck T, Grummt T, Hanslik L, Huppertsberg S, Jekel M, Knepper TP, Krais S, Müller YK, Pittroff M, Ruhl AS, Schmieg H, Schür C, Strobel C, Wagner M, Zumbülte N, Köhler H-R. Relevance of nano-and microplastics for freshwater ecosystems: a critical review. TrAC, Trends Anal Chem. 2019;110:375–92.

52. Leslie HA, Brandsma SH, Van Velzen MJM, Vethaak AD. Microplastics en route: field measurements in the Dutch river delta and Amsterdam canals, wastewater treatment plants, North Sea sediments and biota. Environ Int. 2017;101:133–42.
53. Wang W, Ndungu AW, Li Z, Wang J. Microplastics pollution in inland freshwaters of China: a case study in urban surface waters of Wuhan, China. Sci Total Environ. 2017;575:1369–74.
54. Hitchcock JN. Storm events as key moments of microplastic contamination in aquatic ecosystems. Sci Total Environ. 2020;734: 139436.
55. Sharma S, Chatterjee S. Microplastic pollution, a threat to marine ecosystem and human health: a short review. Environ Sci Pollut Res. 2017;24(27):21530–47.
56. Phuong NN, Duong TT, Le TPQ, Hoang TK, Ngo HM, Phuong NA, Sempere R. Microplastics in Asian freshwater ecosystems: current knowledge and perspectives. Sci Total Environ. 2021; 151989.
57. Wu P, Huang J, Zheng Y, Yang Y, Zhang Y, He F, Gao B. Environmental occurrences, fate, and impacts of microplastics. Ecotoxicol Environ Saf. 2019;184: 109612.
58. Andrady AL. Microplastics in the marine environment. Mar Pollut Bull. 2011;62(8):1596–605.
59. Kooi M, Besseling E, Kroeze C, van Wezel AP, Koelmans AA. Modelling the fate and transport of plastic debris in freshwaters: review and guidance. In: Wagner M, Lambert S, editors. Freshwater microplastics: emerging environmental contaminants? Springer International Publishing; 2017. p. 125–52.
60. Kooi M, Nes EHV, Scheffer M, Koelmans AA. Ups and downs inthe ocean: effects of biofouling on vertical transport of microplastics. Environ Sci Technol. 2017;51(14):7963–71.
61. Lusher AL, Burke A, O'Connor I, Officer R. Microplastic pollution in the Northeast Atlantic ocean: validated and opportunistic sampling. Mar Pollut Bull. 2014;88:325–33.
62. Galloway TS, Cole M, Lewis C. Interactions of microplastic debris throughout the marine ecosystem. Nat Ecol Evol. 2017;1:116.
63. Lagarde F, Olivier O, Zanella M, Daniel P, Hiard S, Caruso A. Microplastic interactions with freshwater microalgae: Hetero-aggregation and changes in plastic density appear strongly dependent on polymer type. Environ Pollut. 2016;215:331–9.
64. Rummel CD, Jahnke A, Gorokhova E, Kühnel D, Schmitt-Jansen M. Impacts of biofilm formation on the fate and potential effects of microplastic in the aquatic environment. Environ Sci Technol Lett. 2017;4(7):258–67.
65. Woodall LC, Sanchez-Vidal A, Canals M, Paterson GL, Coppock R, Sleight V, Thompson RC. The deep sea is a major sink for microplastic debris. Roy Soc Open Sci. 2014;1(4): 140317.
66. Danso D, Chow J, Streit WR. Plastics: environmental and biotechnological perspectives on microbial degradation. Appl Environ Microbiol. 2019;85(19):e01095-e1119.
67. Gewert B, Plassmann MM, MacLeod M. Pathways for degradation of plastic polymers floating in the marine environment. Environ Sci Process Impacts. 2015;17(9):1513–21.
68. Alimi OS, Farner Budarz J, Hernandez LM, Tufenkji N. Microplastics and nanoplastics in aquatic environments: aggregation, deposition, and enhanced contaminant transport. Environ Sci Technol. 2018;52(4):1704–24.
69. Wang F, Wong CS, Chen D, Lu X, Wang F, Zeng EY. Interaction of toxic chemicals with microplastics: a critical review. Water Res. 2018;139:208–19.
70. do Sul JAI, Costa MF. The present and future of microplastic pollution in the marine environment. Environ Pollut. 2014;185:352–64.
71. Urbanek AK, Rymowicz W, Mirończuk AM. Degradation of plastics and plastic-degrading bacteria in cold marine habitats. Appl Microbiol Biotechnol. 2018;102(18):7669–78.
72. Du H, Xie Y, Wang J. Microplastic degradation methods and corresponding degradation mechanism: research status and future perspectives. J Hazard Mater. 2021;418: 126377.
73. Guo X, Wang J. The chemical behaviors of microplastics in marine environment: a review. Mar Pollut Bull. 2019;142:1–14.
74. Worm B, Lotze HK, Jubinville I, Wilcox C, Jambeck J. Plastic as a persistent marine pollutant. Annu Rev Environ Resour. 2017;42(1):1–26.
75. Malankowska M, Echaide-Gorriz C, Coronas J. Microplastics in marine environment: a review on sources, classification, and potential remediation by membrane technology. Environ Sci: Water Res Technol. 2021;7(2):243–58.
76. Cai L, Wang J, Peng J, Wu Z, Tan X. Observation of the degradation of three types of plastic pellets exposed to UV irradiation in three different environments. Sci Total Environ. 2018;628:740–7.
77. Zhang Y, Pu S, Lv X, Gao Y, Ge L. Global trends and prospects in microplastics research: a bibliometric analysis. J Hazard Mater. 2020;400: 123110.
78. Fisner M, Majer A, Taniguchi S, Bícego M, Turra A, Gorman D. Colour spectrum and resin-type determine the concentration and composition of Polycyclic Aromatic Hydrocarbons (PAHs) in plastic pellets. Mar Pollut Bull. 2017;122(1–2):323–30.
79. Haider TP, Völker C, Kramm J, Landfester K, Wurm FR. Plastics of the future? The impact of biodegradable polymers on the environment and on society. Angew Chem Int Ed. 2019;58(1):50–62.
80. Raddadi N, Fava F. Biodegradation of oil-based plastics in the environment: existing knowledge and needs of research and innovation. Sci Total Environ. 2019;679:148–58.
81. Koelmans AA, Bakir A, Burton GA, Janssen CR. Microplastic as a vector for chemicals in the aquatic

environment: critical review and model-supported reinterpretation of empirical studies. Environ Sci Technol. 2016;50(7):3315–26.
82. Fu L, Li J, Wang G, Luan Y, Dai W. Adsorption behavior of organic pollutants on microplastics. Ecotoxicol Environ Saf. 2021;217: 112207.
83. Vieira Y, Lima EC, Foletto EL, Dotto GL. Microplastics physicochemical properties, specific adsorption modeling and their interaction with pharmaceuticals and other emerging contaminants. Sci Total Environ. 2021;753: 141981.
84. Yu Y, Mo WY, Luukkonen T. Adsorption behaviour and interaction of organic micropollutants with nano and microplastics–a review. Sci Total Environ. 2021;797: 149140.
85. Sun Y, Yuan J, Zhou T, Zhao Y, Yu F, Ma J. Laboratory simulation of microplastics weathering and its adsorption behaviors in an aqueous environment: a systematic review. Environ Pollut. 2020;265: 114864.
86. Wang Z, Lin T, Chen W. Occurrence and removal of microplastics in an advanced drinking water treatment plant (ADWTP). Sci Total Environ. 2020;700: 134520.
87. Shen M, Zeng Z, Wen X, Ren X, Zeng G, Zhang Y, Xiao R. Presence of microplastics in drinking water from freshwater sources: the investigation in Changsha, China. Environ Sci Pollut Res. 2021;28 (31):42313–24.
88. Wang G, Lu J, Tong Y, Liu Z, Zhou H, Xiayihazi N. Occurrence and pollution characteristics of microplastics in surface water of the Manas River Basin, China. Sci Total Environ. 2020;710: 136099.
89. Cherniak SL, Almuhtaram H, McKie MJ, Hermabessiere L, Yuan C, Rochman CM, Andrews RC. Conventional and biological treatment for the removal of microplastics from drinking water. Chemosphere. 2022;288: 132587.
90. Johnson AC, Ball H, Cross R, Horton AA, Jurgens MD, Read DS, Svendsen C. Identification and quantification of microplastics in potable water and their sources within water treatment works in England and Wales. Environ Sci Technol. 2020;54 (19):12326–34.
91. Pulido-Reyes G, Magherini L, Bianco C, Sethi R, von Gunten U, Kaegi R, Mitrano DM. Nanoplastics removal during drinking water treatment: laboratory-and pilot-scale experiments and modeling. J Hazard Mater. 2022;436: 129011.
92. Wu J, Zhang Y, Tang Y. Fragmentation of microplastics in the drinking water treatment process-a case study in Yangtze River region, China. Sci Total Environ. 2022;806: 150545.
93. Ball H, Cross R, Grove E, Horton A, Johnson A, Jürgens M, Svendsen C. Sink to river–river to tap: a review of potential risks from nanoparticles and microplastics. London: UK Water Industry Research Limited; 2019.
94. Pivokonsky M, Pivokonská L, Novotná K, Čermáková L, Klimtová M. Occurrence and fate of microplastics at two different drinking water treatment plants within a river catchment. Sci Total Environ. 2020;741: 140236.
95. Zhang Y, Diehl A, Lewandowski A, Gopalakrishnan K, Baker T. Removal efficiency of micro-and nanoplastics (180 nm–125 μm) during drinking water treatment. Sci Total Environ. 2020;720: 137383.
96. Ma B, Xue W, Hu C, Liu H, Qu J, Li L. Characteristics of microplastic removal via coagulation and ultrafiltration during drinking water treatment. Chem Eng J. 2019;359:159–67.
97. Ma B, Xue W, Ding Y, Hu C, Liu H, Qu J. Removal characteristics of microplastics by Fe-based coagulants during drinking water treatment. J Environ Sci. 2019;78:267–75.
98. Arenas LR, Gentile SR, Zimmermann S, Stoll S. Nanoplastics adsorption and removal efficiency by granular activated carbon used in drinking water treatment process. Sci Total Environ. 2021;791: 148175.
99. Jung JW, Kim S, Kim YS, Jeong S, Lee J. Tracing microplastics from raw water to drinking water treatment plants in Busan, South Korea. Sci Total Environ. 2022;825: 154015.
100. Li Y, Li W, Jarvis P, Zhou W, Zhang J, Chen J, Tian Y. Occurrence, removal and potential threats associated with microplastics in drinking water sources. J Environ Chem Eng. 2020;8(6): 104527.
101. Barchiesi M, Chiavola A, Di Marcantonio C, Boni MR. Presence and fate of microplastics in the water sources: focus on the role of wastewater and drinking water treatment plants. J Water Process Eng. 2021;40: 101787.

Pesticides in Surface Water Resources: Occurrence, Fate, Modeling and Treatment

Sehnaz Sule Kaplan-Bekaroglu, Filiz Dadaser-Celik, and Nuray Ates

Abstract

Pesticides are chemicals used for agricultural, industrial, and household purposes to increase crop yield and enhance product quality, avoid pests, plant diseases, weeds and repel pests. Pesticides are classified as herbicides, insecticides, fungicides and rodenticides based on target pests. Pesticides are transported to the environment through surface runoff, seepage, evaporation, drift, biotic and abiotic processes and cause serious risks to aquatic flora and fauna, and human health. Atrazine, metolachlor, diuron, dimethoate, terbutryn, diazinon, chlorpyrifos and imidacloprid are most frequently detected in high concentrations in surface waters. To understand the fate and behaviour of pesticides in the environment, 4 groups of models have been developed: (a) field scale, (b) basin scale, (c) static or dynamic river water quality, and (d) ecological risk assessment models. In order to protect the environment and human health, WHO (World Health Organization), USEPA (US Environmental Protection Agency) and EU (European Union), and individual countries have set standards for pesticides. As many pesticides (e.g., organochlorides, organophosphates) are highly resistant to degradation, they can persist in the environment and water bodies for long time. Since conventional treatment plants are not specifically designed to remove them, pesticides that remain in the water without treatment eventually end up in the human body. Mostly recommended processes for effective treatment of pesticides are advanced oxidation processes, adsorption, membrane filtration, ion exchange, and their hybrid combinations.

Keywords

Pesticides · Emerging contaminants · Fate · Modelling · Toxicity · Surface water

1 Introduction

Pesticides are commonly used to increase crop yield and to enhance the crop quality, to avoid pests, plant diseases, weeds, and to control vectors of public health [1, 2]. Environmental Protection Agency (USEPA) defines the pesticides as chemical substances used for controlling and repelling of pests [3]. Plant diseases cause great damage to the agricultural economy and when

S. S. Kaplan-Bekaroglu
Environmental Engineering Department, Suleyman Demirel University, Isparta, Turkey
e-mail: sulebekaroglu@sdu.edu.tr

F. Dadaser-Celik · N. Ates (✉)
Environmental Engineering Department, Erciyes University, Kayseri, Turkey
e-mail: nuraya@erciyes.edu.tr

A. M. Bahadir et al. (eds.), *Hydrology and Urban Water Supply*, Water and Wastewater Management,
https://doi.org/10.1007/978-3-031-72589-0_18

pesticides are not applied, the yield of basic food products such as rice, wheat, maize, and potatoes decrease by about 10–15% [4]. The European Commission (EC) database for pesticides lists 1472 types of pesticides, including active substances, safeners, and synergists and only 453 of which are approved [5].

The categorization of pesticides bases on several criteria including their target pests, chemical structures, and active ingredients, sources of origin, pesticide formulation, and toxicity of pesticides [6, 7]. The active ingredients of pesticides can be organic and inorganic. Organic pesticides can be produced from natural sources (biopesticides) or synthetic chemicals [4, 8]. Herbicides, insecticides, rodenticides, fungicides, various other substances are the groups of pesticides based on target pests [2, 9].

The chemical composition and active ingredient based classification of pesticides divides them into mainly five groups as organochlorines, organophosphorus, carbamates, pyrethrins and pyrethroids [10]. Organochlorine pesticides especially used as insecticides are lipophilic and are resistance to decay, therefore they tend to bioconcentrate in tissues and persist in soil and water bodies [9]. Carbamate and organophosphate are the classes of pesticides that cause death as a result of the effect on the nervous system of insects [3]. Pyrethroids are synthesized to mimic the effective role of pyrethrins, which are natural insecticides, to increase their potency and to reduce their sensitivity to light and air. Therefore, pyrethroids have longer stability and persistence in the environment than natural form pyrethrins [9, 11].

Pesticide formulations consist of a mixture of active ingredients aimed at controlling target pests and inert ingredients that make pesticides safer, more effective, and easier to apply. In addition, one type of pesticide can be mixed with another non-pesticide compound to increase the effect of pesticides on target pests or for the controlled use of multiple pesticides with a single dose application. Pesticides in liquids, powders, granules, baits, and dust formulations are the most common forms [10].

The toxicity of pesticides is classified by the World Health Organization (WHO) into four categories. These toxicity categories were determined based on the estimation of median lethal dose (LD50) for rats and other laboratory animals acute, oral as well as dermal. The toxicity classes from the lowest to the highest (I-IV) are in the order as extremely, highly, moderately, slightly hazardous [12].

Herbicides (47.5%) and insecticides (29.5%) account for most of the total global pesticide utilization followed by fungicides (17.5%) and the others (5.5%) [2, 7]. Based on the 2019 data, on a global basis, the total pesticide use is 4.2 million tons, and 2.7 kg of pesticide is applied per each hectare cultivated area on average [13, 14]. The highest amount of pesticide is consumed in the Eastern Asia (12.66 kg/ha), South America (5.8 kg/ha), and Western Europe (4.39 kg/ha) countries in the world [15].

2 Occurrence of Pesticides in Surface Water Resources

Pesticide concentration in water resources is linked to product type and density and agricultural management practices in the watershed [16]. Pesticide monitoring studies have shown that much higher concentrations and more variety of pesticides have been detected in surface water resources in developing countries than in developed countries [6]. Since organophosphorus compounds are widely applied to control various pests (insects, plants, nematodes, fungi, rodents etc.), their residues are detected at higher concentrations in surface and groundwater than organochlorine pesticides [17]. In the USA, 11 herbicides, 4 insecticides, and 1 fungicide were observed in higher concentrations than chronic reference level for aquatic life in 39 major rivers [18]. The studies revealed that the concentration of pesticides varied from 7 to 121,222 ng/L in rivers and lakes, from 20 to 1,060 ng/L in groundwater, from 141 to 14,629 ng/L in drinking waters, and from 23 to 3,172 ng/L in wastewaters [19].

The most frequently observed herbicides were atrazine in 18% of a total of 86 water samples in the USA [20] and metolachlor in 86% of collected samples from agriculturally impacted water sources [21]. Because of their low degradation rates, they can persist in environment for years. The maximum atrazine concentrations were detected 1.65 μg/L in USA [21], 1.73 μg/L in China [22] and 1.40 μg/L in Argentina [23].

Due to its high stability and higher usage, diuron is an herbicide compound, detected at the highest concentration of 22.77 μg/L in rivers in Costa Rica [24], 0.49 μg/L in drinking waters, and 0.79 μg/L in surface waters in Brazil [25]. Because diuron is also used as an anti-fouling agent in paints in marine navigation besides crop protection (e.g., fruit, cotton, sugar cane and wheat), it was identified in more than 3 out of 4 samples in the Mediterranean coastal waters [25]. Terbutryn, a selective herbicide with a triazine structure, were detected at the maximum concentrations of 4.76 μg/L in Costa Rica [24] and 1.69 μg/L in China [22].

Among the insecticides, dimethoate is the one observed the most commonly and at the highest concentrations in water resources worldwide [6, 24]. Dimethoate, an acaricide organophosphorus compound, was detected with the maximum concentration of 61.20 μg/L in 135 water samples collected from 14 sampling sites in Costa Rica [24]. The other commonly detected insecticides at high concentrations are chlorpyrifos with 0.785 μg/L in Ghana [26] and 3.1 μg/L in Bangladesh [27], diazinon with 768.91 μg/L in Iran [28], 24.67 μg/L in USA [21] and 9.0 μg/L in Bangladesh [27], malathion with 1.8 μg/L in Bangladesh [27] and imidacloprid with 1.89 μg/L in China [22]. The most frequently identified fungicides in water resources are tebuconazole and carbendazim [6]. The maximum concentration of acetochlor was detected with the amount of 0.68 μg/L and 0.46 μg/L in China and USA, respectively [21].

The EC released the first Union-wide watch list in 2015, which includes acetamiprid, clothianidin, imidacloprid, thiacloprid, thiamethoxam, methiocarb, oxadiazon, and tri-allate. The changes made in the second list published in 2018 are the addition of metaflumizone and the excluding of oxadiazone and tri-allate from the list. As a result of the studies conducted in 21 countries, imidacloprid, acetamiprid, and thiamethoxam are frequently detected, while others are more likely to be identified in water sources because of their limited use [29]. Pesticide monitoring studies conducted across European countries have revealed the existence of numerous pesticides in surface waters and groundwater at levels that might cause ecological risk to aquatic organisms.

3 Fate and Modeling of Pesticides Transport in the Environment

The fate and behavior of pesticides in the environment (soil, surface water, and groundwater) is affected by many processes. These processes include transport processes that control how pesticides move from the point of application to surface water; transfer processes that regulate how pesticides are distributed among water, soil, and biota; and transformation processes that include biological, physical and chemical reactions [30]. Transport processes move pesticides away from their point of application through air and water. Pesticides can enter into air through spray drift and evaporation and can be found in solid, gaseous, and liquid forms [31]. In water, pesticides move either in dissolved form or attached to soil particles. Transfer processes include volatilization, surface runoff, leaching, adsorption, or plant uptake (Fig. 1), which control the distribution of pesticides between solids, liquids, and gases [30, 32].

Transfer processes are affected by pesticide characteristics, such as water solubility, volatility, soil adsorption, and persistence and soil characteristics such as texture, permeability, depth, pH, organic matter content [30]. Transformation processes refer to conversions by microorganisms or plants or occur as chemical and photochemical reactions [33]. Both environmental conditions and characteristics of the pesticide would affect the type and degree of transformation [33].

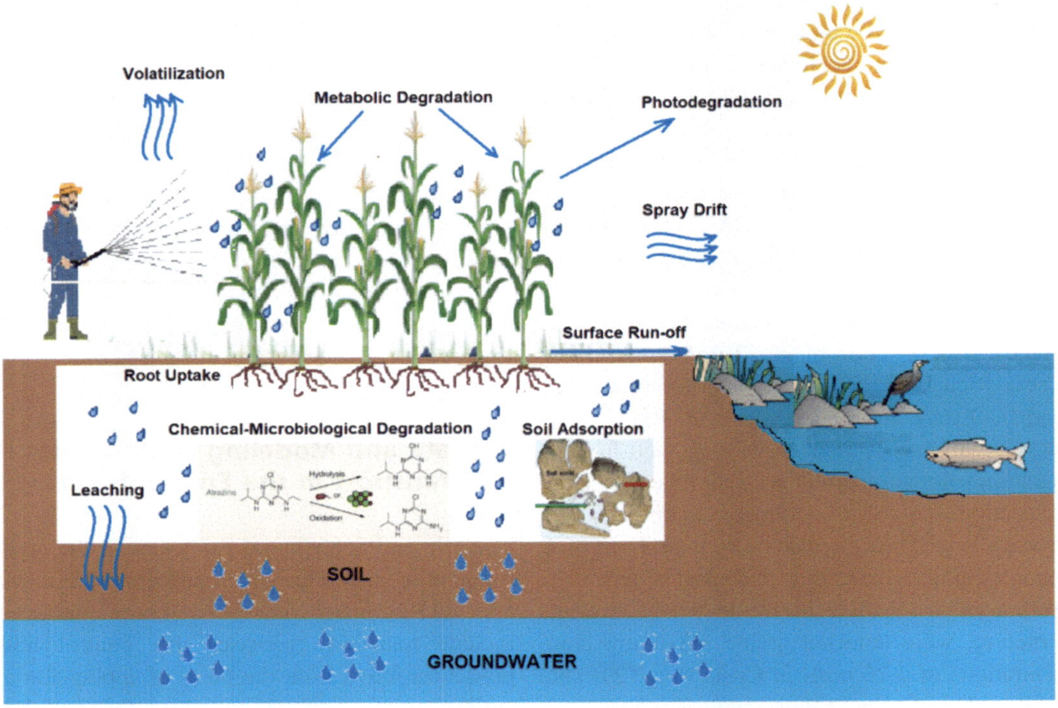

Fig. 1 Major fate and transport mechanisms of pesticides in the environment

Monitoring of pesticide occurrence and fate in the environment requires extensive monitoring network and expensive and labor-intensive monitoring campaigns. Models can be used to understand the fate and behavior of pesticides in the environment and can expand scarce data collected in the field to larger spatial and temporal scales. Modelling utilizes mathematical assumptions and simplification for understanding and predicting real world conditions. All models contain some degree of uncertainty due to the uncertainties in input parameters and model structures [34, 35]. Still, models are useful tools for quantitative assessment of pesticide occurrence and transport. Models can also be used as a risk assessment tool and can help evaluate ecological risks associated with pesticides. One of the most important aspects of models is that they can help evaluate the effectiveness of pollution mitigation strategies.

In general, pesticide fate models can be categorized into four as (1) field-scale models, (2) basin scale models, (3) steady-state or dynamic river water quality models, and (4) ecological risk assessment models [36].

Field scale models focus on pesticide losses in the plant root zone. They can simulate plant growth in addition to movement of water and pesticides in the soil system and major physical, chemical, and biological processes that control pesticide fate in the soil. These models can make assessments about field conditions. However, these models focus on much smaller spatial scales and they cannot fully capture watershed processes. Some field-scale models are RZWQM (Root Zone Water Quality Model), PRZM (Pesticide Root Zone Model) [37], OPUS (Smith 1992), and GLEAMS (Groundwater Loading Effects of Agricultural Management Systems) [38]. RZWQM simulates water and chemical transport and physical, chemical, and biological processes, particularly in the unsaturated zone. PRZM is one-dimensional model that can simulate pesticide fate in the plant root zone. The first version of PRZM was published in 1985 and the model gained new capabilities until the latest version, PRZM5 (Young and Fry 2014). GLEAMS is an extension to the Chemicals, Runoff and Erosion from Agricultural Management Systems (CREAMS) model. These models have been applied in different field conditions [39]. Several studies compared the performance of different field scale models. To example a few, Willian et al. [40] compared PRZM and GLEAMS models to predict the movement of two herbicides in soil. Marin-Benito et al. [41] compared the ability of PRZM and MACRO to simulate soil water content and pesticide concentrations. Fox et al. [42] compared RZWQM and PRZM. Field-scale models were also integrated with basin-scale models such as SWAT to provide a linkage between watershed processes and soil processes [43, 44].

Basin-scale models focus on the hydrology and transport processes at the watershed scale. Compared to the field-scale models, pesticide modelling is more challenging in the basin-scale due to difficulties in definition of transport and degradation processes in different spatial and temporal scales and complexity of basins and hydrologic processes [45]. It can be difficult to represent spatial and temporal variability in soil, weather, and pesticide applications based on scarce point data collected in the field [46]. However, basin-scale models can help evaluate pesticide transport at larger spatial scales. Examples of basin-scale models include AGNPS (Agricultural Non-Point Source Pollution Model) [47], SWAT (Soil and Water Assessment Tool [48], and DWSM (Dynamic Watershed Simulation Model) [49], HSPF (Hydrologic Simulation Program-Fortran) [50], MIKE SHE (Système Hydrologique Européen) [51]. AGNPS was developed for simulating sediment, nutrient, and pesticide transport from agricultural watersheds. The use of AGNPS in pesticide modeling has been limited compared to other basin-scale models. SWAT is a watershed-scale model, widely used for understanding hydrologic processes and nutrient transport and also among the most used models for modeling pesticide fate and transport at the watershed scale. Wang et al. (2019) reviewed pesticide modelling applications with SWAT and identified 50 studies from the SWAT literature database. They further categorized these studies into five groups as (1) sensitivity analysis; (2) exposure modeling for fate and transport; (3) mitigation strategy by using best management practices; (4) Algorithm improvement; and (5) advanced implementations. The number of studies conducted with SWAT has increased in recent years [52, 53]. DWSM is a model developed to simulate rainfall-runoff, soil erosion, and sediment transport processes. It was used for modeling pesticide losses [54]. HSPF is another basin-scale model used widely for pesticide modelling. HSPF can simulate hydrology and water quality for both conventional and toxic organic pollutants, including pesticides. HSPF has been used in agricultural watersheds [55, 56]. MIKE SHE is another basin-scale model that can simulate water and solute transport in addition to pesticide processes. Fauser et al. [57] estimated maximum concentrations in drain water, the time for reaching this concentration, and the time interval for exceeding the limit using MIKE SHE simulations for 37 pesticides. In some other studies, different basin-scale models were compared. Parker et al. [46] evaluated the performance of SWAT, the Nonpoint Source Model (NPSM), a modified version of the HSPF, and the Pesticide Root Zone Model-Riverine Water Quality (PRZM-RIVWQ) model in simulating concentrations of the pesticides atrazine, metolachlor, and trifluralin in surface water.

Steady-state or dynamic river water quality models simulates pesticide transport and behavior in rivers. Examples include QUAL2E (advanced stream water quality model) [58], WASP (Water Quality Analysis Simulation Program—Water Quality Analysis Simulation Program) [59], TOXSWA (TOXic substances in Surface Waters—Toxic substances in surface water) [60], Receiving Water Quality Model [RWQM1] [61]. QUAL2E was used as part of the basin-scale model SWAT. Sumon et al. [62] used the TOXSWA model to calculate pesticide exposure in surface waters in Bangladesh. Adriaanse et al. [63] used TOXSWA to estimate the effect of runoff in pesticide concentrations in runoff. Focks et al. [64] used CASCADE-TOXSWA fate model, to estimate the fate of pesticides in an interconnected system of water bodies.

Ecological risk assessment models aim to determine the risks originating from pesticides. Examples include DREAM [65] and PERPEST [66]. Sumon et al. [62] used a simple first tier risk assessment based on threshold concentrations derived from single species toxicity tests to assess the ecological risk in the form of risk quotients. The PERPEST model was used to refine the ecological risks when the first-tier assessment indicated a possible risk. Sumon et al. [27] evaluated the risks of the pesticides in surface water and sediment using a first-tier risk quotient (RQ) approach and the higher-tier PERPEST model was used to refine the ecological risks of pesticides when RQ indicated a potential risk. Ramo et al. [67] evaluated the toxicity risk to aquatic organisms using the PERPEST model in River Madre de Dios. Many other studies used different risk assessment methods [64]. Several studies used more than one risk assessment model and compared the results with each other [67].

4 Environmental Impacts and Health Effects of Pesticides

The application of pesticides in agricultural activities enhances the crop yield; there is a 45–50% yield loss when pesticides are not applied [68]. However, pesticides cause serious environmental problems when they enter into an ecosystem, especially through water [69]. Pesticide pollution has adverse effects on the lives of aquatic organisms, wildlife, animals, and human health. The entrance of pesticides from both end-of pipe discharge or surface runoff from agricultural, forestry lands, and such areas leads pesticide accumulation in the water bodies, and it negatively impacts wildlife ecosystem, animals, aquatic organisms, and human health [2].

Information on the toxic and ecotoxicological risks of pesticides can be added to this section. In general, human may be exposed to the pesticides into 3 routes through dermal contact during production or application (mostly occupational exposure), ingestion (food and water), and inhalation (contaminated air) as well as bioconcentration in food chain [11]. Pesticide exposure causes acute and chronic effects on human health [70]. According to worldwide investigation, around 11,000 of approximately 385 million incidents of unintentional acute pesticide poisoning result in death, and 44% of farmers (about 860 million) are exposed to pesticide poisoning each year [71].

It has also been reported that pesticides are carcinogenic, mutagenic, and teratogenic in nature [72]. The increase in the incidence of cancer and other chronic diseases in the last 50–60 years is attributed to the increase in pesticide residues reaching water resources due to intensive pesticide use and the long-term persistence of some pesticides [2]. Many health problems and disorders have been associated with pesticides residues and their metabolites including cardiovascular diseases, pulmonary system hormone disruption, intelligence reduction, and reproductive distortion. Moreover, pesticides are considered significantly toxic due to their ability to bioaccumulate in organism tissue and migrate to higher organisms [6].

Based on the data of the EC, it is estimated that pesticide residues in the products harvested and pesticide contaminated air in 24 European Union (EU) countries have caused an average of 2.6 h of life loss per person [73]. Pesticide uptake from fresh food was calculated using application rates from pesticide-related data sources and average human lifespan lost was estimated to be 4.2 min in the USA and 3.2 min in Switzerland [73, 74].

5 Regulation and Legislations on Pesticides

In United States, regulation on pesticides under the name of the Federal Insecticide, Fungicide, and Rodenticide Act was first passed through the commission in 1947 [75]. The Federal Environmental Pesticide Control Act, which was amended in 1972, aims to protect human and environment health from the negative impacts of pesticides by regulating the labeling, distribution, sale and use of pesticides [76]. The Safe Drinking Water Act of 1974 was updated in 1986 and 1996, and federal standards were set for only 22 pollutants. These acts also required the development of standards for a total of 83 specific pollutants, including highly water-soluble pesticides. The law requires national monitoring of linuron, metolachlor, cyanazine, and metribuzin, among many pollutants. The Safe Drinking Water Act of 1996 required the EPA to publish a "candidate pollutant list" every five years [77]. In the pollutant list (CCL 3) published in 2009, 43 pesticides, including alachlor, metolachlor, and their metabolites and acetochlor were included [78].

At the beginning of 2000, a strategy was determined with Directive 2000/60/EC in order to identify high-risk priority substances for aquatic ecosystems. In the scope of water policy, a list of 33 priority substances/substance groups was created by the EC in 2008 with the 2008/105/EC Directive. Based on available data on acute and chronic impacts on the aquatic environment and human health, environmental quality standards have been established for 33 priority substance/substance groups and 8 other pollutants, expressed as annual average value and/or maximum allowable concentration. The EC have updated the water framework policy with the Directives 2000/60/EC and 2008/105/EC dated 12 August 2013 and numbered 2013/39/EC and included 45 priority substances/ substance groups, among which 12 organochlorine and 3 organophosphorus pesticides [79].

Besides, the regulation of Surface Water Quality Management (Türkiye) defines Environmental Quality Standards for 250 pollutants in rivers, lakes, coastal and transitional waters. Of these pollutants, 45 were listed as priority substance/substance groups including 19 pesticides [80].

6 Control Strategies and Treatment Processes for Pesticides

Some pesticides containing persistent organic compounds are highly resistant to degradation and can remain in the environment for long-years and become bioconcentrated up to 70.000-fold in the food chain [8]. Conventionally, water treatment plants are designed for removing suspended solids, macro molecules, and inorganics present in the water but not micropollutants, including pesticides. Therefore, combining conventional treatment processes with advanced treatment processes is a feasible approach for removing of pesticides from drinking water [6, 81]. Pesticide removal from water is mainly affected by pesticide properties (solubility, polarity, volatility, acidity, lipophilicity, etc.), water characteristics (organic matter content, pH, type and concentrations of background ions, temperature, etc.), and the method used for treatment [82]. For effective removal of pesticides from water sources, advanced oxidation processes, adsorption, membrane filtration, ion exchange, and their hybrid combinations can be recommended as advanced treatment methods [83, 84].

Pesticides removal applications in the literature by advanced oxidation processes (AOPs) include chemical (Fenton, peroxonation, persulfate), photo-chemical (UV/H_2O_2, UV/O_3, UV/Fenton), sono-chemical (ultrasound/H_2O_2, ultrasound/O_3), photocatalysis (TiO_2, ZnO, SnO_2, WO_3) electro-chemical (electrochemical or anodic oxidation, electro-fenton), and other processes such as γ-radiation; pulsed electric field, electrolyzed water, high pressure processing, ultrasounds, and non-thermal plasma [82]. While the biodegradability of endosulfan was increased 50% by photo-fenton after 90 min [85], UV-based process combining with $S_2O_8^{2-}$, HSO_4^-, H_2O_2 removed endosulfan with performances of 91%, 86%, and 64%, respectively [86]. Ormad et al. [87] reported ~23% pesticide (DDT dicofol, dieldrin, endrin, HCH, lindane, heptachlor, isodrin, endosulfan) removal by O_3 treatment. Although lower removal yields were observed by O_3/H_2O_2 and O_3/TiO_2 applications,

$O_3/H_2O_2/TiO_2$ increased the pesticide removal up to 36%. Dicofol was degraded 86% within 60 min by ultrasound process with acoustic power of 375 W [88].

Adsorption is a promising process for pesticides removal because it is well-known, low-cost and highly effective; and activated carbon, agricultural/industrial waste, polymeric, inorganic, bio-adsorbents are commonly used adsorbents [84]. Ali et al. [89] reported 95% removal of atrazine by 2.5 g/L of iron nano-composite adsorbent dose within 30 min. Besides, 84% of diazinon was removed by silica-coated magnetic nanoparticles after 30 min and the adsorption kinetic obeyed pseudo-second-order model [90]. Pectin having high number of hydroxyl groups in banana peel structure makes it an effective adsorbent that removed atrazine with 98% from treated water and 39–57% from river water [91]. The attributed adsorption mechanisms are hydrogen bonding or electrostatic interaction for fludioxonil on clay, chemisorption for deltamethrin on modified bentonite, bridge formation through the H-bonding for malathion on modified bentonite, H- bonding, electrostatic interactions, van der Waals' forces for triazine, atrazine and imidacloprid on biochar [84]. Among the various adsorbents, montmorillonite, activated carbon cloth, eucalyptus bark-derived biosorbent and acid-treated patellidae shells provided the highest adsorption capacities for 2,4-D herbicide (1067 mg/g), aldicarb (421.6 mg/g), atrazine (936.1 mg/g) and aldrin (208.3 mg/g), respectively [92].

The major parameters affecting rejection of pesticides by membrane process are membrane characteristics (molecular weight cut-off, desalting degree, porosity, membrane material), pesticide properties (molecular size and weight, polarity, acid dissociation constant, hydrophobicity/hydrophilicity), and feed water composition (pH, ionic strength, presence of organic matter) [93, 94]. In many studies, various atrazine rejection performances have been obtained depending on the molecular weight cut-off (MWCO) of the membrane. The atrazine rejection performances were about 10% by UF membrane with 500 Da MWCO, 31% by NF45 and 39% by NF200 membrane with 300 Da MWCO [95]. On the other hand, atrazine, diuron, and diazinon were removed by NF70 (200–300 Da MWCO) and NF270 membranes (200–400 Da MWCO) with 86–93% [96, 97] and 65–85% [98, 99] removal efficiency, respectively. Fini et al. [100] reported that the highest rejections for dimethoate (85%) and atrazine (95%) were obtained by NF90 among four tested nanofiltration (NF) membranes. Ates et al. [101] reported 95–99% rejection performances for tributyl phosphate, flutriafol, irgarol, and dicofol by RO membranes (SW30-XLE, BW30-LE, GE-AD).

Martins et al. [102] reported 55% diuron removal by SA/SB resin even if it has no charge, moderate removal of diuron by ion exchange resin was attributed to the secondary amine groups in its structure. The removal performances of atrazine varied from 5 to 35% by polyacrylic styrenic resins, respectively. Combining ion exchange process with activated carbon adsorption enhances pesticide removal that both of atrazine and isoproturon were removed 90–100% by the AMBERSORB® and IRA938® resins [103]. Increasing the resin dose has a favorable effect on pesticide removal, while increasing initial pesticide concentration decreases the removal performance. In the study of Liu et al. [104], removal of bentazone was improved from 90% to 98.8% with an increase in MIEX dose from 1 to 10 mL/L of MIEX dose, respectively. On the other hand, removal performance for bentazone was decreased from 99.0% at 0.01 mmol/L to 61.9% at 1.8 mmol/L of initial concentration. Additionally, lower uptake of glyphosate, non-ionic compound, shows MIEX resin preferentially remove pesticides which are hydrophobic ionizable compounds and ion exchange is the dominant mechanism for removal of bentazone by MIEX.

7 Conclusion

Pesticides are widely used all over the world to increase agricultural production yield and to repel pests and vectors from industrial and household environment. Pesticides reaching water resources through various processes and routes seriously threaten human health and the lives of aquatic organisms. Particularly, some type of pesticides that are resistant to biodegradation remain in the water and soil environment for years, and they can be transferred to the humans and animals by becoming more concentrated in the food chain with bioconcentration/biomagnification. The fate and transport of pesticides to water resources and their presence and amounts in aquatic environments generally depend on the properties of pesticides, characteristics of its environment, and climatic conditions. Since conventional water and wastewater treatment plants are not specifically designed for pesticide removal, advanced processes should be integrated to the treatment train. Advanced oxidation processes, adsorption, membrane filtration, ion exchange, and their hybrid combinations are commonly applied treatment processes for pesticide removal from water resources. Long-term and intensive use of pesticides make pests resistant to them. Therefore, to ensure long-term sustainable pest management, replacement of synthetic pesticides with biopesticides with less environmental impact should be considered together with proper application of pesticides in the agricultural activities. In addition, strategies should be developed to manage pests and insects in the urban, agricultural, and industrial environments, providing long-term prevention of pests by natural means.

Acknowledgements The authors thank to the Scientific and Technological Research Council of Turkey (Project No: 118Y402) and the Research Fund of Erciyes University (Project No: FBA-2018-8032) for their support of this study.

References

1. FAO/WHO (Food and Agriculture Organization of the United Nations/World Health Organization). International code of conduct on pesticide management: guidelines on highly hazardous pesticides, Rome; 2016.
2. Rajmohan KS, Chandrasekaran R, Varjani S. A review on occurrence of pesticides in environment and current technologies for their remediation and management. Indian J Microbiol. 2020;60(2):125–38.
3. Rani L, Thapa K, Kanojia N, Sharma N, Singh S, Grewal AS, Srivastav AL, Kaushal J. An extensive review on the consequences of chemical pesticides on human health and environment. J Clean Prod. 2021:283.
4. Chengala L, Singh N. Botanical pesticides—a major alternative to chemical pesticides: a review. Int J Life Sci. 2017;5(4):722–9.
5. EC (European Commission). Active substances, safeners and synergists. https://ec.europa.eu/food/plant/pesticides/eu-pesticides-database/active-substances/ (2022). Accessed 28 May 2022.
6. de Souza RM, Seibert D, Quesada HB, Bassetti FD, Fagundes-Klen MR, Bergamasco R. Occurrence, impacts and general aspects of pesticides in surface water: a review. Process Saf Environ. 2020;135:22–37.
7. Syafrudin M, Kristanti RA, Yuniarto A, Hadibarata T, Rhee J, Al-onazi WA, Algarni TS, Almarri AH, Al-Mohaimeed AM. Pesticides in drinking water: a review. Int J Environ Res Public Health. 2021;18(2):468.
8. Kim KH, Kabir E, Jahan SA. Exposure to pesticides and the associated human health effects. Sci Total Environ. 2017;575:525–35.
9. Raffa CM, Chiampo F. Bioremediation of agricultural soils polluted with pesticides: a review. Bioeng-Basel. 2021;8(7).
10. Yadav IC, Devi NL. Pesticides classification and its impact on human and environment. Environ Sci Eng. 2017;6.
11. Samsidar A, Siddiquee S, Shaarani SM. A review of extraction, analytical and advanced methods for determination of pesticides in environment and foodstuffs. Trends Food Sci Tech. 2018;71:188–201.
12. WHO (World Health Organization). The WHO recommended classification of pesticides by hazard and guidelines to classification 2019. Geneva, Switzerland; 2020.
13. FAO (Food and Agriculture Organization of the United Nations). Pesticides use, pesticides trade and pesticides indicators 1990–2019; 2021.

14. Sabzevari S, Hofman J. A worldwide review of currently used pesticides' monitoring in agricultural soils. Sci Total Environ. 2022;812.
15. FAO (Food and Agriculture Organization of the United Nations). Pesticides indicators. https://www.fao.org/faostat/en/#data/EP (2022). Accessed 09 April 2022.
16. Anderson TA, Salice CJ, Erickson RA, McMurry ST, Cox SB, Smith LM. Effects of landuse and precipitation on pesticides and water quality in playa lakes of the southern high plains. Chemosphere. 2013;92(1):84–90.
17. Pirsaheb M, Hossini H, Asadi F, Janjani H. A systematic review on organochlorine and organophosphorus pesticides content in water resources. Toxin Rev. 2017;36(3):210–21.
18. Stone WW, Gilliom RJ, Ryberg KR. Pesticides in U.S. Streams and Rivers: Occurrence and Trends during 1992–2011. Environ Sci Tech Lett. 2014;48 (19):11025–30.
19. Mojiri A, Zhou JL, Robinson B, Ohashi A, Ozaki N, Kindaichi T, Farraji H, Vakili M. Pesticides in aquatic environments and their removal by adsorption methods. Chemosphere. 2020;253.
20. Battaglin WA, Smalling KL, Anderson C, Calhoun D, Chestnut T, Muths E. Potential interactions among disease, pesticides, water quality and adjacent land cover in amphibian habitats in the United States. Sci Total Environ. 2016;566:320–32.
21. Glinski DA, Purucker ST, Van Meter RJ, Black MC, Henderson WM. Analysis of pesticides in surface water, stemflow, and throughfall in an agricultural area in South Georgia, USA. Chemosphere. 2018;209:496–507.
22. Peng Y, Fang WD, Krauss M, Brack W, Wang ZH, Li FL, Zhang XW. Screening hundreds of emerging organic pollutants (EOPs) in surface water from the Yangtze River Delta (YRD): occurrence, distribution, ecological risk. Environ Pollut. 2018;241:484–93.
23. De Geronimo E, Aparicio VC, Barbaro S, Portocarrero R, Jaime S, Costa JL. Presence of pesticides in surface water from four sub-basins in Argentina. Chemosphere. 2014;107:423–31.
24. Carazo-Rojas E, Perez-Rojas G, Perez-Villanueva M, Chinchilla-Soto C, Chin-Pampillo JS, Aguilar-Mora P, Alpizar-Marin M, Masis-Mora M, Rodriguez-Rodriguez CE, Vryzas Z. Pesticide monitoring and ecotoxicological risk assessment in surface water bodies and sediments of a tropical agro-ecosystem. Environ Pollut. 2018;241:800–9.
25. Caldas SS, Arias JLO, Rombaldi C, Mello LL, Cerqueira MBR, Martins AF, Primel EG. Occurrence of pesticides and PPCPs in surface and drinking water in Southern Brazil: data on 4-year monitoring. J Brazil Chem Soc. 2019;30(1):71–80.
26. Affum AO, Acquaah SO, Osae SD, Kwaansa-Ansah EE. Distribution and risk assessment of banned and other current-use pesticides in surface and groundwaters consumed in an agricultural catchment dominated by cocoa crops in the Ankobra Basin, Ghana. Sci Total Environ. 2018;633:630–40.
27. Sumon KA, Rashid H, Peeters ETHM, Bosma RH, Van den Brink PJ. Environmental monitoring and risk assessment of organophosphate pesticides in aquatic ecosystems of north-west Bangladesh. Chemosphere. 2018;206:92–100.
28. Fadaei A, Dehghani MH, Nasseri S, Mahvi AH, Rastkari N, Shayeghi M. Organophosphorous pesticides in surface water of Iran. B Environ Contam Tox. 2012;88(6):867–9.
29. Pietrzak D, Kania J, Malina G, Kmiecik E, Wator K. Pesticides from the EU first and second watch lists in the water environment. Clean-Soil Air Water. 2019;47(7).
30. Gavrilescu M. Fate of Pesticides in the Environment and its Bioremediation. Eng Life Sci. 2005;5 (6):497–526.
31. Gil Y, Sinfort C. Emission of pesticides to the air during sprayer application: a bibliographic review. Atmos Environ. 2005;39(28):5183–93.
32. Mottes C, Lesueur-Jannoyer M, Le Bail M, Malézieux E. Pesticide transfer models in crop and watershed systems: a review. Agron Sustain Dev. 2014;34(1):229–50.
33. Fenner K, Canonica S, Wackett LP, Elsner M. Evaluating pesticide degradation in the environment: blind spots and emerging opportunities. Science. 2013;341(6147):752–8.
34. Montanari A, Shoemaker CA, van de Giesen N. Introduction to special section on uncertainty assessment in surface and subsurface hydrology: an overview of issues and challenges. Water Resourc Res. 2009;45(12).
35. Cai Y, Yue W, Xu L, Yang Z, Rong Q. Sustainable urban water resources management considering lifecycle environmental impacts of water utilization under uncertainty. Resour Conserv Recycl. 2016;108:21–40.
36. Wu L, Chang H, Ma X. A modified method for pesticide transport and fate in subsurface environment of a winter wheat field of Yangling, China. Sci Total Environ. 2017;609:385–95.
37. Carsel RF, Mulkey LA, Lorber MN, Baskin LB. The pesticide root zone model (PRZM): a procedure for evaluating pesticide leaching threats to groundwater. Ecol Model. 1985;30(1):49–69.
38. Leonard RA, Knisel WG, Davis FM. Modelling pesticide fate with GLEAMS. Eur J Agron. 1995;4 (4):485–90.
39. Alvarez J, Guirao J, Herguedas A, Atienza J. Evaluation of the PRZM2 model for transport of metamitron in undisturbed soil monoliths. In: 4th international conference on water pollution, Lake Bled, Slovenia; 1997, p. 67–74.
40. Willian WT, Mueller TC, Hayes RM, Bridges DC, Snipes CE. Comparison of PRZM and GLEAMS computer model predictions with field data for fluometuron and norflurazon behavior in soil. Weed Technol. 1999;13(3):561–70.

41. Marin-Benito JM, Mamy L, Carpio MJ, Sanchez-Martin MJ, Rodriguez-Cruz MS. Modelling herbicides mobility in amended soils: Calibration and test of PRZM and MACRO. Sci Total Environ. 2020;717.
42. Fox GA, Sabbagh GJ, Chen WL, Russell MH. Uncalibrated modelling of conservative tracer and pesticide leaching to groundwater: comparison of potential Tier II exposure assessment models. Pest Manag Sci. 2006;62(6):537–50.
43. Cheng Y, Zhou JY, Liao JH, Mao DZ, Chen WL, Shan ZJ. Coupled modeling using PRZM/RICEWQ and SWAT for the North Tiaoxi Watershed. Environ Sci Pollut Res. 2020;27(11):12635–45.
44. Gali RK, Cryer SA, Poletika NN, Dande PK. Modeling pesticide runoff from small watersheds through field-scale management practices: Minnesota watershed case study with Chlorpyrifos. Air Soil and Water Res. 2016;9:113–22.
45. Payraudeau S, Gregoire C. Modelling pesticides transfer to surface water at the catchment scale: a multi-criteria analysis. Agron Sustain Dev. 2012;32 (2):479–500.
46. Parker R, Arnold JG, Barrett M, Burns L, Carrubba L, Neitsch SL, Snyder NJ, Srinivasan R. Evaluation of three watershed-scale pesticide environmental transport and fate models 1. JAWRA J Am Water Resourc Assoc. 2007;43(6):1424–43.
47. Young RA, Onstad CA, Bosch DD, Anderson WP. AGNPS (agricultural non-point source pollution model): a watershed analysis tool. USDA ARS, Washington, DC.; 1987.
48. Arnold JG, Srinivasan R, Muttiah RS, Williams JR. Large area hydrologic modeling and assessment—part 1: model development. J Am Water Resour Assoc. 1998;34:73–89.
49. Borah DK, Xia R, Bera M. Chapter 5: DWSM—a dynamic watershed simulation model. In: Mathematical models of small watershed hydrology and applications; 2002.
50. Bicknell BR, Imhoff JC, Jr Kittle JL, Jr Donigian AS, Johanson RC. U.S. environmental protection agency, national exposure research laboratory. Hydrological Simulation Program—Fortran: User's manual for version 11. Athens, Ga, USA; 1997.
51. Abbott MB, Bathurst JC, Cunge JA, O'Connell PE, Rasmussen J. An introduction to the European Hydrological System—Systeme Hydrologique Europeen, "SHE", 2: structure of a physically-based, distributed modelling system. J Hydrol. 1986;87(1–2):61–77.
52. Zhang B, Zhang Q-Q, Zhang S-X, Xing C, Ying G-G. Emission estimation and fate modelling of three typical pesticides in Dongjiang River basin, China. Environ Pollut. 2020;258: 113660.
53. Purnell S, Kennedy R, Williamson E, Remesan R. Metaldehyde prediction by integrating existing water industry datasets with the soil and water assessment tool. Water Res. 2020;183: 116053.
54. Borah DK, Bera M, Shaw S, Keefer L. Illinois state water survey. Dynamic modeling and monitoring of water, sediment, nutrients, and pesticides in agricultural watersheds during storm events, IL, USA; 1999.
55. Laroche A-M, Gallichand J, Lagacé R, Pesant A. Simulating atrazine transport with HSPF in an agricultural watershed. J Environ Eng. 1996;122 (7):622–30.
56. Tsihrintzis VA, Fuentes HR, Gadipudi RK. Modeling prevention alternatives for nonpoint source pollution at a wellfield in Florida. JAWRA J Am Water Resourc Assoc. 1996;32(2):317–31.
57. Fauser P, Thomsen M, Sørensen PB, Petersen S. Predicted concentrations for pesticides in drainage dominated catchments. Water Air Soil Pollut. 2008;187(1):149–56.
58. EPA. U.S. (United States Environmental Protection Agency). The enhanced stream water quality models QUAL2E and QUAL2E-UNCAS: documentation and user model; 1997.
59. Wool TA, Ambrose RB, Martin JL, Comer EA. United States environmental protection agency. Water quality analysis simulation program (WASP), Version 6.0 User's manual. Atlanta, Georgia; 2001.
60. Adriaanse PI, Boesten JJTI, Crum SJH. Estimating degradation rates in outdoor stagnant water by inverse modelling with TOXSWA: a case study with prosulfocarb. Pest Manag Sci. 2013;69 (6):755–67.
61. Reichert P, Borchardt D, Henze M, Rauch W, Shanahan P, Somlyody L, Vanrolleghem P. River water quality model no. 1 (RWQM1): II. Biochemical process equations. Water Sci Technol. 2001;43 (5):11–30.
62. Sumon KA, Rico A, Ter Horst MMS, Van den Brink PJ, Haque MM, Rashid H. Risk assessment of pesticides used in rice-prawn concurrent systems in Bangladesh. Sci Total Environ. 2016;568:498–506.
63. Adriaanse PI, Van Leerdam RC, Boesten J. The effect of the runoff size on the pesticide concentration in runoff water and in FOCUS streams simulated by PRZM and TOXSWA. Sci Total Environ. 2017;584:268–81.
64. Focks A, ter Horst M, van den Berg E, Baveco H, van den Brink PJ. Integrating chemical fate and population-level effect models for pesticides at landscape scale: New options for risk assessment. Ecol Model. 2014;280:102–16.
65. Van-Wendel-de-Joode B, Brouwer D, Vermeulen R, Van Hemmen J, Heederik D, Kromhout H. DREAM: a method for semi-quantitative dermal exposure assessment. Ann Occup Hyg. 2003;47 (1):71–87.
66. Van den Brink PJ, Roelsma J, van Nes EH, Scheffer M, Brock TCM. Perpest model, a case-based reasoning approach to predict ecological risks of pesticides. Environ Toxicol Chem. 2002;21 (11):2500–6.

67. Ramo RA, van den Brink PJ, Ruepert C, Castillo LE, Gunnarsson JS. Environmental risk assessment of pesticides in the River Madre de Dios, Costa Rica using PERPEST, SSD, and msPAF models. Environ Sci Pollut R. 2018;25 (14):13254–69.
68. Oerke EC, Dehne HW, Schönbeck F, Weber A. Crop production and crop protection: estimated losses in major food and cash crops. Elsevier; 2012.
69. Vymazal J, Bfezinova T. The use of constructed wetlands for removal of pesticides from agricultural runoff and drainage: a review. Environ Int. 2015;75:11–20.
70. Recena MCP, Pires DX, Caldas ED. Acute poisoning with pesticides in the state of Mato Grosso do Sul, Brazil. Sci Total Environ. 2006;357(1–3):88–95.
71. Boedeker W, Watts M, Clausing P, Marquez E. The global distribution of acute unintentional pesticide poisoning: estimations based on a systematic review. BMC Public Health. 2020;20(1).
72. Sharma A, Shukla A, Attri K, Kumar M, Kumar P, Suttee A, Singh G, Barnwal RP, Singla N. Global trends in pesticides: a looming threat and viable alternatives. Ecotox Environ Safe. 2020;201.
73. Maggi F, Tang FHM, Black AJ, Marks GB, McBratney A. The pesticide health risk index: an application to the world's countries. Sci Total Environ. 2021;801.
74. Juraske R, Mutel CL, Stoessel F, Hellweg S. Life cycle human toxicity assessment of pesticides: comparing fruit and vegetable diets in Switzerland and the United States. Chemosphere. 2009;77 (7):939–45.
75. USEPA (United States Environmental Protection Agency). Federal Insecticide, Fungicide, and Rodenticide Act (FIFRA). Public Law 104. 61 Statutes at Large Chapter 125. Washington, DC.; 1947.
76. Rauch-Williams T, Snyder S, Drewes J, Dickinson E. Current and proposed paradigms to control CECs in the United States and internationally states and internationally phase 1 report. Denver, CO, USA; 2016.
77. Kroehler CJ. Potable water quality standards and regulations: a historical and world overview. Potable Water-Emerg Glob Prob Solut. 2014;30
78. USEPA. Final contaminant candidate list 3 chemicals: classification of the PCCL to CCL. Agency U. S.E.P. United States Environmental Protection Agency: EPA 815-R-09–008, Washington, DC; 2009.
79. EU. Directive 2013/39/EC of the European Parliament and of the Council of 12 August 2013 Amending Directives 2000/60/EC and 2008/105/EC as regards priority substances in the field of water policy text with EEA relevance Off J L. Union E. European Union: 226; 2013. p. 1–17.
80. MofFWA. Surface water quality regulation. Affairs M.o.F.a.W. Official Gazette: 28483, Ankara, Turkey, 30 Nov 2012; 2012.
81. Guillossou R, Le Roux J, Mailler R, Vulliet E, Morlay C, Nauleau F, Gasperi J, Rocher V. Organic micropollutants in a large wastewater treatment plant: What are the benefits of an advanced treatment by activated carbon adsorption in comparison to conventional treatment? Chemosphere. 2019;218:1050–60.
82. Vagi MC, Petsas AS. Recent advances on the removal of priority organochlorine and organophosphorus biorecalcitrant pesticides defined by Directive 2013/39/EU from environmental matrices by using advanced oxidation processes: an overview (2007–2018). J Environ Chem Eng. 2020;8(1): 102940.
83. Luo YL, Guo WS, Ngo HH, Nghiem LD, Hai FI, Zhang J, Liang S, Wang XCC. A review on the occurrence of micropollutants in the aquatic environment and their fate and removal during wastewater treatment. Sci Total Environ. 2014;473:619–41.
84. Saleh IA, Zouari N, Al-Ghouti MA. Removal of pesticides from water and wastewater: chemical, physical and biological treatment approaches. Environ Technol Innov. 2020;19.
85. Kenfack S, Sarria V, Wethe J, Cisse G, Maiga AH, Klutse A, Pulgarin C. From laboratory studies to the field applications of advanced oxidation processes: a case study of technology transfer from Switzerland to Burkina faso on the field of photochemical detoxification of biorecalcitrant chemical pollutants in water. Int J Photoenergy. 2009.
86. Shah NS, He XX, Khan HM, Khan JA, O'Shea KE, Boccelli DL, Dionysiou DD. Efficient removal of endosulfan from aqueous solution by UV-C/peroxides: a comparative study. J Hazard Mater. 2013;263:584–92.
87. Ormad MP, Miguel N, Lanao M, Mosteo R, Ovelleiro JL. Effect of Application of ozone and ozone combined with hydrogen peroxide and titanium dioxide in the removal of pesticides from water. Ozone-Sci Eng. 2010;32(1):25–32.
88. Debabrata P, Sivakumar M. Sonochemical degradation of endocrine-disrupting organochlorine pesticide Dicofol: investigations on the transformation pathways of dechlorination and the influencing operating parameters. Chemosphere. 2018;204:101–8.
89. Ali I, ALOthman ZA, Al-Warthan A. Sorption, kinetics and thermodynamics studies of atrazine herbicide removal from water using iron nanocomposite material. Int J Environ Sci Tech. 2016;13 (2):733–42.
90. Naeimi Bagheini A, Saeidi M, Boroomand N. Removal of diazinon pesticide using amino-silane modified magnetite nanoparticles from contaminated water. Int J Nanosci Nanotechnol. 2018;14 (1):19–32.
91. Silva CR, Gomes TF, Andrade GC, Monteiro SH, Dias AC, Zagatto EA, Tornisielo VL. Banana peel as an adsorbent for removing atrazine and ametryne from waters. J Agr Food Chem. 2013;61(10):2358–63.

92. Ighalo JO, Adeniyi AG, Adelodun AA. Recent advances on the adsorption of herbicides and pesticides from polluted waters: performance evaluation via physical attributes. J Ind Eng Chem. 2021;93:117–37.
93. Taghizade Firozjaee T, Mehrdadi N, Baghdadi M, Nabi Bidhendi GR. Application of nanotechnology in pesticides removal from aqueous solutions-a review. Int J Nanosci Nanotechnol. 2018;14(1):43–56.
94. Plakas KV, Karabelas AJ. Removal of pesticides from water by NF and RU membranes—a review. Desalination. 2012;287:255–65.
95. Devitt EC, Ducellier F, Cote P, Wiesner MR. Effects of natural organic matter and the raw water matrix on the rejection of atrazine by pressure-driven membranes. Water Res. 1998;32(9):2563–8.
96. Van der Bruggen B, Schaep J, Maes W, Wilms D, Vandecasteele C. Nanofiltration as a treatment method for the removal of pesticides from ground waters. Desalination. 1998;117(1–3):139–47.
97. Chen SS, Taylor JS, Mulford LA, Norris CD. Influences of molecular weight, molecular size, flux, and recovery for aromatic pesticide removal by nanofiltration membranes. Desalination. 2004;160(2):103–11.
98. Ahmad AL, Tan LS, Shukor SRA. Dimethoate and atrazine retention from aqueous solution by nanofiltration membranes. J Hazard Mater. 2008;151(1):71–7.
99. Kosutic K, Furac L, Sipos L, Kunst B. Removal of arsenic and pesticides from drinking water by nanofiltration membranes. Sep Purif Technol. 2005;42(2):137–44.
100. Fini MN, Madsen HT, Muff J. The effect of water matrix, feed concentration and recovery on the rejection of pesticides using NF/RO membranes in water treatment. Sep Purif Technol. 2019;215:521–7.
101. Ates N, Uzal N, Yetis U, Dilek FB. Removal of pesticides from secondary treated urban wastewater by reverse osmosis. Environ Sci Pollut R. 2022.
102. Martins VL, Ogden MD, Jones MR, Trowsdale SA, Hall PJ, Jensen HS. Opportunities for coupled electrochemical and ion-exchange technologies to remove recalcitrant micropollutants in water. Sep Purif Technol. 2020;239.
103. Humbert H, Gallard H, Suty H, Croue JP. Natural organic matter (NOM) and pesticides removal using a combination of ion exchange resin and powdered activated carbon (PAC). Water Res. 2008;42(6–7):1635–43.
104. Liu ZQ, Yan XM, Drikas M, Zhou DN, Wang DS, Yang M, Qu JH. Removal of bentazone from micro-polluted water using MIEX resin: kinetics, equilibrium, and mechanism. J Environ Sci. 2011;23(3):381–7.

Drinking Water Related Health Risks

Emrah Caylak

Abstract

Water is an indispensable source for the existence and continuity of life. The increase in the world's population, the rapid development of industry and technology, the increase in water consumption, as well as climate change and the inability to adequately settle environmental awareness are causing the amount of potable water to gradually decrease. Over the centuries, natural and anthropogenic activities have caused widespread pollution in the waters and leaded to accumulate different pollutants in ecosystem. These pollutants are mainly categorized as biological, inorganic, and organic contaminants. In this study, the potential health risks of these pollutants in drinking water are reviewed.

Keywords

Biological contaminants · Drinking water · Health risk · Inorganic contaminants · Organic contaminants · Pollutants

1 Introduction

Clean drinking water is an essential source for the existence, growth, and development of living organisms. Humans use water in their lives for drinking, food preparation, and personal hygiene purposes such cleaning, bathing etc. Water covers 72% of the Earth. Unfortunately, we can't use most of that as drinking water. It is known that groundwater is the most important supply of drinking water, and less than 1% of the water on Earth is drinkable. Pollution in water sources due to the increase in the population in all countries of the world minimizes people's access to drinking water. Drinking water resources are polluted by natural and anthropogenic activities, which have led to the accumulation of different pollutants in the ecosystem. The ground and surface waters are contaminated natural (geology and mineralogy of the watershed, forest fires, earthquakes, volcanic eruptions, and changes in rainfall patterns) and anthropogenic (industrial, mining, agricultural activities and soil or water purification and errors in treatment and distribution of water) activities [1]. These pollutants are mainly categorized as biological, inorganic, and organic contaminants.

E. Caylak (✉)
Department of Biochemistry, Health Sciences Institute, Firat University, Elazig, Turkey
e-mail: emrah333@hotmail.com

A. M. Bahadir et al. (eds.), *Hydrology and Urban Water Supply*, Water and Wastewater Management,
https://doi.org/10.1007/978-3-031-72589-0_19

2 Pollutants

2.1 Biological Contaminants

Biological contaminants in drinking water are organisms such as bacteria, viruses, protozoa, and parasites. They initiate the evolution of contagious diseases and some of those are pneumonia, hepatitis, gastric ulcers, influenza, and SARS. The presence of microbiological pollutants in drinking water in urban areas is widely observed. Pathogenic microorganisms are bacteria (e.g., Salmonella typhi, Vibrio cholera and Shigella), viruses (e.g., Poliovirus) and protozoa (e.g. Giardia lambia and Cryptosporidium). Such microorganisms are periodically detected in water samples of river, groundwater and drinking water [2]. According to the World Health Organization (WHO), the microorganisms most found in drinking water sources are *Cryptosporidium*, *Legionella*, *Pseudomonas*, *Giardia* and *E. coli*. The presence of these pathogens in the water poses serious threats not only to human, but also to the entire aquatic ecosystem. Therefore, it is mandatory to minimize the exposure to deadly diseases by putting an end to the presence of pathogens in the waters [3].

The microbial contamination of water may occur accidentally, but most contamination is caused by insufficient attention to wastewater disposal and water sterilization procedures. This group of microbes is transmitted directly through drinking water and indirectly through food consumption. Pathogenic microbes have become the main cause of death in the developing countries of the world. Important microbial diseases caused by water and their factors are as follows: Cholera (*Vibriocholera*), bacillary dysentery (*Shigellas*), typhoid (*Salmonella typhi*), paratyphoid (*Salmonella paratyphi* A, B, C), gastroenteritis (other *Salmonellas*, *Proteose*, *Shigella*), infantile diarrhea (*Escherichia coli*), leptospirosis (*Leptospiras*), tularemia (*Pasteurella tularensis*) [4].

Some protozoa and parasites can also be transferred to the human body directly through the consumption of contaminated water and cause disease. Important protozoa and parasites and their diseases include: amoebic dysentery (*Entameoba histolytica*), giardiasis (*Giardia lamblia*), balantidiasis (*Balantidium coli*), daracunculus (*Deracunculus*), desitomatosis (*Fasciola hepatica* or *Dicrocoelium*) [5].

2.2 Inorganic Contaminants

Inorganic substances that pollute the waters as natural sources (earth crust, rock weathering and volcanic eruption, etc.) or human activities (agriculture, mining, industrial activities, untreated wastewater discharge, etc.). The most important of these pollutants are heavy metals, radionuclides (caesium, phosphorous and uranium, etc.), and plant fertilizers (ammonia, nitrate and phosphate, etc.) [1].

2.2.1 Heavy Metals, Light Metals and Metalloids

Heavy and transition metals ions as well as light metals are inorganic pollutants, which include aluminum (Al), arsenic (As), beryllium (Be), boron (B), cadmium (Cd), chromium (Cr), cobalt (Co), copper (Cu), iron (Fe), lead (Pb), lithium (Li), manganese (Mn), mercury (Hg), molybdenum (Mo), nickel (Ni), selenium (Se), vanadium (V), and zinc (Zn). Metals such as Co, Cu, Fe, Zn, Mo, and Se are necessary for living organisms, especially in biological reactions during human development and metabolism. On the other hand, As, Cd, Cr, Pb, and Hg are high health-related risks and cause different diseases and cancer [1].

2.2.2 Aluminum

Aluminum (Al) is present in excess in the earth's crust. Under acidic conditions (pH 4.5–5) water can seep through rocks and clay and mix with drinking water. Aluminum sulfate is added to the water to induce flocculation and remove organic and other contaminants in the water treatment plants. Acids and bases added regularly to treated water, various chemicals such as chlorine, phosphate, silica, and fluoride can create low pH conditions in the water, causing more dissolved aluminum to mix with the water. In the oral intake of Al in drinking water, only about 0.3% is absorbed via the gastrointestinal tract, while about 95% is excreted via the kidneys. This represents a particular health risk, especially for dialysis patients. United States Environmental Protection Agency (USEPA) has determined a concenztration range of 0.05–0.2 mg/L for Al in drinking water; WHO (2010) has increased the range to 0.1–0.2 mg/L [6, 7]. It is also known that Al leads to dementia and Alzheimer's disease [8, 9].

2.2.3 Arsenic

The contamination of water resources with arsenic (As) is due to the dissolution of rocks, minerals and ores containing As or anthropogenically mining, industrial and agricultural activities (the use of arsenic-based pesticides). WHO has established 10 µg/L as a standard for As [1]. High levels of As have been recorded in drinking water sources in Cankiri province of Turkey. Caylak [10] found high concentration of As (10–30 mg/L) in drinking water samples of Cankiri city and Kursunlu Town. It is concentrated in the groundwater of As seeping from rocks containing very high levels of As due to the mineralization zones associated with the volcanic of this region. The epidemiological studies on As in drinking water reported that water containing As used for drinking and bathing causes skin lesions and diseases and cancers of the lungs, liver, kidneys, and bladder in humans [10, 11].

2.2.4 Beryllium

Sources of beryllium (Be) in drinking water include precipitation of atmospheric Be and decoupling of Be containing rocks and soils, coal burning, and other industries using Be. It is rarely found in drinking water in alarming concentrations. WHO has established a health based value of 12 µg/L for Be in drinking water, and the USEPA has set it at 4 µg/L. Although it was found in low concentrations in drinking water sources and drinking water, drinking water in Saudi Arabia contained an average concentration of 1.05 µg/L. This may have raised concentrations due to some industrial sources, such as metal refineries, coal-burning factories, etc. In humans, the lung is the primary target of inhalation exposure to Be. It can lead to the development of chronic Be disease, which is characterized by the formation of granulomas, and increases in lung cancer mortality. Studies on the potential effects of ingested Be to humans are not available. It can cause lesions of the gastrointestinal tract [12, 13].

2.2.5 Boron

Boron (B) is naturally found element in rocks, soil and water and is a possible trace mineral nutrient for humans. It has provided some evidence that low B intake in animals and humans reduce cellular function (estrogen metabolism and cognitive functions) and the activity of other nutrients (vitamin D and calcium). It is released into the environment from these sources by natural decomposition processes, and used in the production of glass, anti-corrosion and cleaning products, agrochemicals, insecticides. Contamination of waters with B occurs from energy and chemical industrial plants or with fertilizers, herbicides, and industrial wastes [14]. There are rich B mine fields in Turkey. Cöl and Cöl [15] determined B levels ranged 2.05 and 29 mg/L in drinking water of Kutahya-Turkey. Caylak [16] also analyzed B in samples of drinking water of Cankiri-Turkey and found concentrations ranged 130–1050 mg/L. The value of those analysis has been found to be significantly higher than the recommended Turkish (1000 µg/L) and WHO (500 µg/L) limits for drinking water. In animals, chronic low-level B exposure is associated with a decrease in growth, cutaneous disorders, and suppression of male reproductive system function. In humans, it has been reported that B causes liver, gastrointestinal, neurological, and cardiovascular diseases, and anorexia, diarrhea, testicular atrophy, and kidney damage [17].

2.2.6 Cadmium

Cadmium (Cd) is found on earth along with zinc (Zn), lead (Pb) and copper (Cu) ores, or in coal and other fossil fuels, and can pass into ground and surface waters in an acidic environment. Contamination of surface water by industrial waste (metal refineries, waste batteries and paint) or certain fertilizers can also occur. Most often, drinking water is contaminated by corrosion of galvanized pipes and metal fittings used in the water treatment plants or distribution system. WHO has established the guideline value as 3 μg/L for Cd in drinking water [1]. Idrees et al. [18], Mohammadi et al. [19] determined Cd concentrations (0.05–0.07 and 0.43 μg/L) in drinking water analyses in various regions of India and Iran. Cd is a toxic heavy metal that accumulates mainly in the kidneys and causes harmful effects on this organ. It has been found that exposure to Cd with drinking water causes convulsions, liver damage, shock, and kidney failure. Prolonged exposure to Cd disrupts calcium metabolism in the body and causes liver, bone, and kidney damage [20].

2.2.7 Chromium

Chromium (Cr) is widely present in the earth's crust, especially in areas with mafic or ultramafic volcanic or metamorphic rocks, it can be released into groundwater. Anthropogenic sources are caused by wastewater flows from the refinery, iron, leather, and paper industries, as well as paint, fungicides; ceramic and glass waste and corrosion of chrome-plated water networks [21]. The maximum permissible concentration of Cr in drinking water was determined by WHO to be 50 μg/L [1]. The average Cr levels in drinking water in Greece were found to be between 1.4 and 50 μg/L [22]. Long-term exposure to the respiratory and gastrointestinal tract, kidney and liver damage due to acute Cr exposure by ingestion with drinking water in humans causes hyperplasia of the small intestine and increases the risk of liver cancer [23].

2.2.8 Cobalt

Cobalt (Co) is a rare element in the earth's crust, it is necessary in trace quantities, since it is an integral component of the vitamin B12 complex of humans and other mammals. In high concentrations, it is toxic to humans and is found in deposits of nickel-bearing laterite near groundwater, as well as in mafic and ultramafic rocks and deposits of sedimentary Cu deposits. It is widely used in industry in the production of steel, magnets, paints, ceramics, glass, petroleum, batteries, and fertilizers. Anthropogenically, Co is mixed with coal-fired power plants, mining, and industrial wastes, as well as surface waters [24]. Atashi et al. [25] found Co levels in drinking water wells in Zahedan, Iran, (172–204 μg/L), that was quite high when compared to WHO standards (5 μg/L). Beyene and Berhe [26] determined that the maximum Co level in drinking water samples obtained from water sources in the Dowhan and Europe regions of Ethiopia was at the levels of 10 and 20 μg/L, respectively. Co is an essential element in humans and animals and is used to treat anemia because it stimulates red blood cell production, while in high concentrations it is toxic. Chronic exposure to Co lung disease and asthma in humans by inhalation; ingestion exposure through vomiting and diarrhea, liver damage, blood effects (damage, macrophage, increased levels of hemoglobin etc.), and has been reported to cause allergic dermatitis. An increase in deaths due to lung cancer has been detected in workers exposed to Co for a long time [24].

2.2.9 Copper

Copper (Cu) is widely found in the earth's crust, and natural metal erosion, leaching into the groundwater table by dissolution can occur. It is released into water by human activities from mining operations, industrial discharge, wastewater treatment plants and Cu concentrations in drinking water can be greatly increased by leaking through pipes used for the distribution of drinking water. It is an essential element for living things and is responsible for the structural and catalytic properties of many enzymes in bone, blood, and iron metabolism [1]. Pettersson et al. [27] analysed samples of tap water in Sweden and determined an increased concentration of Cu (maximum level 5 mg/L). WHO established the guidelines for drinking water for Cu as 2 mg/L [1]. High levels of Cu intake by drinking water have been reported to cause gastrointestinal mucosal ulcerations and bleeding, acute hemolysis, hemoglobinuria, Wilson's disease, and hepatic jaundice. Prolonged exposure to high levels of Cu in drinking water leads to liver and kidney damage, as well as hemolytic anemia [28].

2.2.10 Iron

Iron (Fe) is a very abundant element in the earth's crust and is essential, which found in the structure of metalloenzyme, porphyrin and cytochromes in living organisms. Fe deficiency can cause anemia. Groundwater resources are polluted by the decomposition of Fe in rocks and minerals or from pipes and storage tanks belonging to water distribution networks, industry and sewage waste related to Fe. WHO has established the aesthetic-based water guideline for Fe as 0.3 mg/L [1]. Ghosh et al. [29] found that iron exceeds the limits in groundwater samples of Bangladesh's Jashore. They determined the Fe concentrations (from 0.02 to 6.2 mg/L) and the average value of Fe (1.4 mg/L) was higher than the permissible limit of WHO. However, prolonged exposure to Fe in Parkinson's disease, Huntington's disease, cardiovascular disease, hyperkeratosis, diabetes mellitus, and pigmentation changes, Alzheimer's disease, kidney, liver, respiratory, and neurological disorders such as may cause adverse health risks [30].

2.2.11 Lead

Lead (Pb) is one of the heavy metals commonly found in the earth's crust. In addition to dissolving lead in tap water from natural sources, it is caused by corrosion of PVC and lead pipes and soldering, and fittings used in water distribution systems and installations. In industry, lead is used in the production of accumulator, solder, alloys, batteries, cables and paints, industrial waste can also pollute water sources. According to the WHO standard drinking water criteria, the highest desired Pb limit in drinking water is 10 μg/L [1]. Ehya and Marbouti [31] determined the Pb levels in groundwater sources in Iran reached 840 μg/L, and they suggested that it was related to the inflow of industry. Turhan [32] tested the Pb concentrations in drinking water samples from the Cappadocia region of Turkey and found that they were higher (10.4 and 12.7 μg/L) compared to the WHO's permissible limit. Lead is a very toxic metal and is associated with various health hazards such as anemia, reproductive and mental development disorders. Long-term exposure to Pb with drinking water leads to hypertension, neurodevelopmental and renal dysfunction, adverse pregnancy problems, and increased mortality from cardiovascular diseases [33].

2.2.12 Lithium

Lithium (Li) is a naturally occurring element and is found in variable amounts in all rocks. It can pass into the water supply by decomposition. It is used in batteries of electric vehicles, laptops, and mobile phones, as well as in the glass and ceramic industry. It is known that Li interacts with neurotransmitters and receptors in the human brain, increases serotonin levels and reduces the production of brain norepinephrine. In recent years, studies have been conducted suggesting that adding low concentrations of Li to drinking water may also increase mental health benefits (low suicide mortality rates) for the general population [1]. Li concentrations in tap water were found 0.6–30.7 μg/L in Denmark [34] and 1–191 μg/L in Italy [35]. It is used as a therapeutic agent in the treatment of chronic neurological diseases such as depressive and bipolar disorders, Parkinson, Huntington, Alzheimer [36]. WHO has not defined standards for Li in drinking water. The overdose limit of the serum Li level is 1.8–2.5 mmol/L, and for the toxic level it is 3–10 mmol/L. Many studies have shown that the constant intake of Li in low doses with drinking water can cause chills, nausea, headache, hypothyroidism, weight gain, hypercholesterolemia, hyperparathyroidism, and hypercalcemia. The effects of an overdose of Li with drinking water are nausea, vomiting, diarrhea, chills, ataxia, dysarthria, nystagmus, kidney damage, confusion, and convulsions [37].

2.2.13 Manganese

Manganese (Mn) is commonly found in soil and natural weathering of rocks and soil and anthropogenic activities, such as mining, steel, batteries, fertilizers, fungicides, industrial paints and cosmetics can increase the levels in water sources. Mn can also be added to water as an oxidizing agent or as an impurity in coagulants during the purification of drinking water. Mn under high acidic level is usually more common in groundwater than in surface waters. The maximum acceptable concentration for total Mn in drinking water is established as 400 μg/L by WHO [38]. Ghosh et al. [29] found that Mn exceeds the limits in about 87% of Bangladesh's Jashore groundwater samples. They determined the Mn concentrations (from 0.016 to 2.108 mg/L) and the average value of Mn (0.298 mg/L) was higher than the permissible limit of WHO. Mn is an indispensable element for humans. It has been shown that neurological effects such as common cognitive, and motor disorders, and reproductive system disorders such as testicular dysfunction, and damage, impaired fertility in the female reproductive system occur because of exposure to high levels of Mn in drinking water. In addition, some epidemiological studies have suggested an association between Mn exposure in drinking water and neurological effects such as learning memory impairment in children [39]. The current studies for Mn so far have not been able to determine the carcinogenicity of Mn and it has not been classified as a human carcinogen [38].

2.2.14 Mercury

Mercury (Hg) is a toxic metal found in the earth's crust. Drinking water is polluted by geogenic inputs (lithosphere and fossil fuels) and industrial waste. The WHO has set a maximum permissible Hg concentration of 1 μg/L in drinking water [1]. Khattak et al. [40] identified Hg concentration (2.0 μg/L) in groundwater (shallower aquifer) of District Swabi, Pakistan exceeded WHO recommended limits. Prolonged exposure to Hg using contaminated drinking water causes diseases of the skin, kidneys, and reproductive and nervous systems, as well as cancer [41].

2.2.15 Nickel

Nickel (Ni) is a naturally occurring element. The main natural source of Ni in drinking water can be caused by mobilization from ultramafic rocks into groundwater. The main anthropogenic sources of Ni in water are primarily electroplating, Ni–Cd batteries, and coins, the production of paints, welds, and some electronic products, the burning of fossil fuels, and the discharge of industrial, and municipal waste. The primary source of Ni in drinking water is from pipes and metal fittings used during treatment and distribution. The Ni content of all samples in the study was higher than the permissible limit of 0.02 mg/L defined by WHO [1]. Khan et al. [42] found that the average values of Ni in Iran, Peshawar City's drinking water samples ranged from 0.097 to 0.250 mg/L. Ni leads to oxidative stress by increasing the number of active oxygen species (ROS) and causes toxic effects on infertility and iron homeostasis. Prolonged exposure to Ni can cause serious damage to the skin, kidneys, and diseases of the digestive, circulatory, and nervous systems [43].

2.2.16 Selenium

Selenium (Se) is widely distributed element in the earth's crust. It can be dissolved by the decomposition of minerals in the soil or rock, and can be found in groundwater. Coal and copper power plants, mining, refining, glass, paint, plastics, and ceramics industries are the main anthropogenic sources of Se in water. It is an essential trace element that must be present in the human diet. It is an integral part of selenoproteins, especially in the thyroid hormones, immune response and antioxidant defense system of organisms. Due to Se deficiency, hypothyroidism, Kashin-Beck disease (characterized by rheumatism) and Keshan's disease (characterized by cardiomyopathy) are observed in humans. The maximum acceptable concentration established by WHO for Se in drinking water is 0.05 mg/L [1]. Golubkina et al. [44] identified elevated levels of Se (1.2–18.6 μg/L) in the drinking water of Dorno-Gobi aimag, Mongolia, exceeding WHO's limits. As a result of chronic exposure to drinking water with high levels of Se characterized by symptoms of selenosis, which are, hair and nail loss or abnormalities, skin abnormalities, and even more serious disorders of the nervous system and liver damage. There is not a relationship with exposure to Se and increase in the incidence of cancer [45].

2.2.17 Vanadium

Vanadium (V) is a ubiquitous metal, and high concentrations of it have been found in groundwater, especially in volcanic areas. It is important for the atomic energy industry, aircraft construction and space technology. It is also widely used in the chemical industry to produce sulfuric acid and plastics. High vanadium levels in water sources are caused by fossil fuels with a high V content with the above-mentioned industrial discharges. Its concentrations in drinking water may vary between 0.2 and 100 μg/L [1]. Concentration of V in drinking water is significantly dependent on geographical location. Arena et al. [46] found higher V in groundwater from Mt. Etna, Italy (ranged 15.6–182 μg/L) than the permissible limits of the California Office of Environmental Health Hazard Assessment (OEHHA). The WHO Guidelines for drinking water quality do not contain a limit value for vanadium. It is listed as a contaminant of concern and is on the US-EPA Contaminant Candidate List 3, and OEHHA has published a proposed notification level for vanadium at 15 μg/L in drinking water. It has been shown that V is an essential element for animals. The lack of V in those contributes to the disruption of reproduction and lipid metabolism and causes a decrease in growth. It has been suggested that V may be a regulatory agent for many enzymatic activities in mammalian tissues. It stimulates adenylate cyclase, inhibits cholesterol biosynthesis and plays a role in the regulation of glucose metabolism. Prolonged exposure to V with drinking water can cause arterial hypertension, cardiac arrhythmia, diabetes mellitus and renal lithiasis, and increase the incidence of papillary thyroid cancer [47].

2.2.18 Zinc

Zinc (Zn) is abundant element in the earth's crust. The main use for Zn is the galvanization of iron and steel. It is used in the manufacture of paints, batteries, rubber, insecticides for plumbing and heating fixtures. The concentration of Zn in tap water can be significantly high due to the leaching of Zn from galvanized pipes, water tanks, and brass fittings from surface water. It is an essential element for all organisms, including humans. It is involved in all aspects of metabolism, including Zn-containing proteins and enzymes. Zn deficiency is common in countries where the primary source of protein is cereal grains because Zn absorption is limited due to the high content of phosphates and phytates in cereals. Due to Zn deficiency in humans, there is delayed growth, skin changes, hypogonadism, mental lethargy, and night blindness. WHO permissible limit of the Zn content of drinking water is 3.0 mg/L [1]. Khan et al. [42] conducted sampling of drinking water in Peshawar, Iran and reported Zn concentrations range of 0.007–0.773 mg/L. Prolonged maximum intake of Zn with drinking water, especially due to the interaction between Fe and Zn, leads to digestive disorders, changes in the level of cholesterol and lipoproteins, impairment of cognitive processes and immunity [48].

2.2.19 Radionuclides

The most common radionuclides found in drinking water are uranium (238U, 235U), radium (226Ra) thorium (232Th), cesium (137Cs) and potassium (40 K). In addition, radon (222Rn) produced by the decay of 238U was also found in groundwater. The presence of radioactivity in drinking water occurs due to the dissolution of radioactive substances in the soil and rocks, as well as the passage of water or the mixing of radionuclides into water due to waste. Anthropogenic activities as uranium mining and processing or medical and industrial applications, accidental release of radionuclides (for example, Chernobyl in 1986 and Fukushima Daiichi in 2011 nuclear power plant accidents), fallout of nuclear weapon testing contaminates water. Radioactivity is a process by which atoms spontaneously disintegration, usually occurs together with the emission of radiation. It is measured in units Becquerel (Bq), which measure radioactive decay per second. For drinking water, the activity concentrations of a radionuclide are expressed in Bq/L. WHO guidance levels for radionuclides in drinking water ranged from 0.1 to 10,000 Bq/L. The main radionuclide limits are 1 Bq/L (226Ra, 228Th, 230Th, 234U, 239Pu, 241Am) or 10 Bq/L (90Sr, 131I, 134Cs, 137Cs, 238U). The effective dose is a measure of the energy/biological damage accumulated in a tissue or organ and is called sieverts (Sv) 'radiation dose', usually used as millisievert (mSv). Reference levels of radionuclides for drinking water are based on a value of 1 mSv per year [49]. Küçükömeroğlu et al. [50] determined the radioactivity level in 15 different water works providing drinking water to Trabzon, Turkey and they found the average 222Rn concentration in drinking water ranged from 0.56–13.3 Bq/L. The mean concentrations of 238U, 232Th, 137Cs, and 40 K radionuclides in drinking water were measured as follows 147, 25, 77, and 12 mBq/L, respectively. When the radon concentrations for drinking water were examined, they reported that it was above in the three samples the limit value (11 Bq/L) allowed by the WHO for drinking water. The radionuclides can harm human health by irradiation of the kidneys, liver, bladder, arch, and bone. Also, radon (222Rn) taken with drinking water leads to the risk of lung cancer over time.

2.2.20 Plant Fertilizers

Plants need nutrients such as nitrogen, phosphorus and ammonia during their growth, and fertilizers containing these substances are intensively used by farmers in agriculture to obtain more crops. These substances from fertilizers that are not used by plants and are not kept in the soil can be washed into surface waters or seeped underground and contaminate drinking water sources. The most common groundwater pollutant from fertilizers and organic soil changes is nitrogen. Drinking water is anthropogenically polluted with chemical fertilizers, sewage and household wastes used in agriculture with nitrate (NO_3^-). NO_3^- is very easily soluble in water and leaches in soils, contaminating groundwater. WHO has set the maximum level of pollutants for nitrate in drinking water at 50 mg/L. The researchers reveal that nitrogen accumulates in soils and forms a long-term source of NO_3^- pollution in ground and surface waters, and that even if farmers stopped using nitrogenous fertilizers, high NO_3^- concentrations in rivers and lakes would remain high for decades [51].

NO_3^- taken with drinking water is easily absorbed by the digestive system in humans and NO_3^- in the human body can be reduced to (NO_2^-). Under acidic conditions in the stomach, NO_2^- can be protonated to nitrous acid (HNO_2) and converted into N-nitrosamines and N-nitrosamides. NO_2^- binds to hemoglobin to form methemoglobin, which interferes with the oxygen-carrying capacity of the blood. NO_2^- contaminated water is dangerous for pregnant women and babies, it causes methemoglobinemia in infants, known as Blue Baby Syndrome. High levels of NO_3^- intake with maternal drinking water during pregnancy can cause intrauterine growth retardation, low birth weight, congenital malformations, and neonatal mortality. The relationship between drinking water NO_3^- and thyroid diseases or colorectal, ovarian and kidney cancers, especially due to nitrosamines, which are carcinogenic, has also been reported [51, 52].

Phosphorus (P) is a common component in commercial fertilizers and is found in biological organisms in the form of phosphate (PO_4^{3-}). High concentrations of P in groundwater can be caused by the natural decomposition of PO_4^{3-} from rocks and minerals, agricultural runoff, sewage waste and industrial discharges. In industry, PO_4^{3-} are widely used in the production of food additives, pharmaceuticals, synthetic resins, detergents, and disinfectants. It is added to drinking water distribution systems to prevent metal leakage from water pipes, mostly lead dissolution. On the other hand, some researchers have found that PO_4^{3-} increases bacterial growth, especially Gram-negative bacteria in drinking water, which is not recommended to use this contaminated water as drinking water. The presence of too much P in surface waters can lead to increased growth of algae and large aquatic plants, which can lead to a decrease in dissolved oxygen levels—a process called eutrophication. High P levels can lead to the production of algal toxins, which can be harmful to human and animal health. This leads to a high consumption of dissolved oxygen in the water. As a result of the decrease in dissolved oxygen in the water body, it leads to the death of aquatic plants and animals. Dead plants and animals rot and cause deterioration of water quality [53]. According to WHO standard, the permissible PO_4^{3-} limit in surface water is 5 mg/L. However, within the permissible limits it is not harmful to human health, but above the permissible limits it can damage the kidney and cause osteoporosis [1].

Ammonia is used in fertilizers in the form of ammonia (NH_3) or ammonium (NH_4^+) as a source of nitrogen. It is used industrially as a refrigerant gas or as an intermediate in the production of plastics, textiles, paper, paints, and other chemicals. Ammonia is also a naturally occurring compound that most organisms need for protein synthesis. It has an important role in acid–base regulation and is used in the biosynthesis of purines, pyrimidines, and non-essential amino acids, and is the end-product of metabolism as the normal biodegradation product of nitrogenous organic matter (protein). It is also released into ground and surface waters from mining, industrial activities, as well as domestic and industrial wastewater treatment plants. The environmental limits for ammonia in the surface water of many countries are set at 0.5 mg/L. No drinking water standard for ammonia has been proposed by WHO. Natural levels in groundwater are usually below 0.5 mg/L per liter. Higher levels in surface waters are found in layers rich in humic substances (fecal pollution) or iron, or in forests. It is also used in purification systems to prolong the effectiveness of disinfection chlorine added to drinking water. The addition of ammonia increases the formation of the carcinogen chloramine. Ammonia exposure can have significant effects on human health, such as irritation by inhalation and dermal exposure, memory, and other brain function problems. High levels of ammonia can cause increased bacterial growth in domestic drinking water. Exposure to ammonia from drinking water from water in the range of 0.5–5 mg/L has no direct importance for health. It does not pose a risk to human health, except for people suffering from enzyme deficiency due to genetic disorders or severe renal or hepatic insufficiency, and its carcinogenic effect has not been shown [54].

2.3 Organic Pollutants

Industrial and urban wastes produced by anthropogenic activities pollute the earth's soil and water resources, causing deterioration of soil and food quality. The most important of these organic pollutants are polycyclic aromatic hydrocarbons (PAHs), pesticides, polychlorinated biphenyls (PCBs) and trihalomethanes (THMs). Pollution of fresh water is a natural process in the passage of the normal water cycle. During the water cycle, various organic substances, both soluble and insoluble, are transported to freshwater streams. Among the natural contribution of organic pollutants, volcanic eruption, forest fires, decomposition of natural organisms are important sources. Biological organic substances are also caused by the feces or waste of living organisms from the municipality. Organic pollutants are released into the environment by anthropogenic activities, which are, accidental, discharge (e.g. fuels, solvents), industrial activities (e.g. chemical, petrochemical), agriculture (e.g. pesticides) and military operations (e.g. explosives, chemical weapons). Organic pollutants are mostly of human origin chemicals that enter natural fresh water and they persist as by-products of the breakdown of other chemicals and for a sufficiently long time and can cause harmful effects on the environment. They tend to accumulate in reservoirs such as water and soil. From these reservoirs, they are moved again by various processes, they can change form or species and become accessible to the biological food chain. They are toxic even at low concentrations and can cause acute and chronic disorders associated with mutagenicity, carcinogenicity, and teratogenicity [1].

2.3.1 Polycyclic Aromatic Hydrocarbons (PAHs)

PAHs are composed of two or more chemically fused aromatic (benzene) rings that share a pair of carbon atoms between them. PAHs are toxic and carcinogenic organic compounds that are not necessary for the growth of plants, animals, or humans. They are the combustion products of common fossil fuels such as coal, oil, and natural gas. They can enter naturally the water system from sources such as volcanic eruptions, forest fires, decomposition of vegetable garbage, the synthesis of bacteria and algae, oil spills and erosion of sedimentary rocks containing PAH. They are mainly used extensively in the production of petrochemicals, pharmaceuticals, agricultural fertilizers, photography, plastics industry, asphalt construction products. In the determination of the quality of drinking water, the following six PAHs are usually examined; floranthene, benzo[a]pyrene, benzo[ghi]perylene, benzo[b]floranthene, benzo[k]florantine and indeno[1,2,3-cd]pyrene. The Turkish Legislation and WHO recommended that total PAHs should not exceed 0.1 and 0.2 μg/L while the concentration for the benzo[a]pyrene concentration should remain below 0.01 and 0.7 μg/L, respectively [1, 55]. Caylak [56] determined the levels of B[a]P (0.518–0.459 μg/L), B[ghi]P (0.582–0.757 μg/L) and total PAHs (0.616–0.792 μg/L) in the water samples Of Cankiri, Turkey and found slightly above compared to Turkish and WHO limit values. The estimated cancer risks (CR) of long-term exposure to PAHs by ingestion have been calculated and CR values of B[ghi]P and total PAHs were higher than 1.00 E-05 for ingestion pathway. It has been revealed that PAHs in water samples of Cankiri city have a slight risk of cancer. Humans receive PAHs by various means, including inhalation, dermal contact, and ingestion. PAHs have high lipophilicity and found in almost all internal organs after ingestion and inhalation, stored in adipose tissue and can be released gradually. In humans, to exposure to mild concentrations of PAHs can cause symptoms such as eye irritation, nausea, vomiting, diarrhea, and confusion. Health effects of chronic exposure to PAHs with drinking water include cataracts, decreased immune function, asthma, kidney and liver damage, and jaundice. Benzo[a]pyrene is the most important PAH that causes cancer in organisms. Studies of workers exposed PAHs have shown an increased risk of skin, lung, bladder, and gastrointestinal cancers [57].

2.3.2 Pesticides

Today, pesticides are intensively used by farmers in agriculture for the cultivation of crops. They are vital for crop protection and are of concern in many parts of the world due to their short- and long-term harmful effects on human health and natural ecosystems. Six to seven million tons of pesticides are produced annually on Earth, and they are used about 0.2–2 kg per hectare on arable land. They are consumed annually for agricultural production to preserve and increase crop yields by controlling fungi, weeds, insects, and other pests. The most used pesticides are classified into four main groups: insecticides, rodenticides, herbicides, and fungicides. Most pesticides tend to accumulate biologically by entering freshwater systems and have devastating effects on the environment and human health. Dichlorodiphenyl trichloroethane (DDT), aldrin, dieldrine, endrin, heptachlor, chlordane, mirex and toxaphene are the most common long half-life pesticides among freshwater organic pollutants. Pesticides constitute the main anthropogenic source of organic pollutants in the freshwater system. Surface and groundwater may be contaminated by seeping pesticides into the soil during the use of agricultural chemical products or directly from storage facilities or through sewage treatment plants. WHO guideline levels for pesticides in drinking water were determined as 1 μ/L for total pesticides [1]. Leong et al. [58] and Elfikri et al. [59] reported the concentration of pesticides in the Selangor and Tengi Rivers (are the sources of the drinking water supply) in Malaysia were ranging between 0.010 to 1.849 μg/L and 0.003 to 4.493 μg/L, respectively. People are mainly exposed to aquatic pesticides through dermal contact and ingestion. The effects of exposure to pesticides on humans can be categorized into acute and chronic health problems. Acute health effects include decreased vision, headache, diarrhea, nausea, vomiting, asthma, bronchitis, gastroenteritis, and even death. Chronic health problems include neurological problems, the onset of Parkinson's disease, short attention span, effects such as memory disorders, reproductive problems, disrupting infant development, birth defects, and cancer [60].

2.3.3 Polychlorinated Biphenyls (PCBs)

PCBs are a group of organochlorine compounds that are among the persistent organic pollutants (POPs) and are synthesized by the catalyzed dechlorination of biphenyl. This leads to 209 possible congeners, but only about 130 of them are found in commercial products. Due to its chemical stability and heat resistance, they were man-made organic compounds and widely used as additives in heat exchange fluids, for transformers and large capacitors components, paints, and plastics. Various commercial mixtures were produced with different trade names, such as Roclor, Kanechlor, Declor and Clophen. In general, PCB28, 52, 101, 118, 138, 153, and 180 are indicative PCBs, and they are often used as indicators of environmental pollution. USEPA has set the maximum permissible concentration of PCBs in drinking water at 0.5 μg/L [61]. Cui et al. [62] analyzed the total PCBs concentrations in drinking water source of the Yangtze River in China, and reported the values ranged between 0.00004 and 0.011 μg/L, and PCB18, and PCB28 levels exceeded 0.5 μg/L set by WHO legislation.

Their physicochemical properties cause them to be permanent in the environment. For these reasons, the production and use of PCBs were banned globally, but large quantities of these compounds are still present in old power capacitors and transformers. PCBs released from these sources can enter the environment through evaporation from solutions, runoff, oil leakage, and waste discharges. Volcanic eruptions have been identified as a very small source of natural PCBs. Inhalation and ingestion of contaminated substances for PCBs and contaminated water are sources of human exposure. PCBs stimulate changes in DNA sequence and cause oxidative stress, and their negative effects on humans include impaired reproduction, liver damage, and immunotoxicity, neuro-developmental effects in infants, endocrine disruption, and cancer [61].

2.3.4 Trihalomethanes (THMs)

Disinfection of drinking water sources is a necessity to prevent the occurrence of enteric diseases that can be life-threatening such as cholera, typhoid fever, amoebic dysentery, giardiasis in the public. Ozone, chlorine, chlorine dioxide, and chloramines are added to drinking water to neutralize microbial contamination. Chlorine is one of the most widely used disinfectants in the World. However, chlorine reacts with natural organic or inorganic substances in water to form haloacetic acid, bromide, and trihalomethanes (THMs) harmful to human health [63, 64]. THMs found in drinking water are chloroform ($CHCl_3$), bromoform ($CHBr_3$), bromodichloromethane (BDCM; $CHBrCl_2$), and dibromochloromethane (DBCM; $CHBr_2Cl$). Swimming pools disinfected by chlorination or bromination, and THMs and chloroform were detected at the highest concentration in all pools which are the source of exposure to THMs. Dermal absorption may also be important, especially after swimming and bathing in chlorinated water. THMs are easily absorbed from the lungs and gastrointestinal tract and very harmful and have carcinogenic and mutagenic effects. In studies conducted on experimental animals of THMs, several negative health effects have been observed, such as kidney and liver damage, reproductive disorders, and cancer. It has been shown that there is an association between long-term exposure to THMs with drinking water and bladder, and colon cancers in humans [64]. A maximum contaminant level of 100 μg/L was established for total THMs in drinking water by WHO [1]. Caylak [56] determined the total THMs levels as 54.75–56.89 μg/L in the purified water samples of Cankiri, Turkey and found slightly below compared to Turkish and WHO criteria. Considering the carcinogenic health risk, the values were higher than 10^{-5} for total THMs ($1.82–2.20 \times 10^{-5}$) for ingestion pathway. These results showed that long-term total THMs exposure via ingestion route in drinking water posed little cancer risk for consumers in Cankiri.

3 Conclusion

Due to the increasing population and industrialization, water demand is also increasing every day. To determine the potential health risks of water pollutants the Water Quality Index (WQI), Hazard Quotient (HQ) and Carcinogenic Risk (CR) Assessments were used. Such risk assessments can help to implement advanced treatment methods. Conventional treatment methods to mitigate the risk are no longer sufficient to relieve water resources accordingly.

References

1. Guidelines for drinking water quality, 4th ed. Incorporating the 1st addendum. World Health Organization Geneva; 2017.
2. Caylak E, Tokar M. Investigating chemical and microbiological contaminants in drinking water of Cankiri Province, Turkey. Environ Earth Sci. 2012;67:2015–25.
3. Guidelines for Drinking-Water Quality, 4th edn. Water Sanitation and Health; 2011.
4. Medema GJ, Bahar M, Schets FM. Survival of *Cryptosporidium parvum*, *Escherichia coli*, *Faecal enterococci* and *Clostridium perfringens* in river water-influence of temperature and autochthonous microorganisms. Water Sci Technol. 1997;35:249–52.
5. Marshall MM, Naumovitz D, Ortega Y, Sterling CR. Waterborne protozoan pathogens. Clin Microbiol Rev. 1997;10:67–85.
6. Guidelines for Canadian Drinking Water Quality. Health Canada, federal-provincial-territorial committee on drinking water. Ottawa; 2010
7. Aluminium in Drinking-Water. Background document for development of WHO guidelines for drinking-water quality. Geneva: WHO; 2010.
8. Rondeau V, Commenges D, Jacqmin-Gadda H, Dartigues JF. Relation between aluminum concentrations in drinking water and Alzheimer's disease: an 8-year follow-up study. Am J Epidemiol. 2000;152:59–66.
9. Martyn CN, Coggon DN, Inskip H, Lacey RF. Young WF Aluminum concentrations in drinking water and risk of Alzheimer's disease. Epidemiology. 1997;8:281–6.
10. Caylak E. Health risk assessment for arsenic in water sources of Cankiri province of Turkey. Clean-Soil Air Water. 2012;40:728–34.
11. Wolz S, Dieter HH, Klein D, Schneider K. Oral exposure to inorganic arsenic: an assessment of its carcinogenicity and non-carcinogenic effects. Crit Rev Toxicol. 2009;39:271–98.
12. Al-Saleh I, Al-Doush I. Survey of trace elements in household and bottled drinking water samples collected in Riyadh, Saudi Arabia. Sci Total Environ. 1998;216:181–92.
13. Caylak E, Aytekin M. An industrial risk: beryllium. J Clin Exp Investig. 2012;3:141–8.
14. Drinking Water Health Advisory for Boron. USEPA; 2008.
15. Cöl M, Cöl C. Environmental boron contamination in waters of Hisarcik area in the Kutahya Province of Turkey. Food Chem Toxicol. 2003;41:1417–20.
16. Caylak E. Health risk assessment for trace metals, polycyclic aromatic hydrocarbons and trihalomethanes in drinking water of Cankiri, Turkey. E-J Chem. 2012;9:1976–91.
17. Bialek M, Czauderna M, Krajewska KA, Przybylski W. Selected physiological effects of boron compounds for animals and humans. J Animal Feed Sci. 2019;28:307–20.
18. Idrees N, Tabassum B, Abd Allah EF, Hashem A, Sarah R, Hashim M. Groundwater contamination with cadmium concentrations in some West U. P. Regions, India. Saudi J Biol Sci. 2018;1365–68.
19. Mohammadi AA, Zarei A, Majidi S, Ghaderpoury A, Hashempour Y, Saghi MH, Alinejad A, Yousefi M, Hosseingholizadeh N, Ghaderpoori M. Carcinogenic and non-carcinogenic health risk assessment of heavy metals in drinking water of Khorramabad, Iran. MethodsX. 2019;6:1642–51.
20. Duruibe JO, Ogwuegbu MOC, Egwurugwu JN. Heavy metal pollution and human biotoxic effects. Int J Phys Sci. 2007;2:112–8.
21. Liu H, Yu X. Hexavalent chromium in drinking water: chemistry, challenges and future outlook on Sn(II)- and photocatalyst-based treatment. Front Environ Sci Eng. 2020;14(5):88.
22. Kaprara E, Kazakis N, Simeonidis K, Coles S, Zouboulis AI, Samaras P, Mitrakas M. Occurrence of Cr(VI) in drinking water of Greece and relation to the geological background. J Hazard Mater. 2015;281:2–11.
23. Amrose S, Cherukumilli K, Wright NC. Chemical contamination of drinking water in resource-constrained settings: global prevalence and piloted mitigation strategies. Annu Rev Environ Resour. 2020;45:195–226.
24. Kim JH, Gibb HJ, Howe PD. Cobalt and inorganic cobalt compounds. Chemical safety team & international programme on chemical safety. World Health Organization; 2006. p. 1–93.
25. Atashi H, Sahebi M, Mansoorkiai R, Akbari F. Cobalt in Zahedan drinking water. J Appl Sci Res. 2009;5:2203–7.
26. Beyene HD, Berhe GB. The level of heavy metals in portable water in Dowhan, Erop Wereda, Tigray, Ethiopia. J Nat Sci Res. 2015;5:190–4.

27. Pettersson R, Rasmussen F, Oskarsson A. Copper in drinking water: not a strong risk factor for diarrhoea among young children. A population-based study from Sweden. Acta Pædiatrica. 2003;92:473–80.
28. Bashir F, Irfan M, Ahmad T, Iqbal J, Butt MT, Sadef Y, Umbreen M, Shaikh IA, Moniruzzaman M. Efficient utilization of low cost agro materials for incorporation of copper nanoparticles to scrutinize their antibacterial properties in drinking water. Environ Technol Innov. 2021;21: 101228.
29. Ghosh GC, Khan MJH, Chakraborty TK, Zaman S, Kabir AE, Tanaka H. Human health risk assessment of elevated and variable iron and manganese intake with arsenic-safe groundwater in Jashore, Bangladesh. Sci Rep. 2020;10:1–9.
30. Kell DB. Towards a unifying, systems biology understanding of large-scale cellular death and destruction caused by poorly liganded iron: Parkinson's, Huntington's, Alzheimer's, prions, bactericides, chemical toxicology and others as examples. Arch Toxicol. 2010;84:825–89.
31. Ehya F, Marbouti Z. Hydrochemistry and contamination of groundwater resources in the Behbahan plain, Iran. Environ Earth Sci. 2016;75:1–13.
32. Turhan S. Potentially toxic element contamination of drinking water consumed in Cappadocia region (Turkey) and health risk assessment. Int J Environ Anal Chem. 2022;1–17.
33. Caylak E, Aytekin M, Halifeoglu I. Antioxidant effects of methionine, α-lipoic acid, N-acetylcysteine and homocysteine on lead-induced oxidative stress to erythrocytes in rats. Exp Toxicol Pathol. 2008;60:289–94.
34. Knudsen NN, Schullehner J, Hansen B, Jorgensen LF, Kristiansen SM, Voutchkova DD, Gerds TA, Andersen PK, Bihrmann K, Gronbæk M. Lithium in drinking water and incidence of suicide: a nationwide individual-level cohort study with 22 years of follow-up. Int J Environ Res Public Health. 2017;14:627.
35. Pompili M, Vichi M, Dinelli E, Pycha R, Valera P, Albanese S, Lima A, De Vivo B, Cicchella D, Fiorillo A. Relationships of local lithium concentrations in drinking water to regional suicide rates in Italy. World J Biol Psychiatry. 2015;16:567–74.
36. Marmol F. Lithium: Bipolar disorder and neurodegenerative diseases. Possible cellular mechanisms of the therapeutic effects of lithium. Progress in Neuro-Psychopharmacol Biol Psychiatr.y 2008;32:1761–71.
37. Oruch R, Elderbi MA, Khattab HA, Pryme IF, Lund A. Lithium: a review of pharmacology, clinical uses, and toxicity. Eur J Pharmacol. 2014;740:464–73.
38. Manganese in drinking water. Background document for development of WHO guidelines for drinking-water quality. WHO; 2011.
39. Roels HA, Bowler RM, Kim Y, Claus Henn B, Mergler D, Hoet P, Gocheva VV, Bellinger DC, Wright RO, Harris MG, Chang Y, Bouchard MF, Riojas-Rodriguez H, Menezes-Filho JA, Tellez-Rojo MM. Manganese exposure and cognitive deficits: a growing concern for manganese neurotoxicity. Neurotoxicology. 2012;33:872–80.
40. Khattak SA, Rashid A, Tariq M, Ali L, Gao X, Ayub M, Javed A. Potential risk and source distribution of groundwater contamination by mercury in district Swabi, Pakistan: application of multivariate study. Environ Dev Sustain. 2021;23:2279–97.
41. Martin S, Griswold W. Human health effects of heavy metals. Environ Sci Technol Briefs Citizens. 2009;15:1–6.
42. Khan SA, Din ZU, Zubair A. Levels of selected heavy metals in drinking water of Peshawar city. Int J Secure Netw. 2011;2:648–52.
43. Outcome of a public consultation on the draft update of the risk assessment of nickel in food and drinking water. EFSA Support Publication 17(11):EN-1940–54. EFSA; 2020.
44. Golubkina N, Erdenetsogt E, Tarmaeva I, Brown O, Tsegmed S. Selenium and drinking water quality indicators in Mongolia. Environ Sci Pollut Res. 2018;25:28619–27.
45. Zwolak I, Zaporowska H. Selenium interactions and toxicity: a review. Cell Biol Toxicol. 2012;28:31–46.
46. Arena G, Chiara, C, Dimartino A, Grasso A, Fallico R, Sciacca S, Fiore M, Ferrante M. Determination of total vanadium and vanadium(V) in groundwater from Mt. Etna and estimate of daily intake of vanadium(V) through drinking water. J Water Health. 2015;13:522–30.
47. Domingo JL. Vanadium: a review of the reproductive and developmental toxicity. Reprod Toxicol. 1996;10:175–82.
48. Flinn JM, Hunter D, Linkous DH, Lanzirotti A, Smith L, Brightwell JJ, Jones BF. Enhanced zinc consumption causes memory deficits and increased brain levels of zinc. Physiol Behav. 2005;83:793–803.
49. Criteria for Radionuclide Activity Concentrations for Food and Drinking Water. IAEA TECDOC No. 1788; 2016.
50. Küçükömeroğlu B, Şen A, Duran SU, Çiriş A, Taskin H, Ersoy H. Determination of radioactivity level of water supply network in Trabzon province, Turkey. Isot Environ Health Stud. 2021;57:610–22.
51. Ward MH, Jones RR, Brender JD, de Kok TM, Weyer PJ, Nolan BT, Villanueva CM, van Breda SG. Drinking water nitrate and human health: an updated review. Int J Environ Res Public Health. 2018;15:1557.
52. Greer FR, Shannon M. Infant methemoglobinemia: the role of dietary nitrate in food and water. Pediatrics. 2005;116:784–6.
53. Chandrajith R, Dissanayake CB. Phosphate mineral fertilizers, trace metals and human health. J Natl Sci Found. 2009;37:153–65.
54. Guidelines for Canadian drinking water quality: guideline technical document—Ammonia; 2013.
55. Icme suyu temin edilen sularin kalitesi ve aritilmasi hakkinda yönetmelik, 30823 sayili Resmi Gazete,

Turkey. (in Turkish) https://www.resmigazete.gov.tr/eskiler/2019/07/20190706-8-1.pdf; 2019.
56. Caylak E. Quality and health risk assessment in drinking water of Cankiri, Turkey. Pollut Eff Community Health. 1;1–13.
57. Abdel-Shafy HI, Mansour MS. A review on polycyclic aromatic hydrocarbons: source, environmental impact, effect on human health and remediation. Egypt J Pet. 2016;25:107–23.
58. Leong KH, Benjamin Tan LL, Mustafa AM. Contamination levels of selected organochlorine and organophosphate pesticides in the Selangor River, Malaysia between 2002 and 2003. Chemosphere. 2007;66:1153–9.
59. Elfikrie N, Ho YB, Zaidon SZ, Juahir H, Tan ESS. Occurrence of pesticides in surface water, pesticides removal efficiency in drinking water treatment plant and potential health risk to consumers in Tengi River Basin, Malaysia. Sci Total Environ. 2020;712: 136540.
60. Kim KH, Kabir E, Jahan SA. Exposure to pesticides and the associated human health effects. Sci Total Environ. 2017;575:525–35.
61. Public health goal for water soluable polychlorinated biphenyls expected to be found in drinking water. OEHHA; 2007.
62. Cui X, Dong J, Huang Z, Huang Z, Liu C, Qiao X, Wang X, Zhao X, Zheng B, Shen J. Polychlorinated biphenyls in the drinking water source of the Yangtze River: characteristics and risk assessment. Environ Sci Eur. 2020;32:29.
63. Public health goals for trihalomethanes in drinking water: chloroform, bromoform, bromodichloromethane, dibromochloromethane. OEHHA; 2020.
64. Caylak E, Tokar M. Investigating chemical and microbiological contaminants in drinking water of Cankiri Province, Turkey. Environ Earth Sci. 2012;67:2015–25.

Arsenic Removal Technologies: A Critical Review

Kemal Aktaş and Ayça Erdem Ünşar

Abstract

Arsenic (As) contamination in water sources is a significant global issue due to its detrimental health effects. Various As removal technologies have been developed to mitigate this problem and ensure safe drinking water for affected populations. This critical review provides an in-depth analysis of As removal technologies, evaluating their effectiveness, limitations, and potential environmental impacts. The review focuses on commonly used methods such as coagulation-filtration, oxidation, lime softening, and adsorption, ion exchange, and membrane processes, while also exploring emerging technologies. The study critically examines the advantages and disadvantages of each technology, considering factors such as efficiency, cost, operational complexity, waste generation, and system scalability. Furthermore, the environmental implications associated with treatment residuals and energy consumption are discussed. The review highlights the importance of selecting appropriate As removal technologies based on site-specific conditions and the need for long-term sustainability. It also emphasizes the need for ongoing research and development to improve existing technologies and explore innovative approaches to As removal. Overall, this critical review provides valuable insights into As removal technologies, facilitating informed decision-making for effective mitigation of As contamination in water sources.

Keywords

Arsenic · Health effect · Removal technologies · Water contamination

1 Introduction

In the last century, increasing population and industrial activities have been affecting environmental resources, especially water resources. The introduction of toxic pollutants such as heavy metals into aquatic environments adversely affects the aquatic life [1]. Arsenic (As), one of these pollutants, is one of the most harmful toxic substances found in drinking water [2]. As is an element found abundantly in the atmosphere, soil, and aquatic environments both naturally and anthropogenically. In addition, more than 100 minerals and ores contain arsenic. In nature,

K. Aktaş
School of Engineering, University of British Columbia, Kelowna, BC, Canada

A. Erdem Ünşar (✉)
Faculty of Engineering, Department of Environmental Engineering, Akdeniz University, Antalya, Turkey
e-mail: ayerdem@akdeniz.edu.tr

A. M. Bahadir et al. (eds.), *Hydrology and Urban Water Supply*, Water and Wastewater Management,
https://doi.org/10.1007/978-3-031-72589-0_20

arsenic is in motion with biological activity, and human activities [3]. As is found in aquatic systems in the forms −III, 0, +III and +V [4] and the most common forms are known as As (III) and As (V). Under natural conditions, very different and high arsenic concentrations can be found in groundwater due to water–rock interactions [5].

As contamination in water sources poses a significant threat to public health worldwide. Exposure to elevated levels of As in drinking water can lead to severe health problems [6], including skin lesions, cancers (skin, liver, bladder cancers), cardiovascular diseases, and developmental issues [7]. The World Health Organization (WHO) and the United States Environmental Protection Agency (USEPA) have determined the amount of As allowed in drinking water as 10 μg/L (ppb). As a result, the development of effective As removal technologies has become a crucial area of research and intervention. This critical review aims to provide a comprehensive analysis of various As removal technologies, evaluating their effectiveness, limitations, and potential environmental impacts.

Over the years, numerous technologies have been developed and implemented to address the challenge of As contamination in water sources. These technologies include a range of treatment methods, including oxidation, co-precipitation, adsorption, ion exchange and membrane technologies. Each method offers unique advantages and drawbacks, making it essential to critically assess their performance and suitability for specific scenarios. Among these methods, removal by adsorption stands out [8]. Activated carbon (AC) is used in many different adsorbent adsorption technologies such as natural materials and modified natural materials. Such adsorbents make high removal at the level of mg/L (ppm) and are feasible in terms of cost.

The primary objective of As removal technologies is to reduce or eliminate As from water, ensuring safe drinking water for affected communities. Effectiveness is a crucial factor in evaluating these technologies, as achieving As levels below the permissible limits set by regulatory authorities is essential for protecting public health. The review will examine the efficiency of each technology in terms of As removal rates and its ability to consistently meet regulatory standards.

Apart from effectiveness, other key considerations include the cost and operational complexity associated with implementing and maintaining these technologies. As removal technologies can vary significantly in terms of upfront investment, operational costs, and ongoing maintenance requirements. It is crucial to assess the economic feasibility and long-term affordability of these technologies, particularly in resource-constrained areas where financial resources may be limited.

Environmental implications are also a significant concern when evaluating As removal technologies. Waste generation and management, as well as the potential for energy consumption, need to be carefully examined. Treatment processes may generate residuals or waste streams that require proper disposal or treatment to prevent environmental contamination. Additionally, some technologies, such as membrane processes, may consume significant amounts of energy, leading to increased carbon footprints and operational costs.

Furthermore, the scalability and adaptability of these technologies must be considered. As populations grow or water demand increases, the ability to expand and upgrade treatment systems becomes crucial. Ensuring the scalability of As removal technologies is essential to accommodate changing needs and to meet the demand for safe drinking water effectively.

This critical review will also explore emerging technologies and innovations in the field of As removal. Ongoing research and development efforts aim to improve existing technologies and explore novel approaches to tackle As contamination more efficiently. The review will highlight promising advancements and their potential implications for future As removal strategies.

In conclusion, addressing the challenge of As contamination in water sources requires a comprehensive evaluation of various As removal technologies. This critical review aims to provide a detailed analysis of the effectiveness, limitations, and potential environmental impacts of these technologies. By critically assessing their performance and considering key factors such as effectiveness, cost, environmental implications, and scalability, this review seeks to inform decision-making processes and facilitate the selection of appropriate As removal technologies for ensuring safe drinking water for affected populations.

2 Arsenic Chemistry and Human Health Effect

As is the 20th most abundant element found in the Earth's crust [9]. This is due to soil erosion, mineral ore formations, and volcanic, geothermal, and hydrothermal activities [10]. As (III) is more stable in reduced and As (V) is more stable oxidized environments. However, they can be found in natural environments regardless of their redox state. Under natural conditions, different and high As concentrations can be found in groundwater due to water–rock interactions. Also, in aquifers, it is difficult to follow As concentration and type due to As activity [5].

As is a problem in natural water resources worldwide. Countries such as Bangladesh, Argentina, Chile, and Taiwan lead the countries that have problems with the problem of As in drinking water [11–13]. After Bangladesh, the West Bengal region of India has the most population exposed to As from groundwater [14]. In Table 1, countries with the highest As concentrations in groundwater worldwide and the populations affected by it are given.

As (V) species (AsO_4^{3-}, $HAsO_4^{2-}$, H_2AsO_4) are stable in aerobic environments and are the dominant species [15]. As (III) is the dominant species in anaerobic environments such as groundwater. Redox potentials and pH of the environment are the most important parameters for the control of As species [16]. In order to accurately predict the movements of As in the ground, it is necessary to know that the water chemistry of As to determine its transitions between liquid and solid phases. The fact that As has different structures in different pH and different redox systems in aerobic and anaerobic environments makes As removal difficult. These different physicochemical properties of As are given in Table 2. In this table, some different types of As in the aquatic environment and the known properties of these species are given.

As is one of the pollutants controlled under inorganic pollutants according to EPA regulations in the USA. In Europe, the Regulation on Water Intended for Human Consumption has determined 10 μg/L in the fourth part, Annex-1-b, Chemical parameters section [17]. These standards are mostly accepted around the world, instead of 50 μg/L. As a semi-metal element, it is

Table 1 Arsenic concentration around the world [12]

Country	Population exposed	Concentration (ppb)
Bangladesh	30.000.000	<1–2500
India	6.000.000	<10–3200
Argentina	2.000.000	<1–9900
Vietnam	>1.000.000	1–3050
Mongolia	100.000–600.000	<1–2400
Chili	400.000	100–1000
Mexico	400.000	8–620
Romania	400.000	<2–176
Taiwan	100.000–200.000	<1–2500
Greece	150.000	–
Spain	>50.000	1–100

Table 2 Arsenic reactions in different conditions [12]

Reactions	pKa
As (V)	
$H_2AsO_4^- + H^+ \rightarrow H_3AsO_4$	2.24
$HAsO_4^{-2} + H^+ \rightarrow H_2AsO_4^-$	6.96
$AsO_4^{-3} + H^+ \rightarrow HAsO_4^{-2}$	11.50
As (III)	
$H_2AsO_3^- + H^+ \rightarrow H_2AsO_3$	9.22
$HAsO_3^{-2} + H^+ \rightarrow H_2AsO_3^-$	12.11
$AsO_3^{-3} + H^+ \rightarrow HAsO_3^{-2}$	13.41

an odorless and tasteless substance. It enters drinking water resources from geological sources (soil structure), agricultural areas or industrial sources. Other effects of As, which causes various types of cancer when exposed to high doses for a long time, are; skin problems, abdominal pain, nausea, vomiting, numbness in the hands and feet, as well as partial paralysis and blindness can occur [18]. As is fatal when exposed to a dose of 60 mg/L [19].

3 Arsenic Removal Technologies

As contamination in drinking water is serious health treat. Many authorities limit the As concentration in drinking water to 10 μg/L. However, As concentration in some regions is higher due to mainly geographical reasons. Table 3 shows typical As removal technologies and illustrates advantages and disadvantages.

3.1 Coagulation

Coagulation is the process of ensuring that thermodynamically unstable, negatively charged suspended solids (SS) that do not settle quickly in water aggregate and precipitate by using a positively charged coagulant. SS first comes out of the steady state and then they start to form flocs. Pollutants in the water whose mass increases as a result of these two steps begin to collapse with the effect of gravity and are purified from the water. The gravitational force and friction are effective on the precipitated particles [32]. When charged inorganic salts are added to water, they neutralize SS and increase flock formation by forming short polymer chains. Inorganic hydroxide salts are metal salts that produce a particularly short polymer chain. Inorganic salts are also the cheapest coagulant per kilogram. In addition, it can be found easily and when added in accordance with the rules, it removes SS with high efficiency. Inorganic coagulants also reduce some of the organic precursors and reduce the formation of chlorine and by-products. Their use in high doses can also engulf bacteria as flocs form and settle to the bottom by gravity. Metal salts are especially used for natural organic matter (NOM) removal. They form water-insoluble complexes with negatively charged NOM positive metal salts. These complexes then precipitate as flocs. Another method is by the precipitation of metal hydroxides by adsorbing NOM. Coagulation efficiency varies depending on the pH of the medium and the coagulant dose [33]. Reaction of different salt coagulants with calcium bicarbonate under low pH and alkalinity conditions.

Tab. 3 Comparison of arsenic removal technologies

Methods	Advantages	Disadvantages	Removal efficiency (%)	References
Coagulation	– Easy to obtain chemicals – Relatively low capital requirement – Easy operation – Wide pH range operation	– High cost – Toxic waste – Low efficiency – Require oxidation	$FeCL_3$: 91 Alum: 96 Polyaluminum chloride: 99.4	[15, 20–22]
Oxidation	– Easy to operate – Low maintenance cost – Oxidize other pollutants	– Require pH control – Low efficiency – Sludge production	As(V): >95	[15, 20–25]
Lime softening	– Easy to operate	– Require pH control – Toxic waste – Very low efficiency	<10–45	[15]
Ion exchange	– Well-tailored system – Wide pH range operation – High speed process – Less toxic sludge production – Successful for As (V) removal	– High cost – Hard to operate – Require regeneration and produce waste – Low As (III) removal – Limited resin capacity	As (V): 95	[15, 20, 21, 25–27]
Membrane filtration	– Simple operation – High removal – No toxic waste – No chemical use – No effect on taste and smell – Removal of other pollutants – Low energy consumption	– High capital and operation cost – Membrane sediment formation – Toxic retentate – Require preconditioning – Influence of unwanted ions on removal efficiency	Total As: 96	[15, 20, 21, 26, 28, 29]
Adsorption	– Low cost – Low maintenance – Low by-products – No sludge production – Effective removal of As (V)	– Influence of organic substances and salts on removal efficiency – Require additional treatment (oxidation and/or filtration) – Difficulty removing heavy flocs – Difficulty in As (III) removal	As (III): 30–60 As (V): <90	[20, 21, 23–26]
Chemical precipitation	– Wide application – Simple process – Extensive removal of heavy metals – Effective removal As (V)	– High sludge production – Slow process speed – Require chemicals – Require additional treatment – Low As (III) removal	As (V): 95	[26, 28]

(continued)

Tab. 3 (continued)

Methods	Advantages	Disadvantages	Removal efficiency (%)	References
Nano-photocatalyst	– Low cost – Ability to reuse – High reaction speed – Stability – High efficiency – Easy accessibility	– Accumulation occurs in solutions – Required sophisticated seperation techniques	Zero-valent-iron (ZVI): total As −98 ZnO: total As −98.3	[25, 30, 31]

- $Al_2(SO_4)_3 + 3Ca(HCO_3)_2$ $2Al(OH)_3 + 3CaSO_4 + 6CO_2$
- $Fe_2(SO_4)_3 + 3Ca(HCO_3)_2$ $2Fe(OH)_3 + 3CaSO_4 + 6CO_2$
- $2FeCl_3 + 3Ca(HCO_3)_2 2Fe(OH)_3 + 3CaCl_2 + 6CO_2$

3.2 Oxidation

Oxidation processes have become widespread with the use of chemical oxidants (chlorine and ozone) in water treatment. With the use of chemical disinfection and sand filters, there has been a great decrease in diseases caused by contaminated water. Although this situation has made the use of chemical oxidants widespread, the proof that by-products from the use of chlorine, especially trihalomethane (THM), cause bladder cancer, has led to the development of other methods in recent years [34]. Removal of As (III) species, which are dominant in an oxygen-free environment, is more difficult when compared to As (V). Therefore, the oxidation of As (III) to As (V) increases the removal efficiency. Oxidation with air is a slow process, so different alternatives are used. Ozone, activated carbon, iron and manganese compounds, microbiological oxidation and photo-chemical oxidation are some of these methods (Table 4).

3.3 Lime Softening

Lime softening is a method often used to remove hardness from water using lime or soda. Hardness occurs in the presence of calcium (Ca^{2+}) and magnesium (Mg^{2+}) ions in water. The typical lime softening process consists of quick mixing, floc formation, sedimentation and filtration after lime is added. Pre-settlement and aeration are used to oxidize As compounds. pH control is one of the most important parameters for the

Table 4 Advantages and disadvantages of some oxidants [32]

Methods	Advantages	Disadvantages
Chlorine	Powerful oxidant, easy to add, resistant residual and in practice for long time	Chlorinated by products, smell, and odor and sensitive to pH changes
Ozone	Powerful oxidant, no by products, no smell and odor problems, additional biological degradation, and low pH impact	Short half-life, should be produced on-site, high energy demand, complex formation can happen, and corrosive,
Chlorine dioxide	Powerful oxidant, resistant residual, no by products, and no pH impact	By products such as ClO and ClO_2, should be produced on site, possible hydrocarbon smell
Potassium permanganate	Easy to add and no by product	Unknown by products, pinkish color wastewater, and can cause precipitation
Oxygen	Easy to add, no by products, additional scraping effect, and not toxic	Low oxidant, corrosive and can cause residue

precipitation of lime. The appropriate pH range for As removal is between 9 and 12. Above pH 10.5, magnesium hydroxide is formed and is removed together with As (V). As removal efficiency increases significantly above pH 10.5 [35].

- $CO_2 + Ca(OH)_2CaCO_3 + H_2O$
- $Ca(HCO_3)_2 + Ca(OH)_2 2CaCO_3 + 2H_2O$
- $Mg(HCO_3)_2 + Ca(OH)_2CaCO_3 + MgCO_3 + 2H_2O$
- $MgCO_3 + Ca(OH)_2CaCO_3 + Mg(OH)_2$

3.4 Ion Exchange

EPA defines the ion exchange process as a physical–chemical process. It occurs by the displacement of ions in the solution phase and the solid resin phase. Ions with strong bonds are replaced by ions with weak bonds in the water environment. There are two types of resins, anion, and cation resins. Anion resins are generally used for chlorine and anionic pollutants. Cation resins are used for sodium, potassium, and cationic contaminants. It is used when both types of contaminants are present in mixed resins.

As (III) should be oxidized to As (V) to improve removal efficiency. Ion exchange implications for softening and nitrate removal can also be implemented for As removal. Multiple colons with strong base anionic resin can remove As under pressured and continuous conditions. Removal efficiency is depended on the presence of other anions. Most important anions are sulfate and nitrate. Resin anion removal sequence is given below [36]:

- $SO_4{}^{-2} > HAsO_4^{-2} > NO^{-3}, CO_3^{-2} > NO^{-2} > Cl^-$

3.5 Membrane Filtration

The membrane is a thin and selective barrier layer between two phases (liquid, gas, and vapour). This layer can be in the form of a film, layer, or sheet. Membrane floats are interfaces that act as barriers guiding the transfer of species between two adjacent phases [37]. There are porous and non-porous types of membranes. In porous membranes, separation is achieved by passing a mixture of multiple components through the membrane. The passing part is called flux (permeate) and the leaving part is expressed as concentrated waste (retentate). Membranes are divided into four groups according to their pore size: microfiltration (MF), ultrafiltration (UF), nanofiltration (NF) and reverse osmosis (RO). NF membranes are not as tight and less permeable as RO membranes. The molecular weight separation limit (MWCO) for NF membranes is between 200 and 1000 daltons. They are made to eliminate doubly charged cations. They are commonly known as softening membranes and can also remove many of the disinfection byproducts. Since they are not as impermeable as RO membranes, they can operate at relatively low pressures between 5 and 9 bar. In this case, the use of NF membranes instead of RO reduces operating and investment costs. UF membranes and MF membranes are generally used for particle and pathogen removal. The MWCO value of UF membranes varies between 1000 and 500,000 daltons. UF membranes with an MWCO of 1000 daltons can largely remove disinfection by-products. MF membranes have an average of ten times larger pores than UF membranes. MF membranes are designed to remove micrometer level contaminants. It has a pore structure between 0.05 and 5 μm. While UF membranes can purify viruses, viruses can pass through the pores of MF membranes [38].

3.6 Adsorption

Adsorption is the mechanism resulting from unsaturated molecular forces and unsaturated regions found on solid surfaces. When such solid surfaces come into contact with liquid or gas, the adsorption process begins. In the adsorption process, besides the physical forces such as dipole moment, polarization forces and repulsive forces, there are chemical forces created by the redistribution of electrons in the valence gaps. Besides these forces, there are two types of adsorptions as physical adsorption and chemical adsorption. Physical adsorption is bound to the adsorbate surface by weak van-der-Waals forces. Chemical adsorption involves the exchange or sharing of electrons between the adsorbate and the solid surface and results in a chemical reaction. Adsorption processes are low-cost and simple operating processes. It does not create by-products and hazardous solid waste. Adsorption is considered as the best available technology (BAT) by the EPA [39].

4 Advantages and Disadvantages of Arsenic Treatment Methods

As treatment methods are designed to reduce or remove As contamination from water sources. While these methods can be effective, they also have advantages and disadvantages.

4.1 Advantages of Arsenic Treatment Methods

- *Effectiveness*: One of the key advantages of As treatment methods is their effectiveness in removing As from water sources. Technologies such as coagulation-filtration, adsorption, and ion exchange have proven to be highly efficient in reducing As levels below permissible limits set by regulatory authorities. These treatment methods are capable of achieving significant removal rates, ensuring that treated water meets stringent safety standards [21, 22, 28].
- *Versatility and Applicability*: As treatment methods are versatile and can be applied to various water sources, including groundwater and surface water. This versatility makes them suitable for different regions and communities facing As contamination challenges. Whether in urban or rural areas, treatment methods can be customized to address specific water quality conditions and As concentrations, providing tailored solutions that meet local requirements [26, 28, 40, 41].
- *Technological Advancements and Affordability*: Over the years, As treatment technologies have evolved, becoming more affordable and accessible. Advances in research and engineering have led to the development of cost-effective treatment methods, making them economically viable for implementation in communities affected by As contamination. These advancements have played a crucial role in expanding the reach of treatment methods, enabling a greater number of people to access safe drinking water [21, 23–26].
- *Long-Term Solution*: As treatment methods offer a long-term solution to the problem of As contamination. When properly implemented and maintained, these methods can ensure the continued safety of drinking water sources, protecting public health over an extended period. By effectively removing As from water, treatment methods mitigate the risks associated with long-term exposure to this toxic element, reducing the incidence of As-related health issues in affected communities [25, 28, 31].
- *Environmental Benefits*: In addition to their impact on public health, As treatment methods provide environmental benefits. By removing As from water sources, these methods prevent the discharge of contaminated water back into the environment. This reduces the risk of As contamination in ecosystems and minimizes the potential harm to aquatic life. Furthermore, proper

management and disposal of treatment residuals and waste generated during the treatment process can prevent environmental contamination and ensure sustainable water management practices [26, 28, 29].

4.2 Disadvantages of Arsenic Treatment Methods

Complexity: Many As treatment methods, such as reverse osmosis and electrodialysis, require complex equipment and skilled operators for their proper functioning. The setup and operation of these systems can be technically demanding and may require specialized training and expertise. This complexity can pose challenges, especially in resource-constrained communities, where the availability of skilled personnel and maintenance capabilities may be limited [42, 43].

High Energy Consumption: Certain As treatment methods consume significant amounts of energy. Technologies like membrane filtration and electrodialysis require electricity to drive the treatment process, leading to higher operational costs and environmental impact. The reliance on energy-intensive processes can limit the affordability and sustainability of treatment methods, particularly in areas with limited access to reliable and affordable energy sources [25, 28, 44].

Waste Generation: As treatment processes can generate waste streams containing concentrated As. Proper disposal or treatment of these waste products is crucial to prevent potential environmental contamination. The management of As-rich waste streams can be challenging, requiring appropriate infrastructure and disposal mechanisms to ensure that the waste is safely handled and does not pose a risk to ecosystems and human health [21, 22].

Treatment Residuals: Some As treatment methods produce residuals during the treatment process. These residuals may include sludge, spent media, or filter cartridges that have captured the removed As. The proper management and disposal of these residuals are essential to prevent the release of As back into the environment. Careful handling and disposal measures are required to minimize potential environmental and health risks associated with these treatment residuals [21, 25, 26, 45].

System Scalability: The scalability of As treatment systems can be a significant challenge. As populations grow or water demand increases, treatment systems may need to be upgraded or expanded to meet the requirements. Ensuring the scalability of treatment methods can be complex and costly, requiring careful planning, infrastructure development, and technical expertise. Failing to address scalability issues may limit the effectiveness and long-term sustainability of As treatment efforts [46].

Sustainability and Maintenance: Maintaining As treatment systems over the long term is crucial for sustained effectiveness. Regular monitoring, operation, and maintenance are essential to ensure optimal performance and prevent system failures. However, in resource-constrained areas, the financial and technical resources necessary for ongoing maintenance may be limited. Without proper support and capacity building, treatment systems can deteriorate over time, compromising their effectiveness and putting communities at risk of As exposure [42, 44].

It's important to emphasize that the advantages and disadvantages might vary depending on the specific treatment method, the local context, and the characteristics of the water source. Therefore, a comprehensive assessment of these factors is necessary when selecting and implementing an As treatment method.

5 Discussion and Recommendations

In many regions worldwide, the presence of high levels of As in water has been well-documented. Consequently, ensuring the supply of clean and safe drinking water has emerged as one of the greatest global challenges of the twenty-first century. To address this pressing issue, there is a critical need to develop new technologies, methods and materials that can keep pace with the increasing demand. Given the current situation, the most practical approach to reducing the risk to public health appears to be the removal of As from contaminated water. Several methods are currently being employed for this purpose. However, selecting the optimal procedure is challenging due to the pros and cons associated with each method, as well as the potential risk of secondary As pollution through the by-products of these methods. Therefore, the development of new technologies with the capability of hybrid methodologies is essential to effectively combat the threat of As contamination in the future. These new technologies should prioritize the use of environmentally friendly and sustainable chemicals. Additionally, when selecting a sustainable technology, it is important to take into account additional essential aspects such as the system's adaptability, simplicity, and maintenance convenience. The majority of As treatment processes can be categorized into different types, namely ion exchange, membrane processes, adsorption, and chemical precipitation. However, the capacity of ion exchange techniques to extract As is severely constrained due to competition with other anions present in groundwater. Although membrane techniques can remove As from groundwater, they tend to be expensive. Therefore, current research efforts are focused on investigating affordable and effective methods such as adsorption and chemical precipitation to find sustainable solutions for As removal.

Adsorption has gained popularity because it is efficient, easy to use, cost effective and does not produce toxic byproducts [47]. Nano adsorbents have improved upon limited efficiency of conventional adsorbents, such as the lack of selectivity and active sites [48].

Nanomaterials typically exhibit enhanced performance, often available in powder form. However, their low hydraulic conductivity poses challenges for column-based applications, specifically in terms of convenience. Additionally, the separation and regeneration of specific adsorbents present additional concerns. [49]. Oxidation, photochemical, photocatalytic, and biological reactions are primarily used to oxidize As (III) to As (V), followed by precipitation. Ion exchange technology in drinking water treatment is mostly used for softening and As removal, but also requires pre-oxidation to enhance As removal performance [50]. Different membrane technologies exist for As removal such as nanofiltration (NF) and reverse osmosis (RO). Nanofiltration require sensitive pH adjustment [51] and pre-oxidation [33]. Integrated membrane technologies, such as the combination of RO and membrane distillation, effectively remove total As from water while generating fewer toxic by-products compared to oxidation followed by RO. However, the widespread implementation of membrane technologies is hindered by high energy consumption and the occurrence of membrane fouling, which limits their global applicability. [52]. Coagulation and flocculation are typical conventional methods to remove high concentration of As, however, those processes produce excessive amount of sludge [50].

In summary, despite significant advancements in the development of advanced technology for As removal, the accessibility of these solutions remains a major concern, as they are often expensive and beyond the financial reach of many communities affected by As contamination. There is an urgent need for low-cost devices that can assist underprivileged populations in reducing their exposure to As-contaminated drinking water. Moreover, for these low-cost technologies to be widely accepted and utilized by rural communities, they must demonstrate reliability. The potential for media regeneration, which allows for the reuse or restoration of treatment materials, can significantly enhance the attractiveness and practicality of these technologies. However, addressing these challenges and realizing the potential of low-cost, reliable, and regenerative As removal technologies requires further research and development efforts.

References

1. Jarup L. Hazards of heavy metal contamination. Br Med Bull. 2003;68:167–82.
2. Mandal KB, Suzuki TK. Arsenic round the world: a review. Atlanta. 2002;58:201–35.
3. Cong L. Sorption/desorption of arsenic to nanometer scale magnetite. Doctoral Thesis, Rice University, Houston Texas; 2004. p. 178.
4. Stollenwerk KG. Geochemical processes controlling transport of arsenic in groundwater: a review of adsorption. Netherlands; 2003. p. 67–100.
5. Smedley PL, Kinniburgh DG. A review of the source, behavior, and distribution of arsenic in natural waters. Appl Geochem. 2002;17:517–68.
6. World Health Organization (WHO). Environmental health criteria on arsenic and arsenic compounds. Environmental health criteria series, arsenic, and arsenic compounds, 2nd edn. Geneva; 2001. p. 512.
7. Chakmeng J, Wang J, Shraim A. A global health problem caused by arsenic from natural sources. Chemosphere. 2003;52:1353–9.
8. Negrea A, Popa A, Ciopec M, Lupa L, Negrea P, Davidescu MC, Motoc M, Minzatu V. Phosphonium grafted styrene–divinylbenzene resins impregnated with iron(III) and crown ethers for arsenic removal. Pure Appl Chem. 2014;86(11):1729–40.
9. Woolson EA. Bioaccumulation of arsenicals. In: Arsenical pesticides. American Chemical Society; 1975. p. 97–107.
10. Goldberg S, Johnston CT. Mechanisms of arsenic adsorption on amorphous oxides evaluated using macroscopic measurements, vibrational spectroscopy, and surface complexation modeling. J Colloid Interface Sci. 2001;234:204–16.
11. Mukherjee A, Sengupta MK, Hossain MA, Ahamed S, Das B, et al. Arsenic contamination in groundwater: a global perspective with emphasis on the Asian scenario. J Health Popul Nutr. 2006;24 (2):142–63.
12. Aaklı KOZ. Fabrication of nano and porous materials & their utilization in the purification of water contaminated with arsenic, copper, and lead. Doctoral Thesis, Sabancı University, Istanbul; 2013. p. 160.
13. Shaji E, Santosh M, Sarath KV, Prakash P, Deepchand V, Divya BV. Arsenic contamination of groundwater: a global synopsis with focus on the Indian Peninsula. Geosci Front. 2011;13(3): 101079.
14. Rahaman MS, Mise N, Ichihara S. Arsenic contamination in food chain in Bangladesh: a review on health hazards, socioeconomic impacts and implications. Hyg Environ Health Adv. 2022;2: 100004.
15. Mohan D, Pittman CU. Arsenic removal from water/wastewater using adsorbents—a critical review. J Hazard Mater. 2007;142:1–53.
16. Yan XP, Kerrich R, Hendry MJ. Distribution of arsenic (III), arsenic(V) and total inorganic arsenic in porewaters from a thick till and clay-rich aquitard sequence, Saskatchewan, Canada. Geochim Cosmochim Acta. 2000;64:2637–48.
17. European Union (EU). Council directive 98/83/EC on the quality of water intended for human consumption. Brussels; 1998. https://eur-lex.europa.eu/legal-content/EN/TXT/PDF/?uri=CELEX:01998L0083-20151027&from=EN#:~:text=Objective-,1.,it%20is%20wholesome%20and%20clean.
18. Environment Protection Agency (EPA). 40 CFR Parts 9, 141 and 142. National primary drinking water regulations; arsenic and clarifications to compliance and new source contaminants monitoring; 2001. https://www.govinfo.gov/content/pkg/FR-2001-01-22/pdf/01-1668.pdf.
19. Hurd JJ. Evaluation of three arsenic removal technologies in Nepal. Master Thesis, Massachusetts Institute of Technology, Boston; 2001. p. 65.
20. Pezeshki H, Hashemi M, Rajabi S. Removal of arsenic as a potentially toxic element from drinking water by filtration: a mini review of nanofiltration and reverse osmosis techniques. Heliyon. 2023;9(3): e14246.
21. Siddique T, Dutta NK, Roy Choudhury N. Nanofiltration for arsenic removal: challenges, recent developments, and perspectives. Nanomaterials. 2020;10 (7):1323.
22. Kord Mostafapour F, Bazrafshan E, Kamani H. Survey of arsenic removal from water by coagulation and dissolved air floatation method, Iran. J Health Environ. 2010;3(3):309–18.
23. Teimouri R, Mahdiarfar M. Comparison of safety technologies on arsenic removal from water. J Biosafety. 2017;9(4):79–92.
24. Rahmani AR, Amini S. Removal of arsenic from aquatic environments by amended and unamended oak tree sawdust. J Water Wastewater. 2016;26 (6):42–9.
25. Shiriazar MA, Sepehr E. Methods to correct arsenic contamination in water and soil. Appl Soil Res. 2021;8(4):69–84.
26. Rajendran RM, Garg S, Bajpai S. Economic feasibility of arsenic removal using nanofiltration membrane: a mini review. Chem Pap. 2021;75(9):4431–44.
27. Mohammad-Razdari V, Fanaee SA. Comprehensive review of different types of water desalination. Renew New Energy Quart. 2021;8(1):21–32.
28. Alka S, Shahir S, Ibrahim N, Ndejiko MJ, Vo DVN, Abd Manan F. Arsenic removal technologies and future trends: a mini review. J Clean Prod. 2021;278: 123805.
29. Siddiqui SI, Naushad M, Chaudhry SA. Promising prospects of nanomaterials for arsenic water remediation: a comprehensive review. Process Saf Environ Protect. 2019;126:60–97.
30. Saboori N, Azizi Mobasser J, Asadi A. Removal of arsenic in water environment by reverse osmosis method and zero valent iron nanoparticles. J Water Wastewater. 2018;28(6):39–47.

31. Moafi M, Ardestani M, Mehrdadi N. Use of zinc oxide nano-photocatalyst as a recyclable catalyst for removal of arsenic and lead ions from polluted water. J Health. 2021;12(1):130–43.
32. Kitis M, Yiğit ON, Beyhan M, Sayilgan E. Kimyasal Temel İslemler Ders Notu. Suleyman Demirel University, Isparta; 2011. p. 231.
33. Edzwald JK. Coagulation in drinking water treatment. Water Sci Technol. 1993;27(11):21.
34. Gunten VU. Oxidation processes in water treatment are we on track. Environ Sci Technol. 2018;52 (9):5062–75.
35. Arikan S. Investigation of arsenic adsorption performance of the modified natural materials. PhD Thesis, Dokuz Eylul University, İzmir; 2016. 145 p.
36. Aktas K. Synthesis of nanomaterials and enhancing natural material to remove arsenic from water. PhD Thesis, Akdeniz University, Antalya; 2021. 101 p.
37. Ulbricht M. Advanced functional polymer membranes. Polymer. 2006;47:2217–62.
38. Saleh AT, Gupta KV. An overview of membrane science and technology. Nanomaterial and polymer membranes. Elsevier; 2016. p. 1–23.
39. Albadarin BA, Collins NM, Naushad M, Shirazian S, Walker G, Mangwandi C. Activated lignin-chitosan extruded blends for efficient adsorption of methylene blue. Chem Eng J. 2017;307:264–72.
40. Kundu S, Naskar MK. Perspective of membrane processes for the removal of arsenic from water: an overview. Trans Indian Ceram Soc. 2021;80:28–40.
41. Amini A, Kim Y, Zhang J, Boyer T, Zhang Q. Environmental and economic sustainability of ion exchange drinking water treatment for organics removal. J Clean Prod. 2015;104:413–21.
42. Liu R, Qu J. Review on heterogeneous oxidation and adsorption for arsenic removal from drinking water. J Environ Sci. 2021;110:178–88.
43. Usman M, Katsoyiannis I, Rodrigues JH, Ernst M. Arsenate removal from drinking water using by-products from conventional iron oxyhydroxides production as adsorbents coupled with submerged microfiltration unit. Environ Sci Pollut Res. 2021;28:59063–75.
44. Nguyen TH, Tran HN, Vu HA, Trinh MV, Nguyen TV, Loganathan P, Vigneswaran S, Nguyen TM, Vu DL, Nguyen THH. Laterite as a low-cost adsorbent in a sustainable decentralized filtration system to remove arsenic from groundwater in Vietnam. Sci Total Environ. 2020;699: 134267.
45. Worou CN, Chen ZL, Bacharou T. Arsenic removal from water by nanofiltration membrane: potentials and limitations. Water Pract Technol. 2021;16 (2):291–319.
46. Environmental Protection Agency (EPA). Arsenic: mitigation strategies, documents; 2015. p. 152. https://www.epa.gov/sites/default/files/2015-09/documents/train5-mitigation.pdf.
47. Hassan HR. A review on different arsenic removal techniques used for decontamination of drinking water. Environ Pollutants Bioavailab. 2023;35 (1):2165964.
48. Qu X, Alvarez PJJ, Li Q. Applications of nanotechnology in water and wastewater treatment. Water Res. 2013;47:3931–46.
49. Lata S, Samadder RS. Removal of arsenic from water using nano adsorbents and challenges: a review. J Environ Manage. 2016;166:387–406.
50. Singh R, Singh S, Parihar P, Singh PV, Prasad MS. Arsenic contamination, consequences and remediation techniques: a review. Ecotoxicol Environ Saf. 2015;112:247–70.
51. Zhao S, Zou L, Tang CY, Mulcahy D. Recent developments in forward osmosis: opportunities and challenges. J Membr Sci. 2012;396:1–21.
52. Uddin M, Mozumder M, Islam M, Deowan S, Hoinkis J. Nanofiltration membrane process for the removal of arsenic from drinking water. Chem Eng Technol. 2007;30:1248–54.

SPRINGER NATURE

GPSR Compliance

The European Union's (EU) General Product Safety Regulation (GPSR) is a set of rules that requires consumer products to be safe and our obligations to ensure this.

If you have any concerns about our products, you can contact us on ProductSafety@springernature.com

In case Publisher is established outside the EU, the EU authorized representative is:

Springer Nature Customer Service Center GmbH
Europaplatz 3
69115 Heidelberg, Germany

The manufacturer's authorised representative in the EU is Springer Nature Customer Service Centre GmbH, Europaplatz 3, 69115 Heidelberg, Germany. If you have any concerns regarding our products, please contact ProductSafety@springernature.com

Printed and bound by CPI Group (UK) Ltd, Croydon, CR0 4YY
30/09/2026
02225598-0020